U0022696

Deepen Your Mind

前言

Android 作業系統目前在全球佔據主流，大量網際網路、市政、金融、O2O、叫車平台等公司及部門將業務依靠於 App 的方式交付給最終使用者，這些 App 真的安全嗎？有各類爬蟲軟體對票務、企業個人資訊等 App 內容的瘋狂抓取，因此 App 的安全、逆向工程及自動化利用技術越來越受到 App 開發者的關注。

自從 Frida 於 2014 年年末問世以來，迅速在全球安全社區掀起了 "Frida 熱潮"，借助 Frida 動態修改記憶體的特性實現了快速逆向和演算法呼叫功能，Android 應用安全分析和對抗技術從未像如今這樣成熟和自動化。

作為 Android 應用安全測評工程師，或巨量資料平台擷取工程師，逆向研究員對於 App 的逆向分析研究及其演算法的還原和介面呼叫的熱愛彷彿是刻在骨子裡的。

與逆向技術的發展相對應的是，很多大型軟體和平台的開發者也逐漸把演算法藏得越來越深，越來越難以逆向。這裡面最具有代表性的是強混淆框架 Ollvm 和 Arm 層的虛擬機器保護技術 Vmp，前者注重增加演算法本身的複雜度，後者透過增加一套中間層將演算法保護起來，使得逆向工作變得更加困難，顯然，逆不出中間層也就還原不出演算法。

面對這種情況我們該如何應對呢？解決辦法是採用黑盒呼叫的方式，忽略演算法的具體細節，使用 Frida 把 SO 載入起來，直接呼叫裡面的演算法得到計算結果，構造出正確的參數，將封包傳給伺服器。也可以將呼叫過程封裝成 API 曝露給同事使用，甚至架設計算叢集，加快運行速度，提高運行效率。本書詳細地介紹了基於 Frida 和 Xposed 的演算法批次呼叫和轉發實踐，並舉出了具體的案例分析。

如果 App 對 Frida 或 Xposed 進行了檢測，我們還可以採用編譯 Android 原始程式的方式打造屬於自己的抓取封包沙盒。對系統來說，由於 App 的全部程式都是依賴系統去完成執行的，因此無論是保護 App 在執行時期的脫殼，還是 App 發送和接收資料封包，對系統本身來說 App 的行為都是沒有隱私的。換句話說，如果在系統層或更底層對 App 的行為進行監控，App 的很多關鍵資訊就會曝露在 "陽光" 之下一覽無餘。之後可以直接修改系統原始程式，使用 r0capture 工具為 Hook 的那些 API 中加入一份日誌，即可把處於明文狀態的封包列印出來，從而實現無法對抗的抓取封包系統沙盒。

Frida 以其簡潔的介面和強大的功能迅速俘獲了 Android 應用安全研究員以及爬蟲研究員的芳心，成為逆向工作中的絕對主力，筆者也有幸在 Frida 普及的浪潮中做了一些複習和分享，建立了自己的社群，與大家一起跟隨 Frida 的更新腳步共同成長和進步。

本書充實地介紹了如何安裝和使用 Frida、基本的環境架設、Frida-tools、Frida 指令稿、Frida API、批次自動化 Trace 和分析、RPC 遠端方法呼叫，並包含大量 App 逆向與協定分析案例實戰，此外，還介紹了更加穩定的框架 Xposed 的使用方法，以及從 Android 原始程式開始訂製屬於自己的抓取封包沙盒，打造無法被繞過的抓取封包環境等內容。

本書技術新穎，案例豐富，注重實際操作，適合以下人員閱讀：

- Android 應用安全工程師。
- Android 逆向分析工程師。
- 爬蟲工程師。
- 巨量資料收集和分析工程師。

在本書完稿時，Frida 版本更新到 15，Android 也即將推出版本 12，不過請讀者放心，本書中的程式可以在特定版本的 Frida 和 Android 中成功運行。

Android 逆向是一門實踐性極強的學科，讀者在動手實踐的過程中難免會產生各式各樣的疑問，因此筆者特地準備了 GitHub 倉庫更新和勘誤，讀者如有疑問可以到倉庫的 issue 頁面提出，筆者會盡力解答和修復。筆者的 GitHub：https://github.com/r0ysue/AndroidFridaSeniorBook。

編按：本書之原作者為中國大陸人士，書中許多「手機介面」、「網站」均為簡體中文版介面。為求書籍之完整性，本書中此類介面均維持簡體中文介面，請讀者閱讀時可對照上下文。

陳佳林

目錄

06 Android 原始程式編譯與 Xposed 魔改

07 Android 沙盒之加解密庫「自吐」

Android 逆向環境架設

欲善其事，必先利其器。本章將會介紹筆者在 Android 逆向工程中用到的環境設定，包括主機和測試機的基礎環境設定。一個良好的工作系統能給工作人員在工作過程中帶來很多便利，讓大家不必因為環境問題焦頭爛額，因此在開始逆向工作之前，架設一個良好的環境是非常必要的。

1.1 虛擬機器環境準備

推薦使用虛擬機器而非實機，主要的原因有以下 3 點：

（1）虛擬機器附帶「時光機」功能 ——「快照」，這個特性讓使用者能夠隨時得到一個全新的實機，不會因為一個設定失誤導致系統崩潰，最終只能因為直接重裝系統而懊惱。圖 1-1 所示為筆者在日常工作中開發 FART 脫殼機時建立的諸多虛擬機器快照。

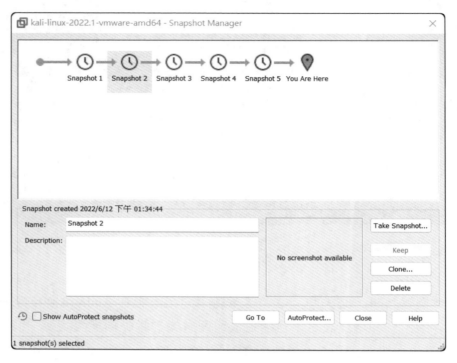

▲ 圖 1-1　帶快照功能的虛擬機器

（2）虛擬機具有良好的隔離特性，做實驗的過程中不會「污染」實機，在分析惡意樣本時，使用虛擬機器能夠極佳地保護物理機的環境不受損壞，是測試全新功能的天然「沙盤」。

（3）虛擬機器環境不受物理機系統限制，無論是 Mac 系統還是 Windows系統，其安裝的虛擬機器系統都能夠任意選擇，包括 Ubuntu、CentOS、Windows 等。

在這裡筆者推薦讀者使用 VMware 出品的系列虛擬機器軟體。VMware具有良好的跨平台特性，可以隨時將已經部署好的環境在不同平台上遷移使用。

對於虛擬機器環境的選擇，筆者更加推薦 Ubuntu 系列的 Linux 作業系統，無論是 Android 原始程式的編譯，還是 Frida、GDB、OLLVM 等後續重要的環境，經過筆者測試，這個系列的系統總是能夠表現出更少被系統環境「拖累」的特性。

在筆者的工作中，主要使用 Kali Linux 這個系統，Kali Linux 是基於 Debian 的 Linux 發行版本，與 Ubuntu 師出同門，是設計用於數位取證的作業系統。Kali Linux 預先安裝了許多滲透測試軟體，包括 Metasploit、BurpSuite、SQLMap、Nmap 等 Web 安全相關軟體，是一套開箱即用的專業滲透測試工具箱。

Kali Linux 附帶 VMware 鏡像版本，下載對應版本後，解壓並按兩下開啟 .vmx 檔案，即可透過 VMware 開啟虛擬機器的系統。

由於筆者這裡選擇的 Kali Linux 版本為 2021.1，在這個版本中 Kali 的預設使用者已經不再是 root/toor，而是 kali/kali。但是筆者建議第一次使用 Kali 使用者登入系統後，使用如下命令設定 root 使用者密碼以重新啟用 root 使用者，這樣在後續工作中便不會因為使用者許可權不夠而出現各種類型的顯示出錯。

```
# sudo passwd root
```

修改使用者完畢後，重新使用 root 使用者登入系統，其介面顯示如圖 1-2 所示。

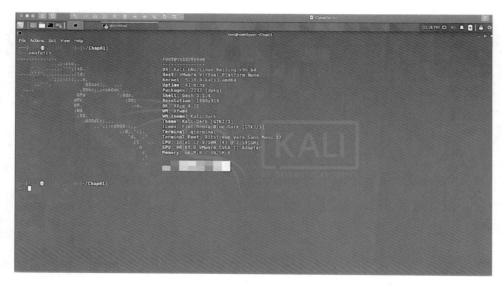

▲ 圖 1-2　Kali Linux 介面

另外，由於虛擬機器本身的時間不是東八區的，在開啟虛擬機器後還需要開啟 Terminal 軟體並輸入如下命令設定時區：

```
┌──(root@vxidr0ysue)-[~/Chap01]
└─# dpkg-reconfigure tzdata
```

```
Current default time zone: 'Asia/Shanghai'
Local time is now:      Sat Apr 17 15:20:35 CST 2021.
Universal Time is now:  Sat Apr 17 07:20:35 UTC 2021.
```

執行命令後，在彈出的視窗選擇 Asia → Taipei 後，就可以設定成標準台北時間，當然不同的讀者可以根據自己所在的地區進行設定。

由於 Kali Linux 在 2020.3 版本後就開始支援中文字型的顯示，而這裡選擇的是 2021.1 版本的 Kali Linux 虛擬機器，因此無須再和 2019 版本的 Kali Linux 一樣另外設定。但要注意的是，一定不要將系統切換為中文環境，中文環境的 Linux 總會出現各種各樣的問題，並且在出現問題後解決起來十分麻煩。

還需要注意的是，Kali Linux 在 2020.3 版本後預設的 Shell 不再是 Bash 而是 Zsh，雖然 Zsh 的自動提示等擴充功能十分強大，但是由於後續 Android 系統的編譯只支援 Bash 終端，因此還需要使用如下命令完成預設 Shell 的切換：

```
# chsh -s /bin/bash
```

在執行完上述命令並重新啟動系統後，再次開啟 Terminal 執行如下命令，會發現預設 Shell 已經回退到 Bash，最終效果如圖 1-3 所示。

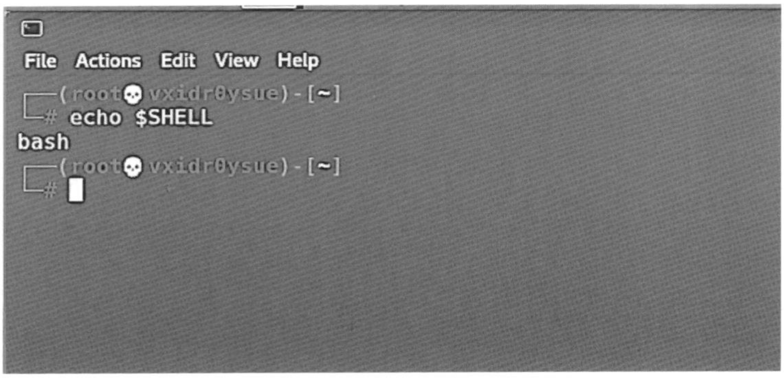

▲ 圖 1-3 預設 Shell

1.2 逆向環境準備

在設定好基礎的系統環境之後，為了進行後面的逆向開發工作，還需要安裝一些基礎的開發工具。

首先，作為 Android 逆向環境開發人員，Android Studio 是一款必不可少的開發工具。在 Eclipse 退出 Android 開發歷史舞臺後，作為 Google 官方的 Android 應用程式開發 IDE，筆者首先推薦這款軟體。在從官網下載和解壓對應 Linux 版本的 Android Studio 後，切換到 android-studio/bin 目錄下，透過執行目前的目錄下的 studio.sh 即可執行 Android Studio。

第一次開啟 Android Studio 會進行 Android SDK 工具的下載，這些工具是後續開發所必需的，因此預設一直點擊 Next 按鈕即可。在這個過程中，可以關注一下 SDK 的保存目錄，預設 SDK 目錄為 /root/Android/Sdk/，這個目錄下存在一些在後續逆向過程中需要的工具，比如 ADB 這個用於與行動裝置進行通訊的工具。這裡將 ADB 工具所在目錄 /root/Android/Sdk/platform-tools/ 加入環境變數，便於在任意目錄下執行 adb 命令。

```
┌──(root@vxidr0ysue)-[~/Chap01]
└─#  adb
bash: adb: command not found

┌──(root@vxidr0ysue)-[~/Chap01]
└─# echo "export PATH=$PATH:/root/Android/Sdk/platform-tools" >> ~/.bashrc
```

在將 ADB 工具加入環境變數後，為了使得設定生效，需要重新開啟 Terminal，再次執行 adb 命令，結果如下：

```
┌──(root@vxidr0ysue)-[~/Chap01]
└─# adb shell

* daemon not running; starting now at tcp:5037
* daemon started successfully
adb: no devices/emulators found
```

回到正題，在下載外掛程式完畢後，Android Studio 的介面如圖 1-4 所示。

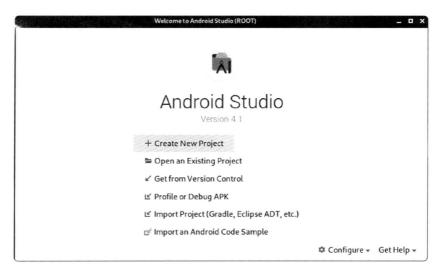

▲ 圖 1-4 Android Studio 的介面

請注意，在第一次建立 Project 時，Android Studio 需要進行一段可能費時很長的同步環節，用於下載 Gradle 建構工具以及其他相關依賴，這個時候只需要去喝杯茶，靜靜等待即可。

在 Android Studio 設定好後，筆者還會推薦一些在日常工作中使用的小工具，這些工具也許不會直接對工作有幫助，但是一旦掌握了這些工具，使用者的日常工作會變得更加得心應手。

首先，推薦 htop 這款加強版 top 工具。與 top 工具相同，htop 可以用於動態查看當前活躍的、佔用高的處理程序，但是比 top 工具的顯示效果更加人性化，具體效果如圖 1-5 所示。這個工具在編譯 Android 原始程式時非常好用，當我們執行 make 命令系統開始編譯 Android 原始程式之後，透過 htop 工具可以發現記憶體 Mem 以肉眼可見的速度跑到底之後，開始侵佔 Swp 的進度指示器。另外，htop 中的 Uptime 後顯示的是開機時間；Load average 是指平均負載，比如虛擬機器被分配了四核心 CPU，那麼平均負載跑到 4 的時候説明系統已經滿載。圖 1-5 中左側 1、2、3、4 的進度指示器表示對應 CPU 當前的負載狀態，其餘 htop 操作指南讀者可以自行去網上搜索。

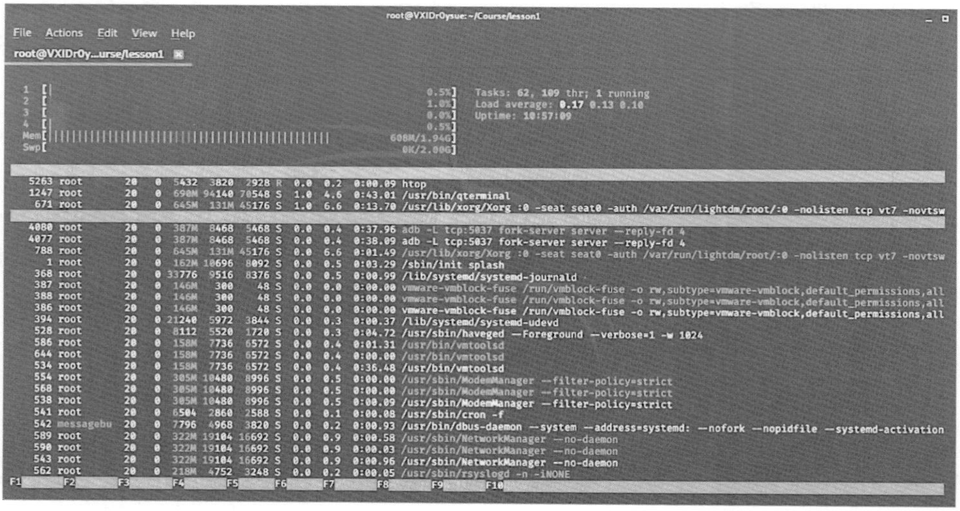

▲ 圖 1-5 htop 介面

另外，筆者還要推薦一款即時查看系統網路負載的工具 jnettop，在安裝和使用軟體（比如 Frida）的過程中，可以利用 jnettop 工具即時查看對應的下載速度和對應的 IP，甚至讀者在 AOSP 編譯時開啟 jnettop，會觀察到編譯過程中出現連接國外的伺服器下載依賴套件等行為。除此之外，值得一提的是，在抓取封包時開啟這個工具往往會有奇效，比如能夠即時查看對方的 IP 等。jnettop 介面顯示如圖 1-6 所示，可以看到主機連接的遠端 IP、通訊埠、速率以及協定等內容。

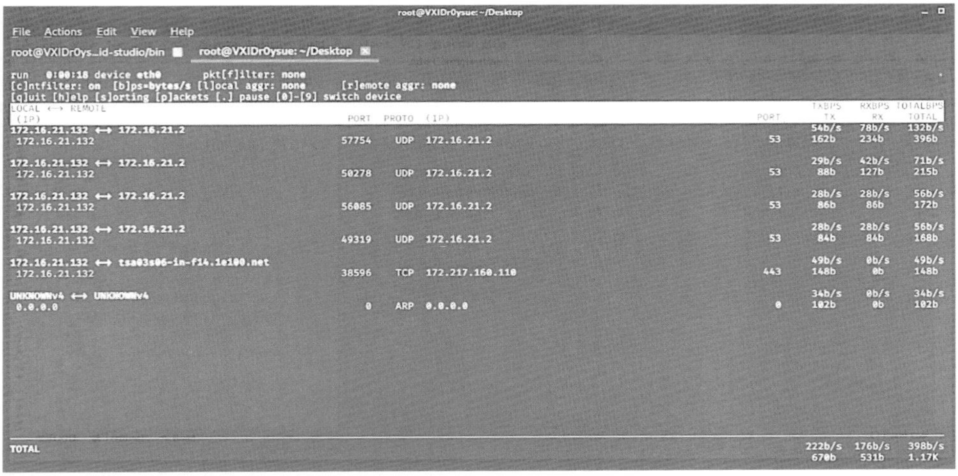

▲ 圖 1-6 jnettop 介面

在過去筆者經常會被問到如圖 1-7 所示因為視窗大小限制導致 jnettop
工具無法執行的問題（Too small terminal (detected size: 79×34)），
真讓人哭笑不得，筆者在這裡統一回答這個問題。實際上 jnettop 工具
本身在執行時期對終端大小是有所要求的，否則 jnettop 工具就無法開
啟，minimum required size: 80×20 這個提示表明終端長和寬至少為
80×20。

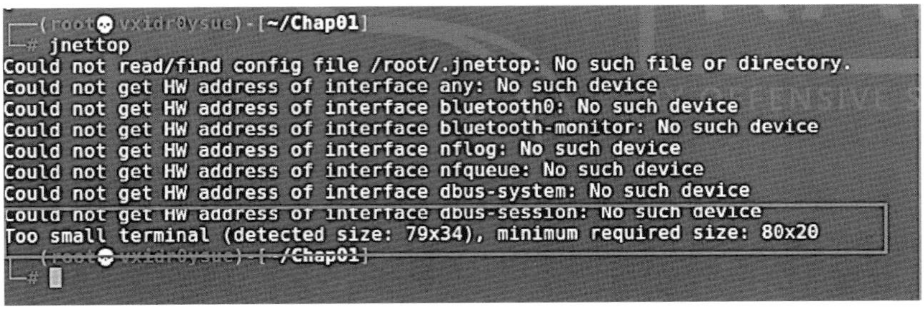

▲ 圖 1-7 jnettop 因終端視窗過小導致無法開啟

1.3 行動裝置環境準備

1.3.1 更新韌體

在 Android 逆向的學習中，提及基礎一定不能錯過更新韌體，而在更新韌體之前，一定要準備一台測試機，這裡筆者推薦 Google 官方的 Nexus 系列和 Pixel 系列的測試機。之所以推薦 Google 原生系統，是因為 Google 官方不僅提供了鏡像，而且在對應的原始程式網站上能夠找到對應鏡像的全部原始程式，因此筆者推薦 Google 官方推出的手機。筆者在這本書中選擇了 Nexus 5X，讀者如選擇其他型號的手機，僅供參考。（編按：本小節為原作者手機，使用簡體中文介面演示）

在拿到測試機後要完成更新韌體，首先需要開啟手機的「開發者選項」，具體步驟如下：

步驟 01 進入「設定」頁面，點擊「系統」，然後點擊「關於手機」，進入「關於手機」介面，如圖 1-8 所示。

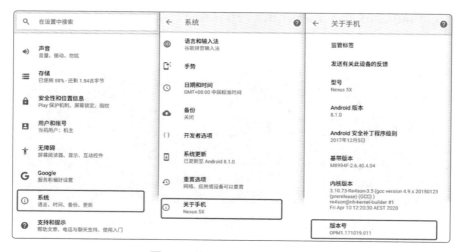

▲ 圖 1-8 進入「關於手機」介面

步驟 02 連續多次點擊「版本編號」所在 View，直到螢幕提示已進入「開發者模式」，如圖 1-9 所示。

步驟 03 在出現頁面提示「已處於開發者模式」後返回上一級目錄，也就是進入「系統」介面，此時會出現「開發者選項」，點擊「開發者選項」，如圖 1-10 所示。

▲ 圖 1-9 開啟「開發者模式」　　▲ 圖 1-10 進入「開發者選項」介面

步驟 04 在進入「開發者選項」介面後，首先開啟「USB 偵錯」。在這個選項開啟後，使用 USB 線連接電腦，手機端就會出現「允許 USB 偵錯嗎？」對話方塊，如圖 1-11 所示。

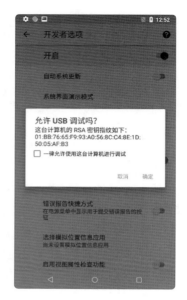

▲ 圖 1-11 請求允許 USB 偵錯

在同意 USB 偵錯之前和之後使用 adb devices 命令的結果如下：

```
┌──(root@vxidr0ysue)-[~/Chap01]
└─#  adb devices # USB偵錯同意前
List of devices attached
0041f34b7d58b939        unauthorized

┌──(root@vxidr0ysue)-[~/Chap01]
└─# adb devices # USB偵錯同意後
List of devices attached
0041f34b7d58b939        device
```

步驟 05 再次回到 Android 測試機上，此時還有一個「OEM 解鎖」選項需要允許，如圖 1-12 所示。這個選項決定了後續能否完成更新韌體，也就是更新韌體中常聽到的 Bootloader 鎖。

步驟 **06** 此時，在電腦的終端上執行命令 adb reboot bootloader 或者將手機關機後同時按住手機電源鍵與音量減鍵，進入 Bootloader 介面。OEM 未解鎖之前的 Bootloader 介面，如圖 1-13 所示。

▲ 圖 1-12 請求允許「OEM 解鎖」

▲ 圖 1-13 OEM 未解鎖之前的 Bootloader 介面

步驟 **07** 保持手機使用 USB 線連接上電腦，再次在電腦終端中執行 fastboot oem unlock 命令，然後測試機就會彈出確認介面，此時按音量減鍵直到選中 YES 選項後按電源鍵，至此，OEM 鎖就成功解鎖了。如圖 1-14 所示為解鎖後的 Bootloader 介面。

```
┌──(root@vxidr0ysue)-[~/Chap01]
└─# fastboot oem unlock
OKAY [170.246s]
Finished. Total time: 170.246s
```

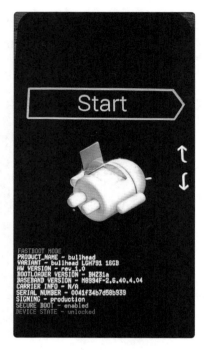

▲ 圖 1-14 OEM 已解鎖的 Bootloader 介面

在 OEM 解鎖後,一個完整的可供更新韌體的手機就準備完成了,此時如果要刷入新的特定系統,就要準備更新韌體套件。這裡的更新韌體套件其實也可以叫作官方鏡像套件,Google 官方提供了一個官方鏡像的網站(網址:https://developers.google.com/android/images),筆者這裡下載 Nexus 5X 的對應更新韌體套件,由於 Android 8.1.0_r1 這個版本的系統支援的裝置比較多,因此在這裡筆者選擇這個版本的系統進行演示。Android 8.1.0_r1 對應代號為 OPM1.171019.011,版本與代號對應關係的 網 址 為 https://source.android.com/setup/start/build-numbers#source-code-tags-and-builds,在找到代號後,再次回到官方鏡像站下載對應版本的鏡像。

在下載完畢後，解壓更新軔體套件並進入更新軔體套件目錄，同時手機進入 Bootloader 介面並使用 USB 線連接上主機，然後直接執行 flash.sh 檔案。對應步驟如下：

```
┌──(root@vxidr0ysue)-[~/Chap01]
└─# unzip bullhead-opm1.171019.011-factory-3be6fd1c.zip
Archive:  bullhead-opm1.171019.011-factory-3be6fd1c.zip
   creating: bullhead-opm1.171019.011/
  inflating: bullhead-opm1.171019.011/radio-bullhead-m8994f-2.6.40.4.04.img
  inflating: bullhead-opm1.171019.011/flash-all.bat
  inflating: bullhead-opm1.171019.011/bootloader-bullhead-bhz31a.img
  inflating: bullhead-opm1.171019.011/flash-base.sh
  inflating: bullhead-opm1.171019.011/flash-all.sh
 extracting: bullhead-opm1.171019.011/image-bullhead-opm1.171019.011.zip
┌──(root@vxidr0ysue)-[~/Chap01]
└─# cd bullhead-opm1.171019.011/
┌──(root@vxidr0ysue)-[~/Chap01/bullhead-opm1.171019.011]
└─# ./flash-all.sh
...
Rebooting                                        OKAY [  0.020s]
Finished. Total time: 213.643s
```

之後，手機系統便會進入初始化介面，在完成語言、WiFi 等相關的設定後，一台「新」的測試機就誕生了。當然，為了方便後續測試，此時還需要再次開啟「開發者選項」以獲取 USB 偵錯許可。

如圖 1-15 所示，在聯網之後會發現測試機系統時間與計算機時間不對應，且頁面提示「此 WLAN 網路無法存取網際網路」。此時可以透過以下命令解決這個問題，在命令執行結束後，待測試機重新開機後便會發現問題消失。

```
┌──(root@vxidr0ysue)-[~/Chap01]
└─# adb shell settings put global captive_portal_http_url https://www.
```

```
google.cn/generate_204
  ┌──(root@vxidr0ysue)-[~/Chap01]
  └─# adb shell settings put global captive_portal_https_url https://www.
google.cn/generate_204
  ┌──(root@vxidr0ysue)-[~/Chap01]
  └─# adb shell settings put global ntp_server 1.hk.pool.ntp.org
  ┌──(root@vxidr0ysue)-[~/Chap01]
  └─# adb shell reboot
```

▲ 圖 1-15　WLAN 網路無法存取網際網路及時間不同步問題

1.3.2　ROOT

上一小節中，我們已經完成了 Nexus 5X 版本的更新韌體工作，此時獲得的是一個全新的沒有做任何操作的新機。開啟測試機的開發者模式後，開啟 USB 偵錯按鈕，此時就可以使用 ADB 連接手機了。在這一節中將演示對 Nexus 5X 進行 Root 的過程。具體步驟如下：

步驟 01 要進行 ROOT，首先需要將 TWRP 刷入 Recovery 分區。

TWRP（Team Win Recovery Project）是一個開放原始程式軟體的訂製 Recovery 映射，供基於 Android 的裝置使用，允許使用者向協力廠商安裝韌體和備份當前的系統，通常在 Root 系統時安裝。而 Recovery 指的是一種可以對 Android 機內部的資料或系統進行修改的模式（類似於 Windows PE 或 DOS），也指 Android 的 Recovery 分區。

由於筆者使用的是 TWRP 的官方鏡像檔案，這裡提供 TWRP 對應的官方網址：https://twrp.me/Devices。在進入該網址後，選擇對應型號的裝置和對應版本的 IMG 鏡像檔案，比如這裡先點擊 LG 進入 LG 廠商的裝置列表，選擇 LG Nexus 5X（bullhead），然後在 Download Link 這裡選擇對應的美版或者歐版，此處選擇美版，也就是 Primary（Americas），具體需要讀者參考自己的手機類型進行選擇。下載完成後，就可以選擇不同版本的 twrp-3.3.0-0-bullhead.img 下載了，這裡選擇 3.3.0 版本的 TWRP。

下載完畢後，如果在 Nexus 系列的手機上，還需要將 TWRP 刷入 Recovery 分區：使裝置進入 Bootloader 介面，並使用 Fastboot 工具將 TWRP 鏡像刷入 Recovery 分區。

```
┌──(root@vxidr0ysue)-[~/Chap01]
└─# adb reboot bootloader
┌──(root@vxidr0ysue)-[~/Chap01]
└─# fastboot flash recovery twrp-3.3.0-0-bullhead.img
Sending 'recovery' (16317 KB)                    OKAY [  1.225s]
Writing 'recovery'                               OKAY [  0.267s]
Finished. Total time: 1.539s
```

步驟 02 在進入 Bootloader 介面後，按音量上下鍵直到頁面出現 Recovery mode 字串後，使用電源鍵確認進入 Recovery 復原模式，這時就進入 TWRP 的介面了。

步驟 03 在進行步驟 04 之前，還需要使用 adb 命令將 Root 工具推送到測試機的 /sdcard 目錄下。Root 工具可以選擇 Magisk 或者 SuperSU，這裡以 Magisk 為例。先從 GitHub 上 Magisk 的倉庫的 Release 中下載新版的 ZIP 檔案，網址為 https://github.com/topjohnwu/Magisk/releases。注意選擇 Magisk，而非 Magisk Manager，筆者寫作本書時，新版為 Magisk-v20.4.zip。

```
┌──(root@vxidr0ysue)-[~/Chap01]
└─# adb push Magisk-v20.4.zip /sdcard/
Magisk-v20.4.zip: 1 file pushed, 0 skipped. 1.9 MB/s (5942417 bytes in
2.996s)
```

步驟 04 如圖 1-16 所示，在進入 TWRP 介面後，首先滑動最下方的按鈕 Swipe to Allow Modifications 進入 TWRP 主介面。然後點擊 Install，此時會預設進入 /sdcard 目錄，滑到最下方就能看到剛剛推送到手機上的 Magisk-v20.4.zip。

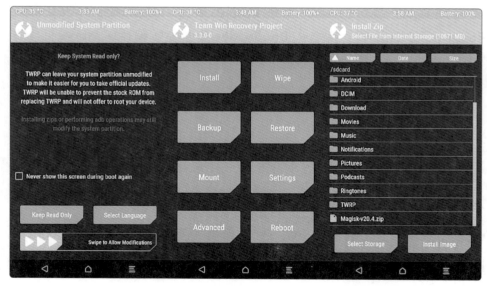

▲ 圖 1-16 TWRP 介面

步驟 05 如圖 1-17 所示，點擊 Magisk-v20.4.zip，進入 Install Zip 介面，滑動 Swipe to confirm Flash 滑動桿，開始刷 Magisk 的流程，然後靜待介面下方出現兩個按鈕，即代表 Root 完畢。

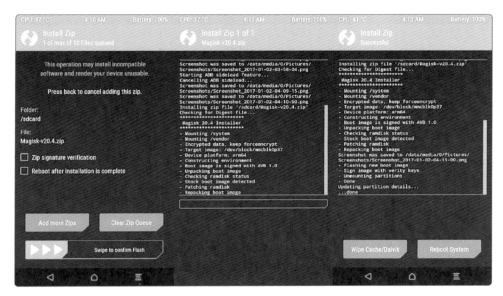

▲ 圖 1-17 刷 Magisk

步驟 06 點擊 Reboot System 按鈕，重新開機系統，會發現手機應用中多了一個 Magisk Manager，此時在 Terminal 中進入 adb shell 終端，輸入 su，會發現手機介面提示 Root 申請，點擊「允許」後，手機的 shell 即可獲得 Root 許可權，如圖 1-18 所示。命令執行結果如下：

```
┌──(root@vxidr0ysue)-[~/Chap01]
└─# adb shell
bullhead:/ $ su
bullhead:/ #
```

至此，就完成了手機的 Root 工作。當然，使用 SuperSU 對裝置進行
Root 的操作也是類似的，僅僅是將 Magisk.zip 換成 SuperSU.zip 而已，
這裡舉出 SuperSU 的官方網址：https://supersuroot.org/。注意 SuperSU
的 Root 和 Magisk 的 Root 是衝突的，在進行 SuperSU 的 Root 之前，
先要將 Magisk 移除掉，這裡的移除不是簡單地移除 Magisk Manger 這
個 App，而是在 Magisk Manger 的主介面點擊「移除」按鈕，從而還原
原廠鏡像，在還原後，就可以愉快地使用 SuperSU 進行 Root 了，如圖
1-19 所示。

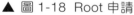

▲ 圖 1-18 Root 申請　　　　　　▲ 圖 1-19 移除 Magisk

1.3.3 Kali NetHunter 更新軔體

為什麼要在刷入官方鏡像且 Root 完成的 Android 測試機上再次刷入 Kali
NetHunter 呢？

正如桌面端的 Kali 是專為安全人員設計的 Linux 定製版作業系統，Kali NetHunter 也是第一個針對 Nexus 行動裝置的開放原始碼 Android 滲透測試平台，刷入這個系統有利於逆向開發人員更加深入地理解 Android 系統，無論是在後面章節中使用 Kali NetHunter 直接從網路卡獲取手機全部流量，還是在刷入 Kali NetHunter 後，逆向人員都可經由 Kali NetHunter 直接執行原本在桌面端 Kali 上可以執行的一切命令。比如 htop、jnettop 等在第 1 章中介紹的所有命令，這些命令原生的 Android 是不支持的。另外，隨著 Kali NetHunter 的刷入，逆向人員便可以憑藉它從核心的層面去監控 App，比如透過 strace 命令直接追蹤所有的系統呼叫，任何 App 都沒有辦法繞過這一方式，畢竟從本質上來說，任何一個 App 都可以當作 Linux 中的一個處理程序。而之所以可以從核心層面去監控 App，是因為安裝的 Kali NetHunter 和 Android 系統共用了同一個核心。可以說，Kali NetHunter 值得每一個 Android 逆向人員所擁有。

另外，由於 Kali NetHunter 對 Android 修改的主要是關於 Android 核心方面的內容，這些修改對平時日常的使用幾乎不會產生任何影響，比如 Xposed 這個 Hook 工具依舊可以在 Kali NetHunter 上正常使用，這大大縮減了逆向人員進行測試的成本。

接下來進入刷入環節。

步驟01 首先，下載 SuperSU 以及調配於 Nexus 5X 版本的 Kali NetHunter，注意這裡的 SuperSU 是 ZIP 格式而非 APK 格式，同時不要使用 SuperSU 官網舉出的新版 SuperSU 工具，而使用 SuperSU-SR5 版。另外，Kali NetHunter 官網舉出的 2020.04 版本的 Kali NetHunter 有 Bug，筆者這裡下載的是 2020.03 版。在官網下載 Kali NetHunter 時，會發現 Nexus 5X 的裝置只支持 Oreo 版本，而 Oreo 是 Android 8 的代號，恰好和之前刷入的手機鏡像一致。

步驟 02 在刷入 Kali NetHunter 之前，還需要對手機進行 Root 操作。這裡由於 Magisk 進行 Root 的方式實際上是「假」Root（讀者有興趣可自行研究），因此筆者選擇 SuperSU 進行 Root。而在安裝 SuperSU 之前，由於 Magisk 和 SuperSU 是不相容的，因此先按照 1.3.1 節的步驟重新刷入一個新的鏡像。

在重新進行更新韌體後，開啟開發者模式與 USB 偵錯功能並確認手機已連接上電腦。然後在主機上使用 adb 命令將 SuperSU-v2.82-201705271822.zip 和下載的 Kali NetHunter 推送到 Android 裝置上。

```
┌──(root@vxidr0ysue)-[~/Chap01]
└─# adb push SuperSU-v2.82-201705271822.zip /sdcard/
SuperSU-v2.82-201705271822.zip: 1 file pushed, 0 skipped. 1.9 MB/s (5903921
bytes in 3.036s)
┌──(root@vxidr0ysue)-[~/Chap01]
└─# adb push nethunter-2020.3-bullhead-oreo-kalifs-full.zip /sdcard/
```

步驟 03 依據 1.3.2 節的步驟重新刷入並進入 TWRP 介面，點擊 Install 按鈕，然後選擇 SuperSU 這個檔案，刷入並重新啟動，從而使得系統再次獲得 Root 許可權，具體操作如圖 1-20 所示。

▲ 圖 1-20 刷入 SuperSU

重新啟動後，進入 Shell 確認獲得 Root 許可權。

```
┌──(root@vxidr0ysue)-[~/Chap01]
└─# adb shell
bullhead:/ $ su
bullhead:/ #
```

步驟 04 最後，重新進入 TWRP，按照同樣的步驟刷入 Kali NetHunter，這個過程可能會很長。最終刷入 Kali NetHunter 並成功重新啟動後，Kali NetHunter 介面展示如圖 1-21 所示。

▲ 圖 1-21 Kali NetHunter 介面展示

此時，不僅桌布發生了變化，開啟設定頁面進入「關於手機」介面，發現 Android 核心也發生了變更，刷之前是 Google 團隊編譯的核心，刷之後變成了 re4son@nh-hernel-builder 編譯的核心，如圖 1-22 所示。

▲ 圖 1-22 Kali NetHunter 刷之前和刷之後的核心對比

從官方文件來看,這個核心是在標準 Android 核心的基礎上系統更新的產物,主要對網路功能、WiFi 驅動、SDR 無線電、HID 模擬鍵盤等功能在核心層面增加支援和驅動,開啟模組和驅動載入支援等。利用這個訂製核心,普通的 Android 手機就可以進行諸如外接無線網路卡使用 Aircrack-ng 工具箱進行無線滲透,模擬滑鼠鍵盤進行 HID BadUSB 攻擊,模擬 CDROM 直接利用手機繞過電腦開機密碼,一鍵部署 Mana 釣魚熱點等功能。

當然,這些與我們進行 AndroidApp 的逆向好像關係不是很大,我們真正關心的是 Kali NetHunter 鏡像的刷入相當於在 Android 手機中安裝了一個完整的 Linux 環境。

在 App 層面，從圖 1-21 可以看到手機上多出了 NetHunter、NetHunter-Kex、NetHunter 終端等 App。

這裡 NetHunter 終端其實就是一個終端程式，可以選擇 ANDROID SU 進入手機的終端或者選擇 Kali 模式。對應之前所説的完整的 Linux 環境，此時透過 NetHunter 終端 App 執行各種 Kali 中可以執行的命令，比如 apt 安裝命令、jnettop 查看網路卡速率、ifconfig 查看 IP 位址等，這裡展示 apt 命令，如圖 1-23 所示。

▲ 圖 1-23 Terminal 命令展示

當然，要使用其他 NetHunter 相關的 App，比如圖 1-22 展示的 NetHunter 終端，需要先開啟 NetHunter App 並允許所有申請的許可權，在 App 進入主介面後，開啟 App 側邊欄，選擇 Kali Chroot Manager 就會自動安裝上 Kali Chroot。在安裝完畢後，點擊 START KALI CHROOT 啟動 Chroot，便可以愉快地使用 NetHunter-Terminal 和 NetHunter-Kex 了，詳細步驟如圖 1-24 所示。

▲ 圖 1-24　Kali Chroot 設定

此時不僅可以透過手機上的 NetHunter 終端執行各種 Android 原本不支持的 Linux 命令，甚至覺得手機介面過小時，可以透過 SSH 連接手機最終在電腦上操作手機。具體關於 SSH 的設定，可以開啟 NetHunter 這個 App，開啟側邊欄，選擇 Kali Services，然後選取 RunOnChrootStart，並且選中 SSH 按鈕來設定，具體操作流程如圖 1-25 所示。這個時候如果電腦和手機在同一內網中，就可以愉快地使用電腦上的終端進行 SSH 連接了。

在開啟 SSH 後，根據筆者手機的 IP 192.168.50.129，最終使用電腦連接手機的效果如下：

```
# 電腦的Shell
┌──(root@vxidr0ysue)-[~/Chap01]
└─# ssh root@192.168.50.129
```

```
root@192.168.50.129's password:

Linux kali 3.10.73-Re4son-3.5 #1 SMP PREEMPT Fri Apr 10 12:20:30 AEST 2020
aarch64
The programs included with the Kali GNU/Linux system are free software;
the exact distribution terms for each program are described in the
individual files in /usr/share/doc/*/copyright.
Kali GNU/Linux comes with ABSOLUTELY NO WARRANTY, to the extent
permitted by applicable law.

# 手機的Shell
root@kali:~# ifconfig
...
wlan0: flags=4163<UP,BROADCAST,RUNNING,MULTICAST>  mtu 1500
        inet 192.168.50.129  netmask 255.255.255.0  broadcast
192.168.50.255
...
```

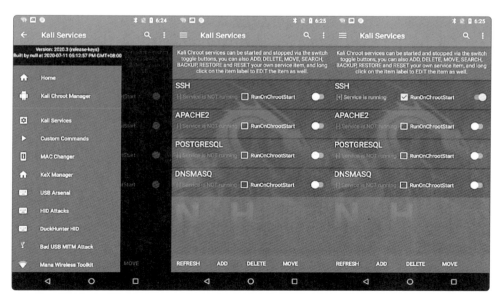

▲ 圖 1-25 NetHunter 開啟 SSH

但可惜的是，Kali NetHunter 僅支援 Nexus 系列以及 OnePlus One 系列的部分手機機型，這實在是一大遺憾。

1.4 Frida 開發環境架設

本小節將開始介紹本書的主角—在 App 逆向工作中常用的逆向工具 Frida。

1.4.1 Frida 介紹

官網對 Frida 的介紹是：Frida 是平台原生 App 的 Greasemonkey，說得專業一點，就是一種動態插樁工具，可以插入一些程式到原生 App 的記憶體空間去動態地監視和修改其行為，這些原生平台可以是 Windows、Mac、Linux、Android 或者 iOS，同時 Frida 還是開放原始碼的。

Greasemonkey 可能看起來比較陌生，其實它是 Firefox 的一套外掛程式系統，透過利用 Greasemonkey 插入自訂的 JavaScript 指令稿可以訂製網頁的顯示或行為方式。換言之，可以直接改變 Firefox 對網頁的編排方式，從而實現想要的任何功能。同時，這套外掛程式還是「外掛」的，非常靈活機動。同樣，Frida 也可以透過將 JavaScript 指令稿插入 App 的記憶體中，對程式的邏輯進行追蹤監控，甚至重新修改程式的邏輯，實現逆向人員想要實現的功能，這樣的方式也可以稱為 Hook。

Frida 目前非常火爆，該框架從 Java 層的 Hook 到 Native 層的 Hook 無所不能，雖然持久化還是要依靠 Xposed 和 Hookzz 等開發框架，但是 Frida 的動態和靈活對逆向及自動化逆向幫助非常大。

Frida 為什麼這麼紅呢？

動靜態修改記憶體實現作弊一直是直接需求，本質上 Frida 做的是跟它一樣的事情。原則上是可以用 Frida 把 CheatEngine 等「外掛」做出來的。當然，現在已經不是直接修改記憶體就可以高枕無憂的年代了。建議讀者也不要這樣做，要知道做外掛的行為是違法的，學安全先學法。

在逆向的工作上也是一樣的道理，使用 Frida 可以「看到」平時看不到的東西。出於編譯型語言的特性，機器碼在 CPU 和記憶體執行的過程中，其內部資料的互動和跳躍對使用者來說是看不見的。當然，如果手上有原始程式，甚至哪怕有帶偵錯符號的可執行檔套件，就可以使用 GDB、LLDB 等偵錯器連上去偵錯查看。

那如果沒有，是純黑盒呢？如果仍舊要對 App 進行逆向和動態偵錯，甚至自動化分析以及規模化收集資訊，此時我們需要的是擁有細微性的流程控制和程式級的可訂製系統以及不斷對偵錯進行動態糾正和可程式化偵錯的框架，Frida 做這種工作可以説是遊刃有餘。

另外，Frida 使用的是 Python、JavaScript 等「膠水語言」，這也是它火爆的一個原因：可以迅速地將逆向過程自動化，並整合到現有的架構和系統中去，為發佈「威脅情報」、「資料平台」甚至「AI 風控」等產品打好基礎。

1.4.2 Frida 使用環境架設

Frida 環境的架設其實非常簡單，官網介紹直接使用 pip 安裝 frida-tools 就會自動安裝新版的 Frida 全系列產品，具體如下：

```
┌──(root@vxidr0ysue)-[~/Chap01]
└─# pip install frida-tools
```

```
Collecting frida-tools
...
Successfully installed colorama-0.4.4 frida-14.1.3 frida-tools-9.0.1
  prompt-toolkit-3.0.8 pygments-2.7.3 wcwidth-0.2.5
┌──(root@vxidr0ysue)-[~/Chap01]
└─# frida --version
14.1.3
```

當然，僅僅在電腦上安裝 Frida 是不夠的，還需要在測試機上安裝並執行對應版本的 Server。例如在 Android 中，需要從 Frida 的 GitHub 首頁的 Release 頁面（https://github.com/frida/frida/releases）下載和電腦上版本相同的 frida-server。

這裡需要注意幾點：第一，frida-server 的版本一定要和電腦上的版本一致，比如筆者前面安裝的 Frida 版本為 14.1.3，那麼 frida-server 的版本也必須是 14.1.3，對應的網址是 https://github.com/frida/frida/releases/tag/14.1.3，可以根據自己主機的 Frida 版本修改網址最後的數字；第二，frida-server 的架構需要和測試機的系統以及架構保持一致，比如這裡使用的 Android 測試機 Nexus 5X 是 ARM64 的架構，就需要下載 frida-server 對應的 ARM64 版本。

可以選擇進入測試機的 Shell 執行如下命令查看系統的架構。getprop 命令是 Android 特有的命令，可用於查看各種系統的屬性。

```
bullhead:/ $ getprop ro.product.cpu.abi
arm64-v8a
```

在下載完 frida-server 後，需要在解壓後將 frida-server 透過 ADB 工具推送到 Android 測試機上。在 Android 中，使用 adb push 命令推送檔案到 data 目錄一般需要 Root 許可權，但是這裡有一個例外，即可以儲存到 /data/local/tmp 目錄，所以 frida-server 一般會被存放在測試機的

/data/local/tmp/ 目錄下。在將 frida-server 存放到測試機目錄下後，使用 chmod 命令指定 frida-server 充分的許可權，這樣 frida-server 就可以執行了。

```
┌──(root@vxidr0ysue)-[~/Chap01]
└─# 7z x frida-server-14.1.3-android-arm64.xz
...
Everything is Ok

┌──(root@vxidr0ysue)-[~/Chap01]
└─# ls
frida-server-14.1.3-android-arm64  frida-server-14.1.3-android-arm64.xz

┌──(root@vxidr0ysue)-[~/Chap01]
└─# adb push frida-server-14.1.3-android-arm64 /data/local/tmp/
frida-server-14.1.3-android-arm64: 1 file pushed, 0 skipped. 18.8 MB/s
 (41309856 bytes in 2.094s)

┌──(root@vxidr0ysue)-[~/Chap01]
└─# adb shell

bullhead:/ $ su
bullhead:/ # cd /data/local/tmp
bullhead:/data/local/tmp # chmod 777 frida-server-14.1.3-android-arm64
bullhead:/data/local/tmp # ./frida-server-14.1.3-android-arm64
```

當然，由於 Frida 迭代更新的速度很快，當讀者看到本書的時候，Frida 版本可能已經不是 14 系列了。一方面，這説明 Frida 的活躍度非常高；另一方面，由於 Frida 迭代更新的速度過快，也會帶來一個弊端：Frida 的穩定性並不能得到有效的保證。故筆者在這裡推薦一款 Python 版本管理軟體 pyenv，透過 pyenv 可以安裝和管理不同的 Python 版本，在不同的 Python 版本上可以安裝不同版本的 Frida 環境，而每一個 pyenv 套件管理軟體安裝的 Python 版本都是相互隔離的。換句話說，無論在這個 Python 環境中安裝了多少依賴套件，對於另一個 Python 版本都是不可見的。

需要注意的是，在安裝 pyenv 之前，建議讀者一定要將虛擬機器進行一次快照。快照是為了防止安裝 pyenv 的最後一步依賴時，導致整個系統無法進入桌面環境。筆者安裝 pyenv 的具體過程如下：

```
┌──(root@vxidr0ysue)-[~/Chap01]
└─# apt update
Get:1 http://kali.download/kali kali-rolling InRelease [30.5 kB]
Get:2 http://kali.download/kali kali-rolling/main amd64 Packages [17.3 MB]
Get:3 http://kali.download/kali kali-rolling/non-free amd64 Packages [202
kB]
Get:4 http://kali.download/kali kali-rolling/contrib amd64 Packages [103
kB]
Fetched 17.6 MB in 1min 8s (259 kB/s)
Reading package lists... Done
Building dependency tree
Reading state information... Done

┌──(root@vxidr0ysue)-[~/Chap01]
└─# git clone https://github.com/pyenv/pyenv.git ~/.pyenv
Cloning into '/root/.pyenv'...
...
done.
Resolving deltas: 100% (12507/12507), done.

┌──(root@vxidr0ysue)-[~/Chap01]
└─# echo 'export PYENV_ROOT="$HOME/.pyenv"' >> ~/.bashrc

┌──(root@vxidr0ysue)-[~/Chap01]
└─# echo 'export PATH="$PYENV_ROOT/bin:$PATH"' >> ~/.bashrc
```

```
┌──(root@vxidr0ysue)-[~/Chap01]
└─# echo -e 'if command -v pyenv 1>/dev/null 2>&1; then\n  eval "$(pyenv
init -)"\nfi' >> ~/.bashrc


┌──(root@vxidr0ysue)-[~/Chap01]
└─# exec "$SHELL"


┌──(root@vxidr0ysue)-[~/Chap01]
└─# apt install -y make build-essential libssl-dev zlib1g-dev \
libbz2-dev libreadline-dev libsqlite3-dev wget \
curl llvm libncurses5-dev libncursesw5-dev xz-utils tk-dev libffi-dev
liblzma-dev python-openssl \
g++ libgcc-9-dev gcc-9-base mitmproxy
```

如果安裝後重新啟動能夠正常進入桌面環境，那麼接下來可以方便地使用 pyenv install 命令安裝不同版本的 Python，在安裝完畢後，還需要執行 pyenv local 命令切換到對應版本。例如安裝 Python 3.8.0 命令如下：

```
┌──(root@vxidr0ysue)-[~/Chap01]
└─# pyenv install 3.8.0
┌──(root@vxidr0ysue)-[~/Chap01]
└─# pyenv local 3.8.0
┌──(root@vxidr0ysue)-[~/Chap01]
└─# python -V
Python 3.8.0
```

在安裝一個新的 Python 環境後，就可以順利進行下一步 Frida 的安裝了。在許多 Frida 版本中，筆者推薦相對穩定的 12.8.0 版本。

> ### 🔍注意
>
> 在安裝自訂版本的 Frida 時，需要先使用 pip 安裝特定版本的 Frida，再安裝對應版本的 frida-tools。12.8.0 版本的 Frida 和對應的 frida-tools 版本對應關係如圖 1-26 所示。
>
> | github.com/frida/frida/releases/tag/12.8.0 | | |
> | --- | --- | --- |
> | frida-v12.8.0-node-v79-darwin-x64.tar.gz | frida-tools | 1/6 |
> | frida-v12.8.0-node-v79-linux-ia32.tar.gz | | |
> | frida-v12.8.0-node-v79-linux-x64.tar.gz | | 19.4 MB |
> | frida-v12.8.0-node-v79-win32-ia32.tar.gz | | 16.4 MB |
> | frida-v12.8.0-node-v79-win32-x64.tar.gz | | 16.7 MB |
> | frida32_12.8.0_iphoneos-arm.deb | | 7.51 MB |
> | frida_12.8.0_iphoneos-arm.deb | | 14.5 MB |
> | python-frida-tools_5.3.0-1.ubuntu-bionic_all.deb | | 70.9 KB |
> | python-frida-tools_5.3.0-1.ubuntu-xenial_all.deb | | 70.8 KB |
> | python-frida_12.8.0-1.ubuntu-bionic_amd64.deb | | 13.8 MB |
> | python-frida_12.8.0-1.ubuntu-xenial_amd64.deb | | 22.6 MB |
> | python2-frida-12.8.0-1.fc28.x86_64.rpm | | 22.8 MB |
> | python2-frida-tools-5.3.0-1.fc28.noarch.rpm | | 76.5 KB |
> | python2-prompt-toolkit-1.0.15-1.fc28.noarch.rpm | | 461 KB |
> | python3-frida-12.8.0-1.fc28.x86_64.rpm | | 22.8 MB |

▲ 圖 1-26　12.8.0 版本的 Frida 對應的 Frida-tools 版本

確定 frida-tools 版本後，即可開始安裝特定版本的 Frida。

```
┌──(root@vxidr0ysue)-[~/Chap01]
└─# python -V
Python 3.8.0


┌──(root@vxidr0ysue)-[~/Chap01]
└─# pip install frida==12.8.0
Collecting frida==12.8.0
...
Successfully built frida
Installing collected packages: frida
Successfully installed frida-12.8.0
┌──(root@vxidr0ysue)-[~/Chap01]
└─# pip install frida-tools==5.3.0
Collecting frida-tools==5.3.0
...
```

```
Successfully built frida-tools
Installing collected packages: wcwidth, six,
  pygments, prompt-toolkit, colorama, frida-tools
Successfully installed colorama-0.4.4 frida-tools-5.3.0 prompt-
toolkit-2.0.10
  pygments-2.7.3 six-1.15.0 wcwidth-0.2.5
┌──(root@vxidr0ysue)-[~/Chap01]
└─# frida --version
12.8.0
```

同樣，再設定好對應版本的 frida-server 後，一個全新的 Frida 就可以投入使用了。

1.4.3 Frida 開發環境設定

相信讀者都知道，在撰寫程式時，一個好的 IDE 會使程式設計工作事半功倍，一個基礎的 IDE 一定要有的功能就是程式的智慧提示；同樣，在使用 Frida 撰寫指令稿時，如果有 Frida 的 API 智慧提示是非常方便的，而 Frida 的作者也非常體貼地提供了一個使得 VSCode、Pycharm 這樣的 IDE 支援 Frida 的 API 智慧提示的方式。具體步驟如下：

步驟 01 安裝 node 和 npm 環境，這裡不要使用 Linux 套件管理軟體 APT 直接安裝，APT 安裝的版本太低，這裡使用 Node.js 官方的 GitHub 提供的方法，具體網址為 https://github.com/nodesource/distributions，這裡根據筆者自己的系統選擇 Debian 版本，並且安裝 Node.js v12.x。

```
┌──(root@vxidr0ysue)-[~/Chap01]
└─# curl -sL https://deb.nodesource.com/setup_12.x | bash -

## Installing the NodeSource Node.js 12.x repo...
...
```

```
┌──(root@vxidr0ysue)-[~/Chap01]
└─# apt-get install -y nodejs
...
The following NEW packages will be installed:
  nodejs
...
Processing triggers for man-db (2.9.0-1) ...
┌──(root@vxidr0ysue)-[~/Chap01]
└─# node -v
v12.19.1
┌──(root@vxidr0ysue)-[~/Chap01]
└─# npm -v
6.14.8
```

步驟 02 使用 git 命令下載 frida-agent-example 倉庫並設定。

```
┌──(root@vxidr0ysue)-[~/Chap01]
└─# git clone https://github.com/oleavr/frida-agent-example.git
...
Resolving deltas: 100% (70/70), done.

┌──(root@vxidr0ysue)-[~/Chap01]
└─# cd frida-agent-example/
┌──(root@vxidr0ysue)-[~/Chap01/frida-agent-example/]
└─# npm install
...

added 244 packages from 208 contributors and audited 245 packages in
82.424s
...

found 0 vulnerabilities

┌──(root@vxidr0ysue)-[~/Chap01/frida-agent-example/]
└─#
```

步驟 03 使用 VSCode 等 IDE 編輯器開啟此專案,此時在子目錄下撰寫 JavaScript 指令稿,就會獲得 Frida API 的智慧提示,如圖 1-27 所示。當然,這裡的 VScode 等 IDE 需要讀者自己下載安裝,這裡不再贅述。

▲ 圖 1-27 VSCode 智慧提示

至此,一個完整的 Frida 逆向開發環境就基本完成了。從下一章開始,我們將介紹 Frida 在逆向工作中的使用方法。

1.5 本章小結

本章主要介紹了筆者在 Android 逆向工作中常用的一些基礎環境設定,包括電腦和手機的基礎環境設定,同時介紹了 Frida 工具的安裝與簡單的指令稿撰寫。當然,這一章雖然沒有過多的技術介紹,但卻是整本書的基石。好的作業環境會讓之後的學習節省很多時間,從而大大提高後續學習的效率,希望讀者能夠「一模一樣」地複現環境,從而保障後面的實踐不會因為環境問題而耽誤寶貴的時間。

02

Frida Hook 基礎與快速定位

第 1 章使用 pyenv 這一 Python 版本管理工具成功安裝了特定版本的 Frida，並且仔細介紹了 Frida 開發環境的架設方式。本章將簡介 Frida 在 Android Hook 上的使用方法，並透過一些開放原始碼專案從兩種角度介紹 Frida 的優勢—快速定位關鍵類別和關鍵方法。要注意的是，雖然本章也介紹了部分基礎知識，但如果讀者有一定的 Frida 基礎，會更容易理解。

2.1 Frida 基礎

2.1.1 Frida 基礎介紹

Frida 存在兩種操作模式：第一，透過命令列直接將 JavaScript 指令稿注入處理程序中，對處理程序進行操作，這種模式稱為 CLI（命令列）模式；第二，使用 Python 指令稿間接完成 JavaScript 指令稿的注入工作，這種模式稱為 RPC[1]（Remote Procedure Call，遠端程序呼叫）模式，這種模式雖然加入了 Python 的包裝，但實際對處理程序進行操作的還是 JavaScript 指令稿。因此本章將重點以 CLI 模式講解 Frida 的使用。

Frida 具體操作 App 的方式有兩種：

一種是 spawn（呼叫）模式，簡而言之就是將啟動 App 的權利交由 Frida 來控制。當使用 spawn 模式時，即使目標 App 已經啟動，在使用 Frida 對程式進行注入時，還是會由 Frida 將 App 重新啟動並注入。在命令列模式中，frida 命令加上 -f 參數就會以 spawn 模式操作目標 App。

另一種是 attach（附加）模式，這種模式是建立在目標 App 已經啟動的情況下，Frida 直接利用 ptrace 原理注入程式進而完成 Hook 操作。在 CLI 模式中，如果不增加 -f 參數，則預設透過 attach 模式注入 App。

這裡需要注意的是，正是由於 Frida 在以 attach 模式注入應用時使用 ptrace 原理完成，因此無法在 IDA 正在偵錯目標應用程式時以 attach 模式注入處理程序中，但是如果先用 Frida 注入程式後再使用 IDA 進行偵錯，則完全沒有任何問題，其中詳細原理讀者可自行研究。

由於 Hook 方案是一種在函數真實執行前對函數執行流程進行動態二進位插樁的方式，因此其時機非常重要：一定要在函數執行前對函數進行 Hook，否則如果在 Hook 之前函數已經執行結束並且不再執行，這樣的 Hook 就沒有意義了。由於 App 中某些函數在啟動時預設只執行一次，因此也就出現了 spawn 和 attach 兩種注入方式。對於只有在 App 啟動早些

---

1 RPC 是一種透過網路從遠端電腦程式上請求服務，而不需要了解底層技術的協定。

時候執行或者只執行一次的方法，通常只有透過 spawn 方式在 App 尚未執行之前就對函數進行 Hook，比如 SO 庫的 .init_array 函數、.init_proc 函數等；而對於頻繁執行的函數或者需要對 App 進行特定操作才執行的函數，則可以在觸發函數執行流程之前以 attach 模式對 App 進行注入。圖 2-1 中左側和右側分別是使用兩種模式 Hook 某 App RegisterNatives 函數的列印結果，可以發現當使用 attach 模式對 RegisterNatives 函數進行 Hook 時，沒有任何資訊列印出來。

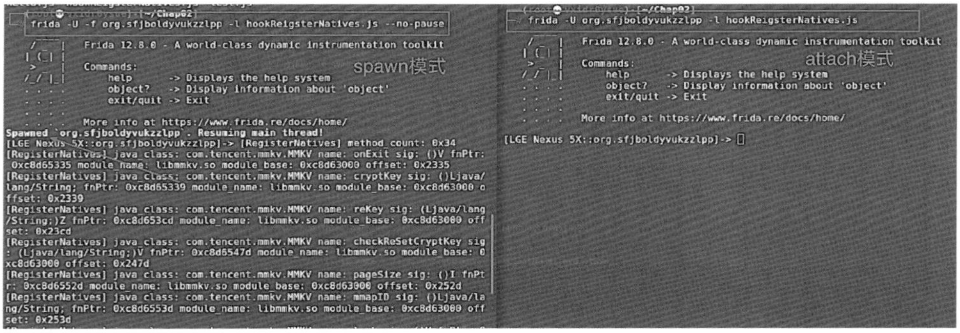

▲ 圖 2-1　spawn 模式與 attach 模式 Hook 結果對比

另外，Frida 通常支援使用兩種模式連接手機：USB 資料線模式和網路模式，當使用 USB 資料線模式連接手機時，手機一定要透過 ADB 協定與電腦相連接，此時在注入 App 時只要加上 -U 參數即可。圖 2-1 就是使用 USB 資料線模式注入應用和 Hook。

當使用網路模式連接手機時，無須保證手機和電腦透過 ADB 協定連接。對應地，frida-server 在執行時期必須使用 -l 參數指定監聽 IP 和通訊埠，主機上的 Frida 則需要透過 -H 參數指定手機的 IP 和通訊埠與手機建立連接。圖 2-2 所示為網路模式下 Frida 注入遠端手機的「設定」應用展示，觀察圖 2-2 可以發現 frida-server 在執行時期透過 -l 參數指定監聽來自任意 IP 8888 通訊埠的連接，而 Frida 則透過 -H 參數指定連接 IP 為 192.168.50.185、通訊埠編號為 8888 的裝置，最終完成對「設定」應用的注入與 Hook 工作。

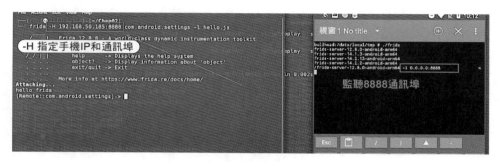

▲ 圖 2-2　在網路模式下使用 Frida

2.1.2　Frida Hook 基礎

相比於 Xposed 使用 Java 程式完成 Hook 模組的撰寫後需要重新啟動才能使得 Hook 程式生效，Frida 更加機動靈活。在每次對目標 App 進行注入時，只需要 frida-server 在測試手機上執行起來，然後使用和 frida-server 版本相同的 Frida 將事先撰寫好的 JavaScript 指令稿注入處理程序即可即時完成對應用的注入和 Hook 工作。除此之外，在注入成功後，哪怕注入指令稿被即時修改，對應的 Hook 效果也能即時生效。那麼 Frida 的 Hook 核心主角—JavaScript 指令稿該如何撰寫呢？本小節將簡介關於 Frida Java 層 Hook 指令稿的語法。

如圖 2-3 所示，以「設定」中「顯示」這個介面為例，Android 版本編號為 8.1.0_r1 的系統中其對應的類別名為 com.android.settings.DisplaySettings。

▲ 圖 2-3「顯示」頁面

在撰寫 Hook 指令稿之前還要介紹的是，當使用者在「顯示」頁面中每次點擊「主動顯示」按鈕時，其對應的函數 int getMetricsCategory() 都會被呼叫。因此，如果想要針對這個函數進行 Hook，其對應的 Frida 指令稿如程式清單 2-1 所示。在 Frida 指令稿成功注入設定應用後，再多次點擊「主動顯示」按鈕，會發現 Frida 的 REPL 介面出現 Hook 日誌資料的列印，其效果如圖 2-4 所示。

◉ 程式清單 2-1　hello.js

```
function hook(){
    Java.perform(function(){
        var settings  = Java.use("com.android.settings.DisplaySettings");
```

```
var getMetricsCategory_func = settings.getMetricsCategory;
getMetricsCategory_func.implementation = function(){
    var result = this.getMetricsCategory() // 執行原函數。
    console.log("getMetricsCategory called",',result =>',result)
    return result
}
})
}
```

```
┌──(root💀wxidl0ysue)-[~/Chap02]
└─ frida -U com.android.settings -l hello.js

    ____
   / _  |   Frida 12.8.0 - A world-class dynamic instrumentation toolkit
  | (_| |
   > _  |   Commands:
  /_/ |_|       help      -> Displays the help system
  . . . .       object?   -> Display information about 'object'
  . . . .       exit/quit -> Exit
  . . . .
  . . . .   More info at https://www.frida.re/docs/home/

[LGE Nexus 5X::com.android.settings]-> getMetricsCategory called ,result => 46
getMetricsCategory called ,result => 46
getMetricsCategory called ,result => 46
getMetricsCategory called ,result => 46
getMetricsCategory called ,result => 46
getMetricsCategory called ,result => 46
getMetricsCategory called ,result => 46
getMetricsCategory called ,result => 46
getMetricsCategory called ,result => 46
[LGE Nexus 5X::com.android.settings]->
```

▲ 圖 2-4 Hook 結果

此時再次觀察程式清單 2-1，會發現圖 2-4 中列印的日誌和指令稿中 console.log() 函數執行的結果相同，説明函數成功被 Hook 了。

這裡還要介紹程式清單 2-1 中幾個比較重要的基礎知識。

第一，所有針對 Java 層函數的 Hook 指令稿必須處於 Java.perform() 的包裝中，Java.perform() 函數的包裝表示將其中的函數注入 Java 執行時期中，那麼如果沒有 Java.perform() 函數的包裹，會發生什麼呢？如圖 2-5 所示，這裡筆者將屬於 Java.perform() 的部分在程式中進行註釋後，再次保存會發現提示錯誤：Current thread is not attached to the Java VM; please move this code inside a Java.perform() callback。

```
[LGE Nexus 5X::com.android.settings] -> Error: Current thread is not attached to the Java VM; please move this code inside a Java.perform() callback
    at frida/node_modules/frida-java-bridge/lib/vm.js:28
    at frida/node_modules/frida-java-bridge/lib/class-factory.js:88
    at hook (/hello.js:3)
    at main (/hello.js:14)
    at frida/runtime/core.js:55
Error: Current thread is not attached to the Java VM; please move this code inside a Java.perform() callback
    at frida/node_modules/frida-java-bridge/lib/vm.js:28
    at frida/node_modules/frida-java-bridge/lib/class-factory.js:88
    at hook (/hello.js:3)
    at main (/hello.js:14)
    at frida/runtime/core.js:55
```

▲ 圖 2-5 沒有 Java.perform() 函數顯示出錯

第二，在使用 Java.use() API 獲取指定類別的 handle 後，這裡以類似於 Java 中呼叫類別靜態方法的方式獲取對應的函數。與 Java 中不同的是，Frida 指令稿中直接以在 "." 連接子後接函數名的方式得到的函數並不一定是我們想要 Hook 的函數。如果函數存在多個多載，此時還需要在函數名後增加 .overload(＜signature＞) 獲取指定函數，比如這裡針對 int getMetricsCategory() 這個無參函數，其對應的 ＜signature＞ 也為空，因此要獲取這個函數，只需要在函數名後接 .overload() 即可。而如果函數存在多個多載，比如 String 字串類別的 subString()，這個用於獲取子字串的函數就存在兩個多載：String substring(int) 和 String substring(int, int)，此時如果只想 Hook substring(int) 函數，在 Frida 中獲取對應函數的 handle，其具體 Hook 程式就必須如程式清單 2-2 中所示的在函數名 substring 之後增加 .overload('int') 關鍵字以獲取特定函數的 handle。

◆ 程式清單 2-2 hookSubString.js

```
function hookSubString(){
  Java.perform(function(){
      var String = Java.use('java.lang.String')
      var subString_int_func = String.substring.overload('int') // 獲取
substring(int)函數的 handle
      subString_int_func.implementation = function(index){
```

```
        var result = this.substring(index)
        console.log("substring called",'index =>',index,',result =>',
result)
        return result
    }
  })
}
```

如果在獲取 subString 函數時不增加 .overload('int')，那麼 Frida 在注入後就會顯示出錯：substring(): has more than one overload, use. overload(＜signature＞)，顯示出錯結果如圖 2-6 所示。

```
[LGE Nexus 5X::com.android.settings]-> hookSubString()
[LGE Nexus 5X::com.android.settings]-> Error: substring(): has more than one overload, use .overload(<signature>) to choose from:
    .overload('int')
    .overload('int', 'int')
    at throwOverloadError (frida/node_modules/frida-java-bridge/lib/class-factory.js:1020)
    at frida/node_modules/frida-java-bridge/lib/class-factory.js:707
    at /hello.js:20
    at frida/node_modules/frida-java-bridge/lib/vm.js:11
    at frida/node_modules/frida-java-bridge/index.js:279
    at hookSubString (/hello.js:21)
    at eval (input:1)
    at eval (native)
    at /hello.js:38
```

▲ 圖 2-6　無 overload 多載顯示出錯

而當按照上述步驟獲取到函數的 handle 後，此時還沒有最終完成對函數的 Hook 效果，真實去完成 Hook 工作的部分實際上是程式清單 2-1 和程式清單 2-2 中的 .implementation 以及在 implementation 後的 JS 函數。在這個 JS 函數中，我們可以執行任意使用者自訂的操作，這裡僅僅是列印一行日誌，並且呼叫原函數獲取了函數的返回值。

由於本書的定位以及篇幅，本小節只簡單介紹了 Frida 函數 Hook 的基礎，且並未展開描述。

2.1.3 Objection 基礎

如果説 Frida 工具提供了各種 API 供使用者自訂使用，在此基礎之上可以實現無數的具體功能，那麼 Objection 可以認為是一個將各種常用的功能整合進工具中並可直接供使用者在命令列中使用的利器，甚至透過 Objection 工具可以在不寫一行程式的前提下完成 App 的逆向分析。

Objection 整合的功能主要支持 Android 和 iOS 兩大行動平台，在對 Android 的支援中，Objection 可以快速地完成諸如記憶體搜索、類別和模組搜索、方法 Hook 以及列印參數、返回值、呼叫堆疊等常用功能，是一個非常方便甚至可以説逆向必備的神器。

Objection 的安裝十分簡單，只需要透過 pip 進行安裝即可，預設安裝 Objection 時會自動安裝新版的 Frida 和 frida-tools。但需要注意的是，如果讀者在安裝特定版本的 Frida 後安裝 Objection，則需要指定 Objection 的版本進行安裝，比如這裡使用 Frida 12.8.0 版本，其對應的 Objection 版本為 1.8.4，因此此時 Objection 的安裝命令如下：

```
# pip install objection==1.8.4
```

當然，如果讀者在未安裝 Frida 的前提下直接使用 pip 命令安裝 Objection，並未下載 Frida，則會自動下載新版的 Frida 和 frida-tools。由於 Frida 不同版本的 API 可能會有一些差異，因此在進行特定版本的 Objection 安裝時還需要注意 Frida、frida-tools 和 Objection 的先後順序（先安裝 Frida 和 frida-tools，再安裝對應版本的 Objection）。

在成功安裝 Objection 後，讓我們來一起了解一下 Objection 的基本使用方式。Objection 的基本命令如圖 2-7 所示。

```
┌──(root㉿kaxxxxysuc)-[~/Chap02]
└─# objection --help
Usage: objection [OPTIONS] COMMAND [ARGS]...

        _     _   _
    ___| |___|_|___|_|___ ___
   | . | . | | -_| --_|  _| . |
   |___|___|_|___|___|_| |___|
       |___|(object)inject(ion)

       Runtime Mobile Exploration
           by: @leonjza from @sensepost

   By default, communications will happen over USB, unless the --network
   option is provided.

Options:
   -N, --network                Connect using a network connection instead of USB.
                                [default: False]

   -h, --host TEXT              [default: 127.0.0.1]
   -p, --port INTEGER           [default: 27042]
   -ah, --api-host TEXT         [default: 127.0.0.1]
   -ap, --api-port INTEGER      [default: 8888]
   -g, --gadget TEXT            Name of the Frida Gadget/Process to connect to.
                                [default: Gadget]

   -S, --serial TEXT            A device serial to connect to.
   -d, --debug                  Enable debug mode with verbose output. (Includes
                                agent source map in stack traces)

   --help                       Show this message and exit.

Commands:
   api           Start the objection API server in headless mode.
   device-type   Get information about an attached device.
   explore       Start the objection exploration REPL.
   patchapk      Patch an APK with the frida-gadget.so.
   patchipa      Patch an IPA with the FridaGadget dylib.
   run           Run a single objection command.
   version       Prints the current version and exists.
┌──(root㉿kaxxxysuc)-[~/Chap02]
└─#
```

▲ 圖 2-7　Objection 命令提示

以「行動 TV」樣本為例，在透過 adb 命令安裝 App 後，首先透過 Jadx 等反編譯工具查看 AndroidManifest.xml 檔案的內容，獲取到套件名為 com.cz.babySister。

在保障對應版本的 frida-server 已經在手機端啟動後，根據圖 2-7 中的命令提示與上面獲取到的 App 套件名，最終得到 Objection 注入「行動

TV」處理程序命令如下：

```
# objection -g com.cz.babySister explore
```

在成功執行注入命令後，如圖 2-8 所示是 Objection 在成功注入處理程序後的 REPL 介面。在這個介面中，我們可以透過 Objection 相關命令對處理程序進行 Hook 等操作。

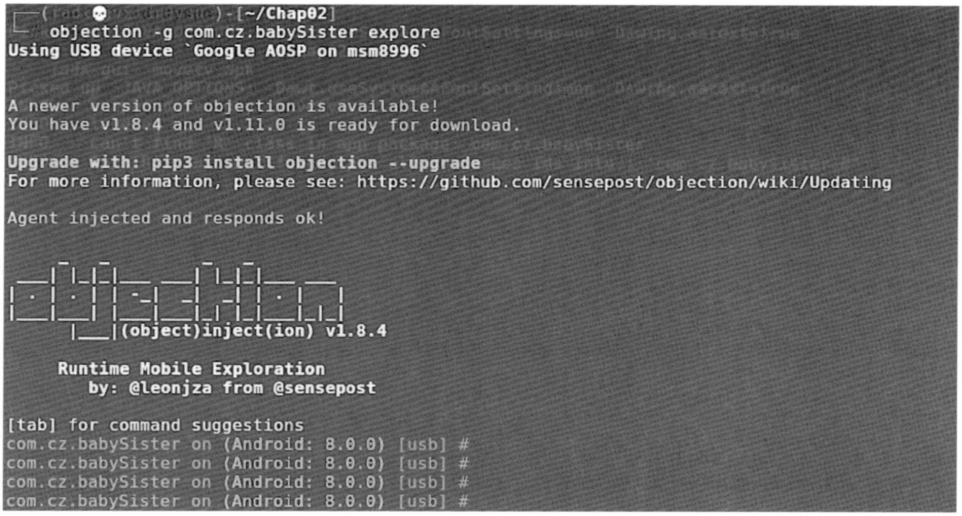

▲ 圖 2-8　Objection REPL 介面

接下來正式介紹 Objection REPL 介面中支援的命令。

（1）記憶體列舉相關命令
Objection 可以快速便捷地列印出記憶體中的各種已載入類別的相關資訊，這對快速定位 App 中的關鍵類別有著關鍵性作用，這裡先介紹幾個常用命令。

① 列舉處理程序記憶體中已載入的類別，其命令格式如下：

```
# android hooking list classes
```

這裡需要注意的是，這裡列出的類別都是處理程序已經載入過的類別，如果處理程序還未載入目標類別，對應類別名是無法被列出的。如圖 2-9 所示是「行動 TV」這個樣本在登入頁面時記憶體中已經載入的類別。

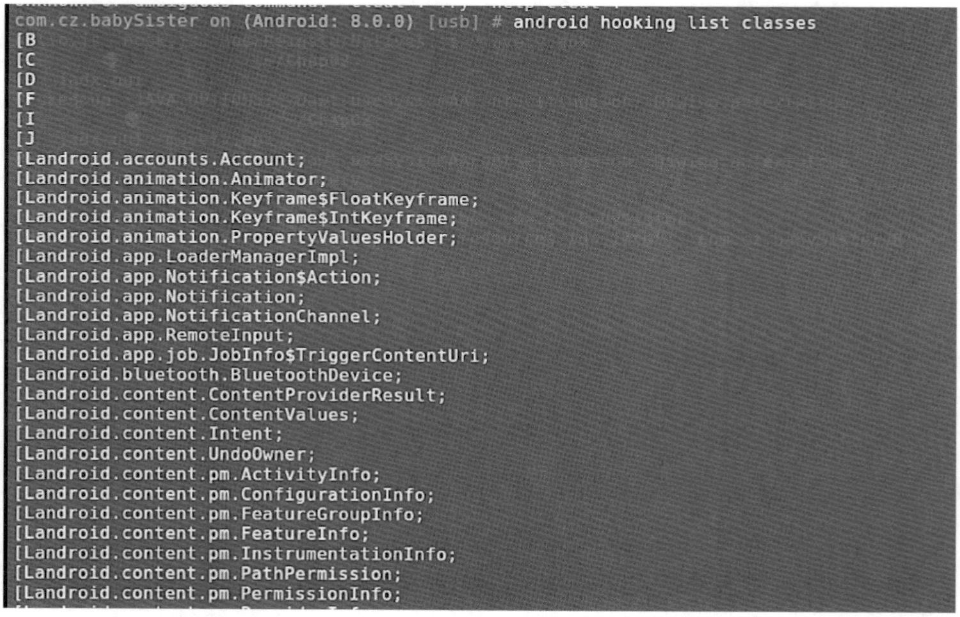

▲ 圖 2-9 列舉處理程序已經載入的類別

② 列舉記憶體已經載入的類別中包含特定字串的類別並列出，其命令格式如下：

```
# android hooking  search  classes  <pattern>
```

這裡的 pattern 可以是任意字串，如圖 2-10 所示是樣本中包含 com.
cz.babySister 字串的類別名。

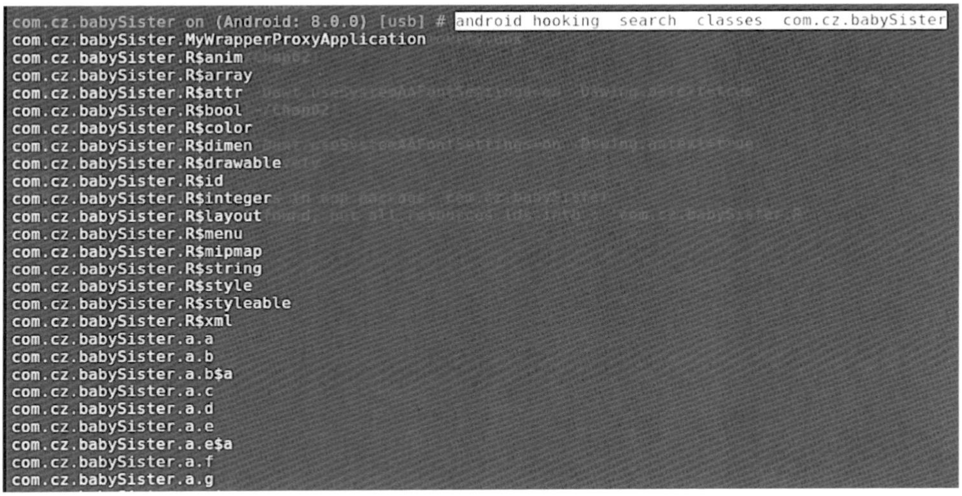

▲ 圖 2-10 搜索包含特定字串的類別

③ 獲取指定類別中所有非建構函數的方法簽名，其命令格式如下：

```
# android hooking list class_methods <class_name>
```

注意，上述命令中，class_name 是包含套件名的完整類別名，比如這裡
要列印 Loading 類別的所有函數，則必須在 Loading 類別名前加上其完
整的套件名 com.cz.babySister.view，並透過 "." 連接子連接。最終列印
Loading 類別所包含函數的效果如圖 2-11 所示。

```
com.cz.babySister on (Android: 8.0.0) [usb] # android hooking list class_methods com.cz.babySister.view.Loading
private void com.cz.babySister.view.Loading.a(android.util.AttributeSet,int,int)
private void com.cz.babySister.view.Loading.a(int)
protected boolean com.cz.babySister.view.Loading.verifyDrawable(android.graphics.drawable.Drawable)
protected void com.cz.babySister.view.Loading.onAttachedToWindow()
protected void com.cz.babySister.view.Loading.onDetachedFromWindow()
protected void com.cz.babySister.view.Loading.onDraw(android.graphics.Canvas)
protected void com.cz.babySister.view.Loading.onMeasure(int,int)
protected void com.cz.babySister.view.Loading.onSizeChanged(int,int,int,int)
protected void com.cz.babySister.view.Loading.onVisibilityChanged(android.view.View,int)
protected void com.cz.babySister.view.Loading.onWindowVisibilityChanged(int)
public float com.cz.babySister.view.Loading.getBackgroundLineSize()
public float com.cz.babySister.view.Loading.getForegroundLineSize()
public float com.cz.babySister.view.Loading.getProgress()
public int com.cz.babySister.view.Loading.getBackgroundColor()
public int[] com.cz.babySister.view.Loading.getForegroundColor()
public void com.cz.babySister.view.Loading.a()
public void com.cz.babySister.view.Loading.setAutoRun(boolean)
public void com.cz.babySister.view.Loading.setBackgroundColor(int)
public void com.cz.babySister.view.Loading.setBackgroundLineSize(float)
public void com.cz.babySister.view.Loading.setForegroundColor(int)
public void com.cz.babySister.view.Loading.setForegroundColor(int[])
public void com.cz.babySister.view.Loading.setForegroundLineSize(float)
public void com.cz.babySister.view.Loading.setLineStyle(int)
public void com.cz.babySister.view.Loading.setProgress(float)

Found 24 method(s)
com.cz.babySister on (Android: 8.0.0) [usb] #
```

▲ 圖 2-11 列印類別中所有方法

（2）Hook 相關命令

作為 Frida 的核心功能，Hook 功能總是繞不過的。同樣，Objection 作為 Frida 優秀的協力廠商工具，也提供了很多激動人心的 Hook 命令。事實上，Objection 在這方面表現得確實令人驚豔。

① Objection 支援 Hook 類別中全部非建構函數的方法，其命令格式如下：

```
android hooking watch class <class_name>
```

與列印特定類別中所有方法的命令相同，這裡 class_name 必須是完整的類別名，同時需要注意 Objection 預設 Hook 類別中的全部函數並不包括類別的建構函數。這裡依舊 Hook Loading 類別的全部方法，其效果如圖 2-12 所示。

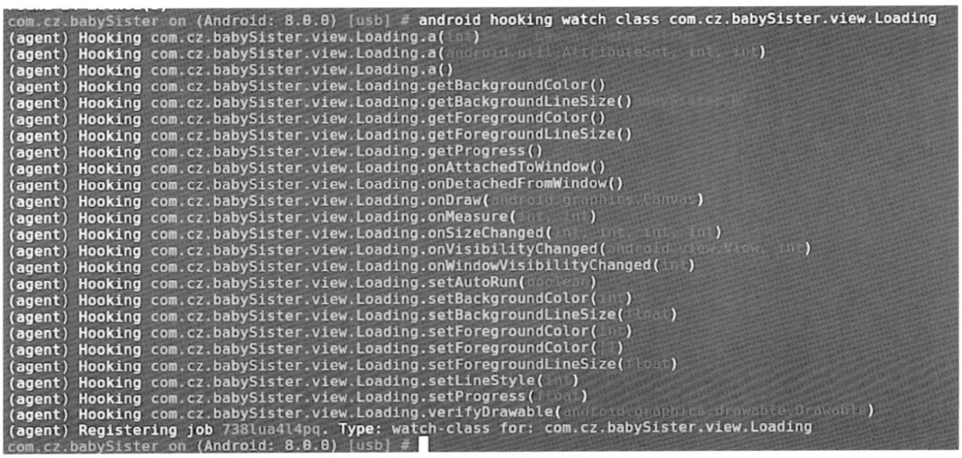

▲ 圖 2-12　Hook 類別中所有方法

② Objection 同樣支援 Hook 指定函數，其命令格式如下：

```
android hooking watch class_method <classMethod> <overload> <option>
```

這裡的 classMethod 與 Hook 類別中全部函數的命令相同，classMethod 必須是完整的類別名加上函數名，並以 "." 連接子連接。而 option 格式支援 3 個參數，其中 --dump-args 參數在被 Hook 函數執行時會列印其參數內容，若加上 --dump-return 參數，則會列印函數返回值，加上 --dump-backtrace 參數列印函式呼叫堆疊，同時這 3 個參數可以組合使用。

與 Frida 不同的是，在 Hook 函數時如果不指定其參數，即這裡的 overload 格式的參數，那麼預設 Hook 所有名稱相同的函數，比如這裡 Hook Loading 類別的 setForegroundColor 方法，觀察圖 2-12 會發現該函數存在兩個多載，此時若不指定參數類型，則其 Hook 效果如圖 2-13 所示。

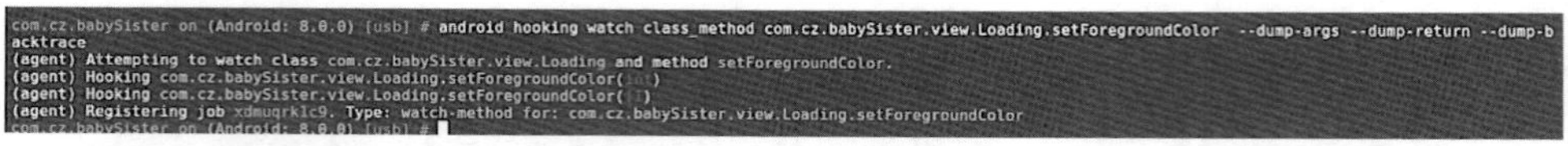

▲ 圖 2-13 Hook setForegroundColor 方法的效果（不指定參數）

如果想要 Hook 指定參數的特定方法，還需要加上函數的參數類型並以雙引號包含，比如這裡想要 Hook 參數類型為 int 的陣列的 setForegroundColor() 方法，那麼最終 Hook 的命令如下：

```
# android hooking watch class_method com.cz.babySister.view.Loading.
setForegroundColor "[I" --dump-args --dump-return --dump-backtrace
```

最終 Hook 效果如圖 2-14 所示。

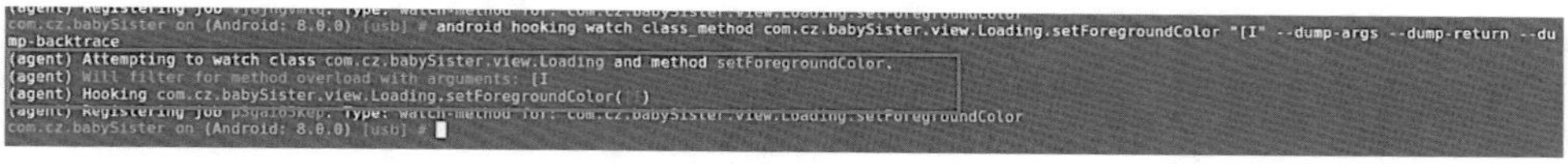

▲ 圖 2-14 Hook setForegroundColor([I]) 方法的效果（指定參數）

Objection 還有很多這裡未介紹的命令，比如 jobs 命令用於 Hook 任務管理，android heap 命令用於操作記憶體中類別的實例等。限於篇幅，這裡僅介紹筆者認為重要的 Objection 命令，直接在 Objection REPL 介面中簡單地透過按空白鍵查看其支援的命令列表。

2.2 Hook 快速定位方案

在逆向分析的過程中，筆者一直推崇「Hook—主動呼叫—RPC」的三段式理論。展開來説，在協定分析的第一步，首先透過 Hook 的方式確定關鍵業務邏輯位置，然後透過主動呼叫實現關鍵業務邏輯的呼叫，最後透過 RPC 遠端程序呼叫的方式進行關鍵業務邏輯的批次呼叫，以期達到後續利用的目的。在 Hook 的過程中，如何在大量的程式中快速定位關鍵業務邏輯的位置，減少逆向人員的工作量是一大重點，Frida 的出現正是將Hook 工作成功地從 Xposed 模組每次編譯都需重新啟動的循環中解放出來的契機，其即時生效的特點大大減少了逆向人員的時間成本，加快了逆向的進度。當然，Frida 在快速逆向中的作用不止於此，本節將介紹基於 Frida 的兩種更加快速的關鍵邏輯定位方式。

2.2.1 基於 Trace 列舉的關鍵類別定位方式

相信有一定基礎的讀者都用過 Android Studio 中附帶的 DDMS 工具，相較於搜索字串這種大海撈針的方法，DDMS 中的 Method Trace 功能能夠讓逆向人員快速得到一段時間內目標 App 執行過的函數記錄，進而快速定位關鍵業務邏輯函數，而實現 DDMS 這項功能的就是筆者在這一小節中要介紹的函數 Trace。

DDMS 是 Google 官方提供的工具，其本意是幫助開發者分析和測試 App 中方法的速度與性能，因此其本身要求 App 的 debuggable 屬性為 true，而這是實際逆向分析中幾乎不可能遇到的應用；此外，由於 DDMS 的 Trace 功能會記錄所有的函數（包括 App 中的函數和系統函數），這樣的操作十分佔用系統性能，其具體效果總是差強人意，正因如此，Trace 方式一直不溫不火，但是 Frida 的出世為函數的 Trace 提供了另一條出路，Frida 不僅無須應用處於 debuggable 狀態，而且支援 Trace 指定類別中的函數，支援 Trace 特定類別中的所有函數，這樣的效果大大縮小了函數 Trace 的範圍，對系統性能的要求有了極大的改善。在這一小節中，我們並不直接撰寫 Frida 指令稿來進行函數的 Trace，而是透過介紹一些基於 Frida 封裝的可用於進行 Trace 定位的工具來介紹基於 Trace 列舉的關鍵類別定位方式所帶來的優勢。

首先不得不提的是前文介紹過的 Objection 工具在 Trace 中的作用。

在 2.1.3 節 中 介 紹 Objection 的 常 用 Hook 命 令 時， 曾 經 介 紹 過 Objection 支持 Hook 一個類別中所有函數的功能，實際上筆者在使用 Objection 的過程中發現 Objection 還支持透過 -c 參數對指定檔案中的所有命令進行執行的功能，如圖 2-15 所示。

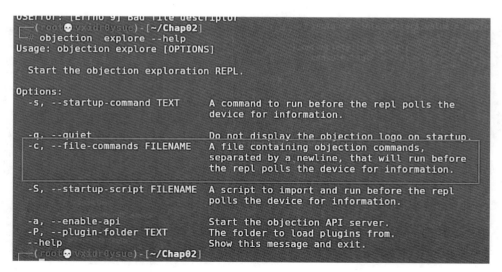

▲ 圖 2-15 Objection Trace 功能

同樣以「行動 TV」樣本為例，其套件名為 com.cz.babySister。在手機上
啟動 frida-server 和樣本應用後，使用 Objection 對樣本進行注入，並使
用如下命令搜索包含套件名的類別：

```
# android hooking search classes com.cz.babySister
```

在獲取到如圖 2-16 所示的所有與套件名相關的類別後，便可以將搜索的
類別保存為檔案，並在每一行行首加上 android hooking watch class 字
串，以使得每一行組成一筆對 Java 類別進行 Hook 的命令，最終組成的
Trace 檔案的部分內容如圖 2-17 所示。當然，這裡在行首增加字串的方
式有很多種，這裡透過 VS Code 編輯器的垂直選擇功能完成內容的補全。

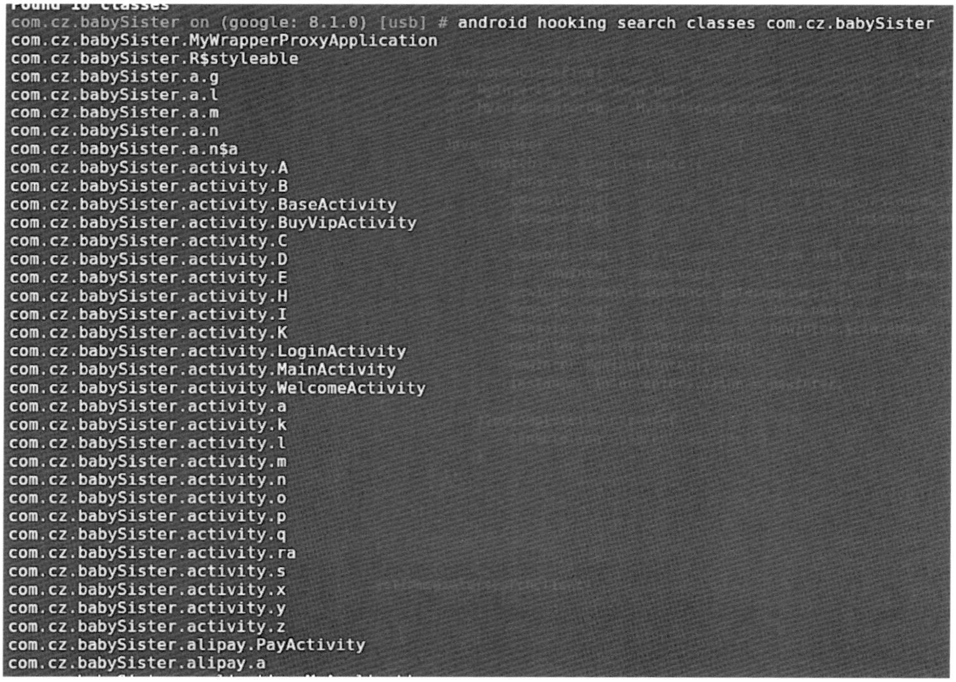

▲ 圖 2-16 套件名相關類別

```
more: cannot open hook: No such file or directory
  (root⦿ vxidr8ysue)-[~/Chap02]
    more hook.txt
android hooking watch class com.cz.babySister.MyWrapperProxyApplication
android hooking watch class com.cz.babySister.R$styleable
android hooking watch class com.cz.babySister.a.g
android hooking watch class com.cz.babySister.a.l
android hooking watch class com.cz.babySister.a.m
android hooking watch class com.cz.babySister.a.n
android hooking watch class com.cz.babySister.a.n$a
android hooking watch class com.cz.babySister.activity.A
android hooking watch class com.cz.babySister.activity.B
android hooking watch class com.cz.babySister.activity.BaseActivity
android hooking watch class com.cz.babySister.activity.BuyVipActivity
android hooking watch class com.cz.babySister.activity.C
android hooking watch class com.cz.babySister.activity.D
android hooking watch class com.cz.babySister.activity.E
android hooking watch class com.cz.babySister.activity.H
android hooking watch class com.cz.babySister.activity.I
android hooking watch class com.cz.babySister.activity.K
android hooking watch class com.cz.babySister.activity.LoginActivity
android hooking watch class com.cz.babySister.activity.MainActivity
android hooking watch class com.cz.babySister.activity.WelcomeActivity
android hooking watch class com.cz.babySister.activity.a
android hooking watch class com.cz.babySister.activity.k
android hooking watch class com.cz.babySister.activity.l
android hooking watch class com.cz.babySister.activity.m
android hooking watch class com.cz.babySister.activity.n
android hooking watch class com.cz.babySister.activity.o
android hooking watch class com.cz.babySister.activity.p
android hooking watch class com.cz.babySister.activity.q
android hooking watch class com.cz.babySister.activity.ra
android hooking watch class com.cz.babySister.activity.s
android hooking watch class com.cz.babySister.activity.x
android hooking watch class com.cz.babySister.activity.y
android hooking watch class com.cz.babySister.activity.z
android hooking watch class com.cz.babySister.alipay.PayActivity
android hooking watch class com.cz.babySister.alipay.a
```

▲ 圖 2-17 檔案 Hook 命令

在得到包含 Hook 命令的檔案後，便可以重新透過 Objection 對應用進行
注入，完成對包含套件名所有類別的 Hook 工作，最終效果如圖 2-18 所
示。而如果此時再去觸發我們所關心的業務邏輯，便能夠快速篩選出關
鍵業務邏輯所在類別的範圍，從而完成關鍵類別的定位工作。

當然，這裡需要注意的是，查看 Objection 1.8.4 版本的原始程式會發現
其 search class 搜索類別的命令是透過先獲取所有已載入類別後再篩選的
方式來得到最終結果的，其獲取所有類別的實現如程式清單 2-3 所示。

▲ 圖 2-18　Objection 批次 Hook

⬥ **程式清單 2-3　getClasses**

```
export const getClasses = (): Promise<string[]> => {
  return wrapJavaPerform(() => {
    return Java.enumerateLoadedClassesSync();
  });
};
```

忽略程式清單 2-3 中外部封裝的部分，會發現其實本質上是使用 Frida 的
API：enumerateLoadedClassesSync() 函數完成類別的獲取。無論是從
API 名稱還是官方的釋義中都能夠發現這個 API 只是列出記憶體中已經載
入的類別，而非應用中所有的類別。因此，若想獲得盡可能完整的類別

清單，需要盡可能多地使用應用後再執行命令。如圖 2-19 所示是筆者在
登入帳戶前後獲取到的套件名相關的類別列表。

```
[tab] for command-suggestions
com.cz.babySister on (google: 8.1.0) [usb] # android hooking search classes com.cz.babySister
com.cz.babySister.MyWrapperProxyApplication
com.cz.babySister.alipay.a
com.cz.babySister.application.MyApplication
com.cz.babySister.application.a
com.cz.babySister.c.a
com.cz.babySister.service.BootReceiver
com.cz.babySister.service.NetService                          第一次
com.cz.babySister.service.TestService
com.cz.babySister.service.b
com.cz.babySister.utils.ParseJson

Found 10 classes
com.cz.babySister on (google: 8.1.0) [usb] # android hooking search classes com.cz.babySister
com.cz.babySister.MyWrapperProxyApplication
com.cz.babySister.R$styleable
com.cz.babySister.a.g
com.cz.babySister.a.g$a
com.cz.babySister.a.l
com.cz.babySister.a.m
com.cz.babySister.a.n
com.cz.babySister.a.n$a
com.cz.babySister.activity.A
com.cz.babySister.activity.B
com.cz.babySister.activity.BaseActivity
com.cz.babySister.activity.BuyVipActivity
com.cz.babySister.activity.C
com.cz.babySister.activity.E                                  第二次
com.cz.babySister.activity.H
com.cz.babySister.activity.I
com.cz.babySister.activity.K
com.cz.babySister.activity.MainActivity
com.cz.babySister.activity.a
com.cz.babySister.activity.s
com.cz.babySister.activity.x
com.cz.babySister.activity.y
com.cz.babySister.activity.z
com.cz.babySister.alipay.PayActivity
com.cz.babySister.alipay.a
com.cz.babySister.application.MyApplication
com.cz.babySister.application.a
com.cz.babySister.b.a
com.cz.babySister.c.a
com.cz.babySister.fragment.CctvFragment
```

▲ 圖 2-19　搜索相關類別對比

如果讀者開啟 Objection 對應原始程式，從 -c 參數解析處進行分析會發
現，該參數只是用於逐一執行檔案中的程式而已，真正用於 Trace 的程式
不過是透過 getDeclaredMethods() 反射相關函數獲取特定類別中所有函
數，並對獲得的函數一一 Hook 而已，其關鍵 Hook 核心程式如程式清單
2-4 所示。

⚫ 程式清單 2-4 Trace 核心程式

```
export const watchClass = (clazz: string): Promise<void> => {
  return wrapJavaPerform(() => {
    const clazzInstance: JavaClass = Java.use(clazz);
    // 獲取指定類別中的所有函數
    const uniqueMethods: string[] = clazzInstance.class.
getDeclaredMethods().map((method) => {
        // ...
    });

    // start a new job container
    ...

    uniqueMethods.forEach((method) => {
      clazzInstance[method].overloads.forEach((m: any) => {

        // get the argument types for this overload
        const calleeArgTypes: string[] = m.argumentTypes.map((arg) =>
arg.className);
        // ...
        // replace the implementation of this method
        // tslint:disable-next-line:only-arrow-functions
        // Hook函數程式
        m.implementation = function() {
          send(
            c.blackBright(`[${job.identifier}] `) +
            `Called ${c.green(clazz)}.${c.greenBright(m.methodName)}(${c.
red(calleeArgTypes.join(", "))})`,
          );

          // actually run the intended method
          return m.apply(this, arguments);
        };
```

```
            // record this implementation override for the job
            ...
          });
        });

        // record the job
        ...
      });
    };
```

另外還需要注意的是，如果 App 本身是保護的應用，在使用 Frida 對應用進行測試時，要儘量選擇 attach 模式進行應用相關類別的 Trace/Hook，否則在應用還沒啟動時，App 真實的類別仍未被殼從記憶體中釋放並載入，此時去 Hook 相關類別會報如圖 2-20 所示的 ClassNotFoundException 異常錯誤。這是由於保護 App 的 ClassLoader 在執行時期的切換問題所導致的，由於 ClassLoader 的原理不屬於本章說明的範圍，這裡不再展開描述，相信有一定基礎的讀者都明白其中的緣由。

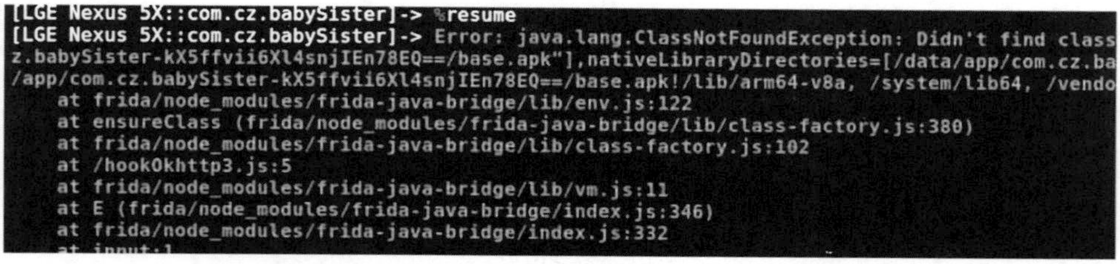

```
[LGE Nexus 5X::com.cz.babySister]-> %resume
[LGE Nexus 5X::com.cz.babySister]-> Error: java.lang.ClassNotFoundException: Didn't find class
z.babySister-kX5ffvii6Xl4snjIEn78EQ==/base.apk"],nativeLibraryDirectories=[/data/app/com.cz.ba
/app/com.cz.babySister-kX5ffvii6Xl4snjIEn78EQ==/base.apk!/lib/arm64-v8a, /system/lib64, /vendo
    at frida/node_modules/frida-java-bridge/lib/env.js:122
    at ensureClass (frida/node_modules/frida-java-bridge/lib/class-factory.js:380)
    at frida/node_modules/frida-java-bridge/lib/class-factory.js:102
    at /hookOkhttp3.js:5
    at frida/node_modules/frida-java-bridge/lib/vm.js:11
    at E (frida/node_modules/frida-java-bridge/index.js:346)
    at frida/node_modules/frida-java-bridge/index.js:332
```

▲ 圖 2-20 ClassNotFoundException 異常錯誤

Objection 的 Trace 功能就暫且介紹到這裡。接下來介紹一款基於 Frida 開發的專門用於 Trace 的工具——ZenTracer，其專案的位址為 https://github.com/hluwa/ZenTracer。

由於 ZenTracer 是一款基於 Frida 和 PyQt5 的工具，因此在執行前先要透過 pip 安裝 PyQt5 和 Frida 的依賴套件，最終成功啟動 ZenTracer 後，其介面如圖 2-21 所示。

▲ 圖 2-21 ZenTracer 主介面

要使用 ZenTracer 首先要點擊圖 2-21 上的 Action 選單，選擇 Match RegEx 或者 Black RegEx 完成類別的過濾工作。顧名思義，Match RegEx 就是 Hook 指定的與輸入正則匹配的類別，而 Black RegEx 是指不 Hook 指定的類別。這裡以 Match RegEx 功能為例，在輸入想要 Hook 的類別前加上 M: 就可以完成對包含指定 pattern 的類別的 Hook 工作，比如這裡想要 Hook 所有包含 com.cz.babySister 的類別，其最終輸入如圖 2-22 所示。若使用 Black RegEx 功能，則只需將 M: 替換為 B: 即可指定不 Hook 對應類別。

▲ 圖 2-22 Hook 包含 com.cz.babySister 的類別

別在確定匹配規則後，再次點擊 Action 選單下的 Start 選項，ZenTracer
就會完成對當前前臺 App（手機頁面上正在顯示的應用）中所有符合匹
配規則的目標類別的 Hook 工作，最終 Trace 效果如圖 2-23 所示。

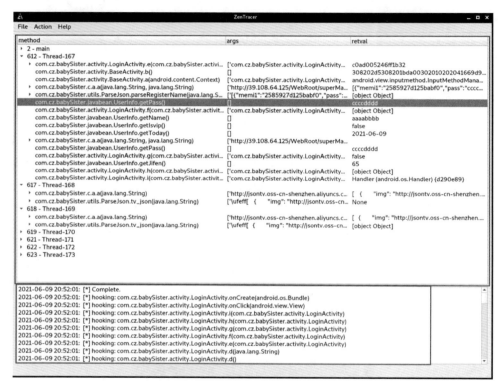

▲ 圖 2-23　ZenTracer Hook 結果

觀察圖 2-23，會發現圖片下方是 Hook 的類別的相關資訊，而圖片上方
顯示的是 Hook 後執行的函數記錄與其參數和返回值資訊。為了更好地
觀察結果與後續分析，ZenTracer 還友善地提供了將 Hook 結果匯出為
JSON 格式檔案的功能，只需要依次點擊 File → Export JSON 即可完成檔
案的匯出。

此時再次查看 ZenTracer Trace 相關程式，會發現其 Trace 關鍵程式同樣和 Objection 類似，只是將遍歷得到的目標類別直接傳遞給對應的 Hook 程式而已。相比於 Objection 需要先導出命令到檔案再 Trace 的方式，ZenTracer 更加方便快捷，其關鍵程式如程式清單 2-5 所示。

● 程式清單 2-5 ZenTracer Trace 關鍵程式

```
// 遍歷記憶體中已經載入的類別
別Java.enumerateLoadedClasses({
    onMatch: function (aClass) {
        for (var index in matchRegEx) {
            // 判斷是否與指定類別匹配
            if (match(matchRegEx[index], aClass)) {
                var is_black = false;
                for (var i in blackRegEx) {
                    // 過濾黑名單中的類別
                    if (match(blackRegEx[i], aClass)) {
                        is_black = true;
                        log(aClass + "' black by '" + blackRegEx[i] + "'");
                        break;
                    }
                }
                if (is_black) {
                    break;
                }
                log(aClass + "' match by '" + matchRegEx[index] + "'");
                // 對目標類別進行Hook的程式，具體程式這裡限於篇幅不再列出
                traceClass(aClass);
            }
        }

    },
    onComplete: function () {
```

```
        log("Complete.");
    }
});
```

相比於 DDMS，Objection 和 ZenTracer Trace 函數的範圍更加靈活，能
夠依據使用者的想法對執行函數進行追蹤，同時 ZenTracer 還支援對函
數參數和返回值的列印工作，這大大縮減了從 Trace 結果中獲取關鍵函
數的時間成本，為逆向工作提供了更多的便利；另一方面，基於 Frida
的 Objection 和 ZenTracer 無須樣本程式處於 debuggable 狀態，減少了
對應用程式另外操作的工作成本。當然，由於 Frida 本身不太穩定以及
Trace 本身對程式的侵入成本，被測試的程式肯定是會經常崩潰的，此
時可以透過切換 Android 和 Frida 版本或者換更高性能的手機的方式來
減少崩潰的頻率。當然，即使按照上述方式提高了穩定性，Frida 還是存
在崩潰的可能性，但相信用過的讀者都會覺得瑕不掩瑜，基於 Frida 的
Objection 以及 ZenTracer 對逆向工作效率的提高都是呈指數級的。

2.2.2 基於記憶體列舉的關鍵類別定位方式

實際上，基於記憶體列舉的關鍵類別定位在 Xposed 時代就出現了：當逆
向分析人員透過分析發現某些類別可能是關鍵類別時，可以透過對關鍵
類別進行 Hook 去驗證分析的結果，只是 Xposed 的 Hook 每次都需要對
手機進行重新啟動以生效，而 Frida 則能夠更加有效率地去 Hook 驗證分
析結果。

當然，筆者認為相比于 Xposed 而言，Frida 的進步不僅僅是提高了 Hook
的效率，Frida 還支援對處理程序記憶體的漫遊功能，能夠透過 Java.
choose() 這個 API 在目標處理程序的 Java 堆中尋找和修改已存在的 Java
物件實例，同時還能修改物件中屬性的值，這樣的功能使得逆向人員從

只能單純地透過 Hook 去獲取和修改物件值的侷限中釋放出來，分析人員不僅能夠對未執行的函數設定 Hook，同時還能夠對已經建立的實例進行操作，這大大拓寬了逆向工作的思路。

接下來以經典的針對 OkHttp 框架的 Hook 抓取封包問題為例進行介紹。

相信對 OkHttp 原理有一定了解的讀者都知道 OkHttp 的核心其實是攔截器。簡單來説，在 OkHttp 中一個完整的網路請求會被拆分成幾個步驟，每個步驟都透過攔截器來完成，可以説透過攔截器就能夠完整地得到每次發送封包和收封包的資料。在筆者對 OkHttp3 的原始程式分析過程中發現，OkHttp 中的攔截器由 okhttp3.OkHttpClient 類別中的 List 成員 _interceptors 陣列管理，這個陣列中包含著對應 Client 中的所有攔截器。那麼是不是意味著我們自己寫一個簡單的只是列印日誌的 LogInterceptor 並增加到 _interceptors 陣列中就可以完成對所有資料封包的抓取呢？答案是肯定的。

筆者在研究過程中透過 Java.choose() 這個 API 從記憶體中搜刮 okhttp3.OkHttpClient 的實例物件，並修改原物件的 _interceptors 陣列內容，最終使得這個攔截器的 List 陣列中包含我們自訂的攔截器 LogInterceptor，從而完成資料的抓取封包，其效果如圖 2-24 所示。

讓我們回過頭來觀察實現修改記憶體中 okhttp3.OkHttpClient 物件的程式內容。觀察程式清單 2-6，發現除了主要的 Java.choose() API 的使用外，還有一些值得關注的地方。

首先，在程式清單 2-6 中，Java.openClassFile() 這個 API 用於開啟自訂的 DEX 檔案，myok2curl.dex 和 okhttplogging.dex 分別用於完成 log 日誌的列印工作和具體攔截器的實現，在 Frida 執行指令稿前，兩個 DEX 檔案已經事先放置於 /data/local/tmp 目錄下並指定其執行許可權。再加上

load() 函數的使用，將開啟的 DEX 檔案載入進記憶體後，便可以在指令
稿中載入原本 App 並不存在的類別和函數。

```
e : 高二 , school_year : 2009 }]}]}
06-10 20:03:28.629 23646 24632 E okhttpGET: <-- END HTTP (1195-byte body)
06-10 20:03:28.632 23646 24859 E okhttpGET: <-- 200 https://xesapi.speiyou.cn/xes/homepage/v1/bro
951431 (180ms, unknown-length body:)
06-10 20:03:28.632 23646 24859 E okhttpGET: server: Tengine
06-10 20:03:28.632 23646 24859 E okhttpGET: content-type: application/json;charset=UTF-8
06-10 20:03:28.632 23646 24859 E okhttpGET: date: Thu, 10 Jun 2021 12:03:28 GMT
06-10 20:03:28.632 23646 24859 E okhttpGET: vary: Accept-Encoding
06-10 20:03:28.632 23646 24859 E okhttpGET: vary: Origin
06-10 20:03:28.632 23646 24859 E okhttpGET: vary: Access-Control-Request-Method
06-10 20:03:28.632 23646 24859 E okhttpGET: vary: Access-Control-Request-Headers
06-10 20:03:28.633 23646 24859 E okhttpGET: x-guider: 6fe4f196c8ee30f55416e3946d07e415
06-10 20:03:28.633 23646 24859 E okhttpGET: currentservertime: 1623326608231
06-10 20:03:28.633 23646 24859 E okhttpGET: x-request-id: dayu_c1e2f9e5e1821dfad9d9b9e1fc87e240
06-10 20:03:28.633 23646 24859 E okhttpGET: via: cache39.l2ea120-8[46,0], vcache10.cn3596[55,0]
06-10 20:03:28.633 23646 24859 E okhttpGET: timing-allow-origin: *
06-10 20:03:28.633 23646 24859 E okhttpGET: eagleid: 755bb11e16233266082085899e
06-10 20:03:28.633 23646 24859 E okhttpGET:
06-10 20:03:28.633 23646 24859 E okhttpGET: {"status":0,"code":0,"msg":"","data":{"liveTitle":"",
eListUrl":""}}
06-10 20:03:28.633 23646 24859 E okhttpGET: <-- END HTTP (135-byte body)
06-10 20:03:28.635 23646 24858 E okhttpGET: <-- 200 https://xesapi.speiyou.cn/v1/grid/list/simple
06-10 20:03:28.635 23646 24858 E okhttpGET: server: Tengine
06-10 20:03:28.635 23646 24858 E okhttpGET: content-type: application/json;charset=UTF-8
06-10 20:03:28.635 23646 24858 E okhttpGET: date: Thu, 10 Jun 2021 12:03:28 GMT
06-10 20:03:28.635 23646 24858 E okhttpGET: vary: Accept-Encoding
06-10 20:03:28.635 23646 24858 E okhttpGET: vary: Origin
06-10 20:03:28.635 23646 24858 E okhttpGET: vary: Access-Control-Request-Method
06-10 20:03:28.635 23646 24858 E okhttpGET: vary: Access-Control-Request-Headers
06-10 20:03:28.635 23646 24858 E okhttpGET: x-guider: 488445969932701f432e845a46e1cf56
06-10 20:03:28.636 23646 24858 E okhttpGET: currentservertime: 1623326608234
06-10 20:03:28.636 23646 24858 E okhttpGET: x-request-id: dayu_9aef34ade38fdb02063c56e207b1decl
06-10 20:03:28.636 23646 24858 E okhttpGET: via: cache55.l2ea120-8[44,0], vcache10.cn3596[56,0]
06-10 20:03:28.636 23646 24858 E okhttpGET: timing-allow-origin: *
06-10 20:03:28.636 23646 24858 E okhttpGET: eagleid: 755bb11e16233266082085900e
06-10 20:03:28.636 23646 24858 E okhttpGET:
06-10 20:03:28.636 23646 24858 E okhttpGET: {"status":0,"code":0,"msg":"","data":[{"gt_name":"幼儿
2025"},{"grid_id":"-9","grid_name":"托班","school_year":"2024"},{"grid_id":"-8","grid_name":"小班"
year":"2022"},{"grid_id":"-6","grid_name":"大班","school_year":"2021"}]},{"gt_name":"小学","grid
rid_id":"2","grid_name":"二 年 级","school_year":"2019"},{"grid_id":"3","grid_name":"三 年 级","schoo
017"},{"grid_id":"5","grid_name":"五年级","school_year":"2016"},{"grid_id":"6","grid_name":"六 年 级
:"7","grid_name":"初 一","school_year":"2014"},{"grid_id":"8","grid_name":"初 二","school_year":"20
name":"高中","grid_info":[{"grid_id":"10","grid_name":"高 一","school_year":"2011"},{"grid_id":"11
e":"高 三","school_year":"2009"}]}]}
06-10 20:03:28.636 23646 24858 E okhttpGET: <-- END HTTP (1195-byte body)
06-10 20:03:28.638 23646 24856 E okhttpGET: <-- 200 https://xesapi.speiyou.cn/xes/homepage/v1/bro
951431 (185ms, unknown-length body:)
06-10 20:03:28.638 23646 24856 E okhttpGET: server: Tengine
```

▲ 圖 2-24 抓取封包效果

自訂 DEX 檔案的載入主要是為了避免將 Java 翻譯成 JavaScript 的複雜工
作，相反可以直接使用 Java 語言撰寫自訂的類別並編譯成 DEX 檔案供後
續使用。

在這個程式中，筆者也實現了使用 JavaScript 撰寫自訂的類別的方式一

Java.registerClass()，觀察這部分實現會發現，相比直接使用簡單的 Java. openClassFile(<dex>).load() 函數來説，registerClass 函數的實現著實有點隔靴搔癢的感覺。

而剩下的主體程式就是使用 Java.choose 函數找到對應的 OkHttpClient 物件，並向 _interceptors 陣列中增加自訂的 MyInterceptor 攔截器物件。這樣的功能毫無疑問 Xposed 是無法實現的。

● 程式清單 2-6　hookOkHttp3.js

```
function searchClient(){
    Java.perform(function(){
        // 載入包含CurlInterceptor攔截器的DEX
        Java.openClassFile("/data/local/tmp/myok2curl.dex").load();
        console.log("loading dex successful!")
        const curlInterceptor =  Java.use("com.moczul.ok2curl.
CurlInterceptor");
        const loggable = Java.use("com.moczul.ok2curl.logger.Loggable");
        var Log = Java.use("android.util.Log");
        var TAG = "okhttpGETcurl";
        //註冊類別——一個實現了所需介面的類別
        var MyLogClass = Java.registerClass({
            name: "okhttp3.MyLogClass",
            implements: [loggable],
            methods: {
                log: function (MyMessage) {
                    Log.v(TAG, MyMessage);
                }}
        });
        const mylog = MyLogClass.$new();
        // 得到所需的攔截器物件
        var curlInter = curlInterceptor.$new(mylog);


        // 載入包含logging-interceptor攔截器的DEX
        Java.openClassFile("/data/local/tmp/okhttplogging.dex").load();
```

```
    var MyInterceptor = Java.use("com.r0ysue.learnokhttp.okhttp3Logging");
    var MyInterceptorObj = MyInterceptor.$new();

    Java.choose("okhttp3.OkHttpClient",{
        onMatch:function(instance){
            console.log("1. found instance:",instance)
            console.log("2. instance.interceptors():",instance.
interceptors().$className)
            console.log("3. instance._interceptors:",instance._
interceptors.value.$className)
            console.log("5. interceptors:",Java.use("java.util.
Arrays").toString(instance.interceptors().toArray()))
            var newInter = Java.use("java.util.ArrayList").$new();
            newInter.addAll(instance.interceptors());
            console.log("6. interceptors:",Java.use("java.util.Arrays").
toString(newInter.toArray()));
            console.log("7. interceptors:",newInter.$className);
            newInter.add(MyInterceptorObj);
            newInter.add(curlInter);
            instance._interceptors.value = newInter;

        },onComplete:function(){
            console.log("Search complete!")
        }
    })
    })
}
setImmediate(searchClient)
```

另外，在筆者的研究過程中還使用了一個 Objection 外掛程式—
WallBreaker，其專案位址為 https://github.com/hluwa/Wallbreaker。
WallBreaker 可以用於快速定位一個類別中所包含的屬性與函數，甚至可
以直接透過物件的控制碼獲取所在類別中的所有屬性的值。這個功能在
筆者研究 OkHttp3 攔截器機制的過程中提供了巨大的幫助。圖 2-25 是筆

者在驗證 OkHttpClient 類別的 _interceptors 成員就是對應 Client 的攔截
器時的截圖。

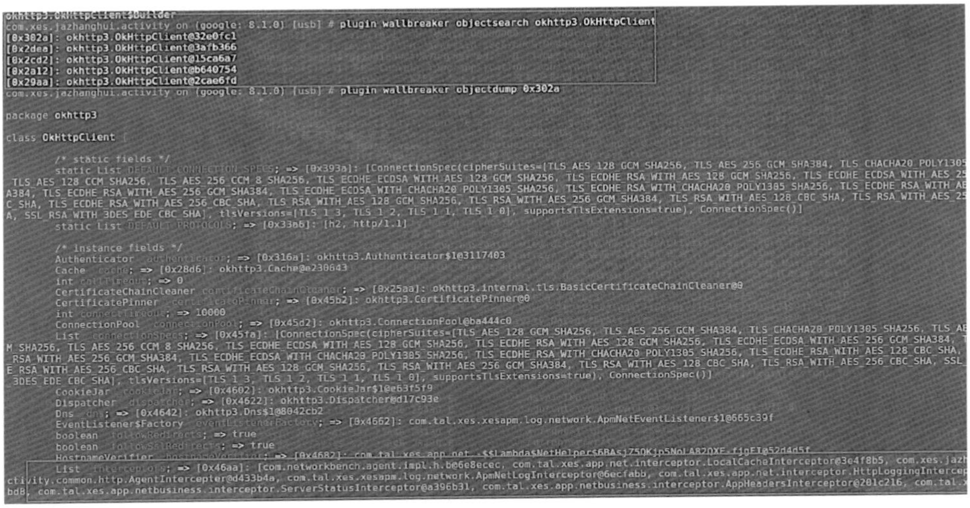

▲ 圖 2-25　OkHttpClient 物件解析

對 WallBreaker 的原始程式進行分析，會發現這個功能也是利用 Java.
choose() 函數實現的對記憶體中類別物件的搜索，如程式清單 2-7 所示。
後續對物件屬性和函數的列印其實是透過物件控制碼值進行反射，從而
獲取對應成員和函數，具體這裡不再分析，如果讀者對其實現感興趣，
可以自行閱讀專案原始程式。

● 程式清單 2-7　objectsearch 功能實現

```
export const searchHandles = (clazz: string, stop: boolean = false) => {
    let result: any = {};
    Java.perform(function () {
        Java.choose(clazz, { // <===
            onComplete: function () {
            },
```

```
            onMatch: function (instance) {
                const handle = getHandle(instance);
                result[handle] = objectToStr(instance);
                if (stop) {
                    return "stop"
                }
            },
        });
    }
);
    return result;
};
```

相比於 Xposed 而言，Frida 在 native 層的 Hook 也是頗有建樹。逆向人員同樣能夠透過在 native 層進行記憶體列舉完成很多工作。以在 Android 安全中困擾著許多安全人員的脫殼問題為例，比如筆者在工作過程中經常使用的 dexdump 專案，其位址為 https://github.com/hluwa/FRIDA-DEXDump，dexdump 的核心原理是在目標處理程序的記憶體空間中遍歷搜索包含 DEX 檔案特徵（dexdump 主要利用檔案表頭是 dex03? 模式的特徵）的資料，並在匹配到符合特徵的 DEX 資料後完成真實 DEX 檔案的dump 工作。這個方法的實現主要是利用 Frida 的 Memory.scanSync() 函數，dexdump 的主要程式如程式清單 2-8 所示。

● 程式清單 2-8 dexdump 核心程式

```
Memory.scanSync(range.base, range.size, "64 65 78 0a 30 ?? ?? 00").
forEach(function (match) {
    if (range.file && range.file.path
        && (// range.file.path.startsWith("/data/app/") ||
            range.file.path.startsWith("/data/dalvik-cache/") ||
            range.file.path.startsWith("/system/"))) {
        return;
```

```
    }

    if (verify(match.address, range, false)) {
        var dex_size = get_dex_real_size(match.address, range.base, range.
base.add(range.size));
        result.push({
            "addr": match.address,
            "size": dex_size
        });

        var max_size = range.size - match.address.sub(range.base);
        if (enable_deep_search && max_size != dex_size) {
            result.push({
                "addr": match.address,
                "size": max_size
            });
        }
    }
});
```

當然，dexdump 僅僅能夠解決一代整體保護的脫殼工作，於是又有大神利用 Frida 撰寫了 frida_fart 用於解決二代取出保護的脫殼問題，其專案位 址 為 https://github.com/hanbinglengyue/FART/blob/ master/frida_fart.zip。frida_fart 中存在著兩種解決脫殼的方案，其中 Hook 版本是透過 Hook 在 App 執行過程中用於載入和執行 DEX 檔案的 ART 虛擬機器中的 native 函數—LoadMethod 函數，並透過這個函數的參數分別獲取載入的 Java 函數所在的 DEX 檔案和函數的真實內容，其主要程式如程式清單 2-9 所示。

◯ 程式清單 2-9 frida_fart_hook.js

```
Interceptor.attach(addrLoadMethod, {
```

```
onEnter: function (args) {
    this.dexfileptr = args[1];
    this.artmethodptr = args[4];
},
onLeave: function (retval) {
    var dexfilebegin = null;
    var dexfilesize = null;
    if (this.dexfileptr != null) {
        dexfilebegin = Memory.readPointer(ptr(this.dexfileptr).
add(Process.pointerSize * 1));
        dexfilesize = Memory.readU32(ptr(this.dexfileptr).add(Process.
pointerSize * 2));
        var dexfile_path = savepath + "/" + dexfilesize + "_loadMethod.
dex";
        var dexfile_handle = null;
        try {
            dexfile_handle = new File(dexfile_path, "r");
            if (dexfile_handle && dexfile_handle != null) {
                dexfile_handle.close()
            }

        } catch (e) {
            dexfile_handle = new File(dexfile_path, "a+");
            if (dexfile_handle && dexfile_handle != null) {
                var dex_buffer = ptr(dexfilebegin).
readByteArray(dexfilesize);
                dexfile_handle.write(dex_buffer);
                dexfile_handle.flush();
                dexfile_handle.close();
                console.log("[dumpdex]:", dexfile_path);
            }
        }
    }
    var dexfileobj = new DexFile(dexfilebegin, dexfilesize);
    if (dex_maps[dexfilebegin] == undefined) {
```

```
            dex_maps[dexfilebegin] = dexfilesize;
            console.log("got a dex:", dexfilebegin, dexfilesize)
        }
        if (this.artmethodptr != null) {
            var artmethodobj = new ArtMethod(dexfileobj, this.artmethodptr);
            if (artmethod_maps[this.artmethodptr] == undefined) {
                artmethod_maps[this.artmethodptr] = artmethodobj
            }
        }
    }
});
```

觀察程式清單 2-9 會發現，frida_fart 針對二代取出殼的解決方案，本質
上使用的是 Frida 在 native 層 Hook 的基礎 API——Interceptor.attach()。
而程式中針對 dexFile 記憶體物件和 ArtMethod 記憶體物件的解析主要是
利用 Frida 中的 Memory 讀取記憶體相關的 API 多記憶體資料進行讀寫入
操作，而這些都是在原生的 Xposed 中無法想像的工作。

2.3 本章小結

本章主要介紹了在逆向工作中一個非常好用的工具 Frida，並介紹了一
些 Frida 指令稿撰寫的基礎，同時還介紹了一款基於 Frida 開發的協力廠
商工具 Objection 及其對應的基本操作。有了 Objection，App 測試過程
中 Java Hook 工作幾乎就被完全取代了，當然 Frida 指令稿還能夠做很多
Objection 中不能執行的操作，比如 native 函數的 Hook、Hook 篡改函
數邏輯的工作等。在介紹完這些基礎內容後，筆者以一種高屋建瓴的角
度介紹了兩種快速定位關鍵類別的思路，區別於簡單的字串搜索，Frida

以豐富的 API 和強大的功能為這兩種快速定位方案開闢了新氣象，雖然
Frida 的穩定性有待商榷，但瑕不掩瑜，相信讀者在本章中都能夠有所收
穫。

Frida 指令稿開發之主動 呼叫與 RPC 入門

在逆向分析過程中，快速定位到演算法的關鍵函數只是第一步。為了詳細分析函數的流程與邏輯，往往還需要反覆地呼叫這個關鍵的函數，而如果每次都依賴程式本身的邏輯，那麼往往傳入的參數每次都是變化的，這非常不利於逆向工作的進展，因此主動呼叫的意義也就出現了－透過傳入可控的參數反覆偵錯目標函數，最終確認其整體的邏輯與細節。當然，由於一些目標 App 的某些函數內部保護措施和邏輯十分複雜，也確實存在著無法在一定時間中得到預期目標的情況，這時如果想要獲得函數的執行結果，使用主動呼叫和 RPC（遠端程序呼叫）結合的方式不失為一種選擇。本章將帶領讀者一起來領略主動呼叫與 RPC 的魅力。

3.1 Frida RPC 開發姿勢

按照筆者推崇的 Frida 理論：先 Hook 定位關鍵邏輯，然後主動呼叫建構參數進行利用，最後透過 RPC 匯出結果進行規模化呼叫。本節本應該先介紹主動呼叫的理論與方法，但考慮到後續知識的連貫性，因此這裡先對 RPC 的一些使用方法進行介紹。

從 frida-python 看 RPC 開發

讀者還記得在安裝 Frida 時使用的是哪種方式進行安裝的嗎？沒錯，就是使用 pip 這一 Python 的套件安裝程式對 Frida 環境進行設定的。事實上，Frida 是一款基於 Python 和 JavaScript 的處理程序級 Hook 框架，其中 JavaScript 語言承擔了 Hook 函數的主要工作，而 Python 語言的角色則相當於一個提供給外界的綁定介面，使用者可以透過 Python 語言將 JavaScript 指令稿注入處理程序中，只是相比於命令列注入的方式，Python 透過程式注入的方式更加優雅。另外，官方也提供了透過 Python 遠端外部呼叫 JavaScript 中的函數的方式，並在對應的專案倉庫（專案位址：https://github.com/frida/frida-python）中提供了一些例子。本節將透過其中的一部分程式來介紹一些基礎的 frida-python 遠端程序呼叫的方式。

在正式介紹官方倉庫的例子前，首先介紹一些透過 Python 實現 Frida 注入的基礎知識：相對於命令列直接指定參數注入透過 USB 連接的手機處理程序，透過 Python 注入 Android 處理程序的方式步驟更加分明。

步驟01 透過 Frida 獲取特定裝置。程式清單 3-1 分別展示了連接 USB 裝置和網路裝置的方式。

🔻 **程式清單 3-1 獲取裝置**

```
import frida
# 獲取USB連接裝置
device = frida.get_usb_device()
# 獲取網路裝置
device = frida.get_device_manager().add_remote_device('192.168.50.96:6666')
```

步驟02 在獲取到裝置 device 後，與 Frida 透過命令列實現處理程序注入的兩種方式──spawn 和 attach 對應，使用 Python 完成處理程序注入的方式同樣也有兩種。以注入「設定」應用為例，其程式如程式清單 3-2 所示。

🔻 **程式清單 3-2 注入處理程序**

```
# -*- coding: utf-8 -*-
import time
import frida

# spawn 方式注入處理程序
pid = device.spawn(["com.android.settings"]) # 注意這裡spawn的參數是一個list
類型的參數
device.resume(pid) # 喚起處理程序，也可以在透過attach函數注入處理程序後再呼叫
time.sleep(1) # 這裡休眠是為了等待處理程序被完全喚起
session = device.attach(pid)

# attach模式注入處理程序
session = device.attach("com.android.settings") # 直接透過指定套件名進行處理
程序注入
```

步驟03 成功注入處理程序後，還需要最後一步：Hook 指令稿的注入。這一步實際上就是將 JavaScript 指令稿作為字串或者位元組流透過 Frida 提供的 API 載入進對應的處理程序 session 中。簡單的 JavaScript hook 指令稿透過 Python 注入的方式如程式清單 3-3 所示。

⬤ **程式清單 3-3 注入指令稿**

```
import frida

script = session.create_script("""
setImmediate(Java.perform(function(){
    console.log("hello python frida");
}))
""") # 讀取Hook指令稿內容
script.load() # 將指令稿載入進處理程序空間中
```

在程式清單 3-3 中，建立指令稿的方式是透過讀取一段代表 Hook 指令稿的字串。在真實地完成指令稿的注入時，推薦將 JavaScript 指令稿和 Python 程式分離，透過讀取檔案的方式將指令稿載入進處理程序中，這樣在撰寫指令稿時有智慧提示，而且可以單獨使用命令列測試指令稿的正確性，筆者通常使用如程式清單 3-4 所示的方式注入指令稿。

⬤ **程式清單 3-4 檔案方式注入處理程序**

```
with open("hook.js") as f:
    script = session.create_script(f.read())
script.load() # 將指令稿載入進處理程序空間中
```

在介紹了透過 Python 注入指令稿的基礎知識後，讓我們來正式了解一下 Frida 官方提供的一些例子。

筆者認為學習 RPC 實際上就是學習一些關於 JavaScript 指令稿和 Python 進行互動的方式。

以 frida-python 倉庫的 example 目錄下的 rpc.py 指令檔為例，這裡將程式修改為適合 Android 應用的形式，具體內容如程式清單 3-5 所示。

◉ 程式清單 3-5 rpc.py

```python
# -*- coding: utf-8 -*-
from __future__ import print_function
from frida.core import Session

import frida
import time

device = frida.get_usb_device() # 透過USB連接裝置
# frida.get_device_manager().add_remote_device('192.168.50.96:6666')

pid = device.spawn(["com.android.settings"]) # spawn方式注入處理程序

device.resume(pid)
time.sleep(1)

session = device.attach(pid)

script = session.create_script("""
rpc.exports = {
  hello: function () {
    return 'Hello';
  },
  failPlease: function () {
   return  'oops';
  }
};
""")
script.load() # 載入指令稿
api = script.exports # 獲取rpc匯出函數
print("api.hello() =>", api.hello()) # 執行匯出函數
print("api.fail_please() =>", api.fail_please()) # 執行匯出函數
```

在確保手機使用 USB 資料線連接上電腦並且測試機上對應版本的 frida-

server 正在執行後，直接透過 python 命令執行 rpc.py 指令稿，其結果如
圖 3-1 所示。

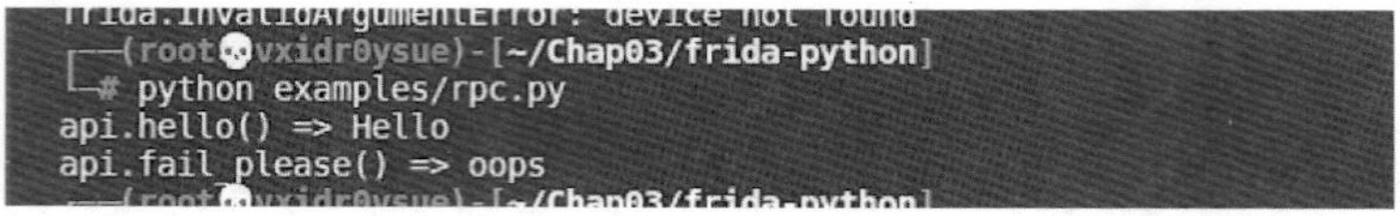

▲ 圖 3-1 rpc.py 執行結果

在程式清單 3-5 這個例子中，會發現如果想要在 Python 中呼叫 JavaScript
中的函數，首先需要在 JavaScript 中將對應函數寫到 rpc.exports 這個
字典中，在撰寫完成後，如果想要在 Python 中進行呼叫，只需先透過
script.exports 獲取對應的匯出函數字典，再透過對應的字典鍵值進行呼
叫即可。

另外，細心的讀者會發現，JavaScript 指令稿中的 failPlease 鍵值在
Python 指令稿中呼叫時，從最初的駝峰命名法（第一個單字以小寫字母
開始，從第二個單字開始以後的每個單字的字首都採用大寫字母）變成
了底線命名法（每個單字用底線隔開並且單字都是小寫）。簡單來說，就
是所有的 JavaScript 指令稿中帶大寫字母的匯出函數鍵值被替換為 "_" 加
上對應小寫字母的方式，對應程式清單 3-5 中 JavaScript 中的 failPlease
匯出函數變成了 Python 中的 fail_please。

如果說 rpc.py 介紹的是在 Python 中遠端主動呼叫 JavaScript 中的函數的
方式，那麼接下來要介紹的就是 JavaScript 主動向 Python 發送資料的方
式。

以 examples 目錄下的 detached.py 檔案為例，其程式在修改為調配於
Android 應用後，具體內容如程式清單 3-6 所示。在執行指令稿後，手動
透過 adb 命令斷開 USB 連接，執行結果如圖 3-2 所示。

● 程式清單 3-6 detached.py

```python
# -*- coding: utf-8 -*-
from __future__ import print_function

import sys
import frida
import time


def on_detached():
    print("on_detached")

def on_detached_with_reason(reason):
    print("on_detached_with_reason:", reason)

def on_detached_with_varargs(*args):
    print("on_detached_with_varargs:", args)

device = frida.get_usb_device()
# frida.get_device_manager().add_remote_device('192.168.50.96:6666')

pid = device.spawn(["com.android.settings"])

device.resume(pid)
time.sleep(1)
session = device.attach(pid)

print("attached")
session.on('detached', on_detached) # 注入分離回應函數
session.on('detached', on_detached_with_reason) # 注入分離回應函數
session.on('detached', on_detached_with_varargs) # 注入分離回應函數
sys.stdin.read()
```

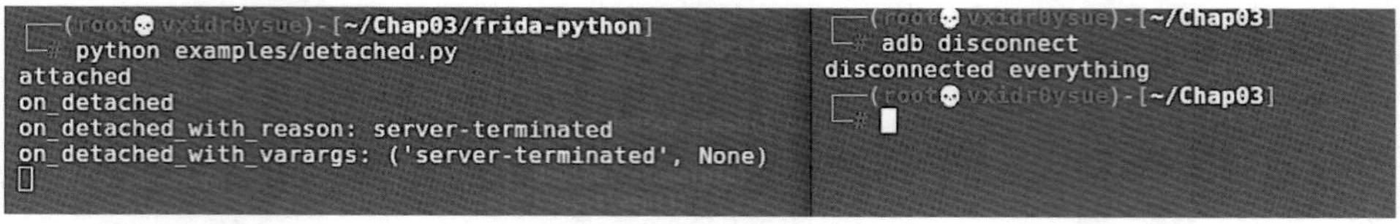

▲ 圖 3-2 執行效果

觀察圖 3-2 會發現，列印出來的資訊是 frida-server 處理程序終止導致的注入分離，而這個實現正是透過 session.on() 這個 API 指定 deatached 行為對應的處理函數列印出來的日誌。

同樣，讀者如果研究 crash_report.py 檔案程式，會發現相對於 detached.py，crash_report.py 只是多了一個針對處理程序崩潰的響應函數而已。但要注意的是，針對處理程序崩潰的回應函數是透過裝置 device 增加的，而非處理程序的 session（device.on() 函數），其中 crash_report.py 中的主要程式如程式清單 3-7 所示。

◯ 程式清單 3-7 crash_report.py

```
def on_process_crashed(crash):
    print("on_process_crashed")
    print("\tcrash:", crash)

...
device = frida.get_usb_device()
# frida.get_device_manager().add_remote_device('192.168.50.96:6666')

pid = device.spawn(["com.android.settings"])

...

device.on('process-crashed', on_process_crashed)  # 處理程序崩潰回應函數

session = device.attach(pid)
```

```
# session = device.attach("Hello")
session.on('detached', on_detached) # 注入分離回應函數
...
```

到這裡，frida-python 中與 RPC 遠端呼叫相關的程式差不多就介紹完畢了，當然 frida-python 的 examples 中遠不止前面介紹的這些程式，比如指令稿 child_gating.py 介紹了子處理程序的注入方式，bytecode.py 介紹了將指令稿編譯為位元組碼後再載入指令稿的方式，inject_library 資料夾中的程式則介紹了手動向處理程序中注入一個動態函數庫檔案的方式，等等。因為這部分程式與本章 RPC 內容的相關性不大，故這裡不再贅述，讀者如果感興趣，可以自行研究。

另外，在上一章中曾介紹過 ZenTracer 在 Trace 快速定位類別方面的應用，但並未深究其程式細節，事實上在獲取到 ZenTracer 全部程式後會發現，整個專案除去部分 UI 相關的程式，真實用於 Hook 的程式只有一個 traceClass() 函數，剩下的部分都是用於在 JavaScript 和 Python 介面之間的資料傳遞，資料的傳遞方式是透過 send 函數將 JavaScript 中的資料傳輸到 Python 用於接收資訊的 FridaReceive 函數中。其具體程式如程式清單 3-8 和 3-9 所示。

◯ 程式清單 3-8　JavaScript 中的資料傳輸到 Python

```
function log(text) {
    var packet = {
        'cmd': 'log',
        'data': text
    };
    send("ZenTracer:::" + JSON.stringify(packet)) // 向Python發送資料，與
Python互動
}

function enter(tid, tname, cls, method, args) {
```

```javascript
    var packet = {
        'cmd': 'enter',
        'data': [tid, tname, cls, method, args]
    };
    send("ZenTracer:::" + JSON.stringify(packet)) // 向Python發送資料，與
Python互動
}

function exit(tid, retval) {
    var packet = {
        'cmd': 'exit',
        'data': [tid, retval]
    };
    send("ZenTracer:::" + JSON.stringify(packet)) // 向Python發送資料，與
Python互動
}
```

◯ 程式清單 3-9 在 Python 訊息中接收訊息

```python
# 訊息訊息回應函式定義
def FridaReceive(message, data):
    if message['type'] == 'send': # 接收JS中send函數傳遞的資料
        if message['payload'][:12] == 'ZenTracer:::':
            packet = json.loads(message['payload'][12:]) # 載入JSON檔案
            cmd = packet['cmd']
            data = packet['data']
            if cmd == 'log':
                APP.log(data)
            elif cmd == 'enter':
                tid, tName, cls, method, args = data
                APP.method_entry(tid, tName, cls, method, args)
            elif cmd == 'exit':
                tid, retval = data
                APP.method_exit(tid, retval)
    else:
        print(message['stack']) # 列印呼叫堆疊
...
```

```
def _attach(pid):
    if not device: return
    app.log("attach '{}'".format(pid))
    session = device.attach(pid)
    session.enable_child_gating()
    source = open('trace.js', 'r').read().replace('{MATCHREGEX}', match_
s).replace("{BLACKREGEX}", black_s) # 透過字串匹配修改Js檔案Hook目標
    script = session.create_script(source)
    script.on("message", FridaReceive) # 訊息訊息回應函數
    script.load()
    scripts.append(script)
```

與程式清單 3-7 中使用 session.on('detached', on_detached) 函數指定處理程序崩潰的回應函數相比,在程式清單 3-9 中則透過 script.on ("message", FridaReceive) 函數註冊用於接收 message 資訊的函數,做到了當 JavaScript 中呼叫 send 訊息函數時,訊息序列被發送到指定的 FridaReceive 函數中。另外,這裡值得一提的是,ZenTracer 中利用 JavaScript 指令稿以字串的方式讀取這一特點,直接透過特定字串匹配後,替換的方式最終做到了對指定函數進行 Hook 的效果。當然其實這裡不大推薦這種方式,筆者更推薦透過 RPC 呼叫的方式傳遞參數完成函數的 Hook,但不排除撰寫程式時存有一些特殊考慮,筆者限於水準並未發現。

ZenTracer 作者的另一個專案—FRIDA-DexDump 同樣利用了很多 RPC 相關的知識,與 ZenTracer 相比,作者在這個專案中頻繁地呼叫 JavaScript 中的一些匯出函數,比如掃描記憶體中符合條件的 DEX 檔案的函數 scandex() 等。這裡不再一一介紹,如果讀者感興趣,可以自行研究。

在 RPC 呼叫中還會有著一些與 send() 相對應的函數,比如 wait() 和 recv() 這兩個函數分別用於在 JavaScript 中阻塞執行緒和接收從 Python 中透過 script.post() 函數傳回的資料。由於這幾個 API 筆者平時使用的頻率並不是很高,這裡不再詳細説明。

3.2 Frida Java 層主動呼叫與 RPC

在上一節中,我們一起學習了關於 Frida RPC 的一些基礎知識。所謂萬事俱備,只欠東風。在這一節中就來完善 Frida 三劍客的最後一環一主動呼叫,並結合之前所學的知識完整地展示三劍客的威力。

這裡以筆者自己撰寫的一個 demoso1 專案為例進行介紹。demoso1 中存在著兩個 native 函數在應用開啟後被迴圈呼叫,其中 method01 是一個靜態函數,用於對輸入參數進行 AES/CBC/PKCS5Padding 加密後並返回,method02 是一個成員函數,用於將以 AES/CBC/PKCS5Padding 加密後的加密解密成明文並返回。二者的 Java 層宣告與具體呼叫如程式清單 3-10 所示。

⬇ 程式清單 3-10 AES 加解密函數

```
@Override
protected void onCreate(Bundle savedInstanceState) {
    ...
    while(true){
        try {
            Thread.sleep(1000);
        } catch (InterruptedException e) {
            e.printStackTrace();
        }
        Log.i("r0addmethod1", method01("roysue"));
        Log.i("r0addmethod2", method02(method01("roysue")));
    }
}
/**
* AES加密, CBC, PKCS5Padding
*/
```

```
public static native String method01(String str);

/**
* AES解密, CBC, PKCS5Padding
*/
public native String method02(String str);
```

有一定基礎的讀者一定知道如果 native 函數是靜態註冊的，其在 so 層最終生成的函數名是有一定規律的（以 Java_ 開頭的匯出函數），而這種靜態註冊方式非常不安全，因此這裡利用 RegisterNatives() 函數對這兩個函數進行了動態註冊以加強安全性，具體動態註冊相關的程式如程式清單 3-11 所示。

❍ 程式清單 3-11 在 JNI_Onload 中動態註冊

```
#define NELEM(x) ((int) (sizeof(x) / sizeof((x)[0])))
...
JNIEXPORT jint JNI_OnLoad(JavaVM* vm, void* reserved){
    JNIEnv *env;
    vm->GetEnv((void **) &env, JNI_VERSION_1_6);
    JNINativeMethod methods[] = {
            ...
            {"method01", "(Ljava/lang/String;)Ljava/lang/String;", (void *)
method01},
            {"method02", "(Ljava/lang/String;)Ljava/lang/String;", (void *)
method02},
    };
    // 動態註冊
    env->RegisterNatives(env->FindClass("com/example/demoso1/
MainActivity"), methods, NELEM(methods));
    return JNI_VERSION_1_6;
}
```

最終專案執行後，App 部分日誌列印如圖 3-3 所示。

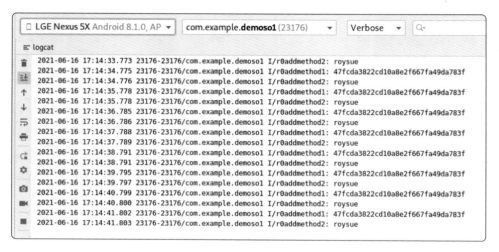

▲ 圖 3-3 執行部分日誌

在這一節中，筆者將從 Java 層和 so 層分別介紹 Frida 的主動呼叫方式與對應的 RPC 使用方式，並對比最終 RPC 的效率。

為了確定 method01 和 method02 函數在記憶體中參數和返回值的類型，這裡首先使用上一章介紹的 Objection 來確認這兩個函數的簽名與呼叫情況。

首先在手機上成功執行 frida-server 和 demoso1 應用並使用 USB 線連接手機，待 Objection 成功注入目標處理程序後，執行如下兩行命令列出並 Hook MainActivity 類別中的所有函數，用於確認在記憶體中兩個目標函數對應的函數簽名和呼叫情況，最終結果如圖 3-4 所示。

```
# android hooking list class_methods com.example.demoso1.MainActivity
# android hooking watch class com.example.demoso1.MainActivity
```

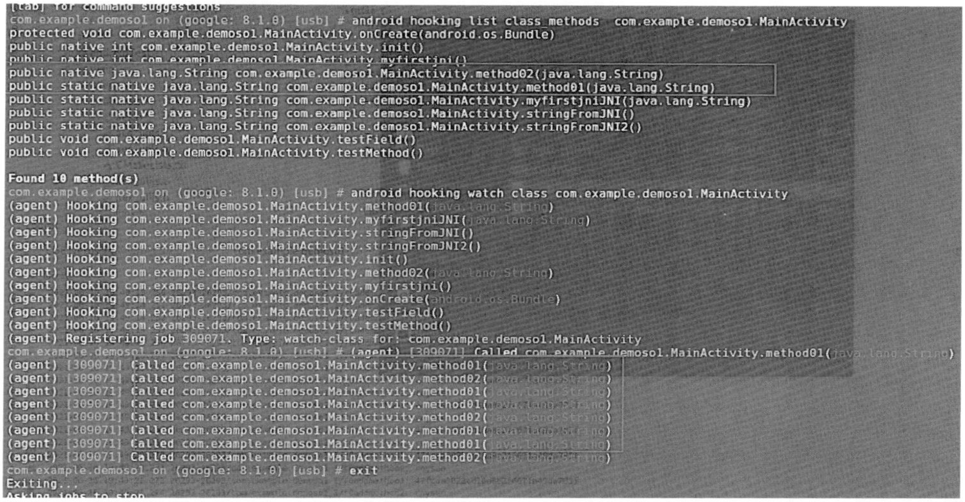

▲ 圖 3-4 列出 MainActivity 類別記憶體中的所有函數

根據圖 3-4 確認記憶體中 method01 函數的簽名後,接下來介紹靜態函數 method01 的主動呼叫方式。

在 Objection 中確認 method01 函數被呼叫後,要透過程式進行函數的主動呼叫,筆者認為首先要透過程式寫一個對應函數的 Hook 指令稿。

之所以這樣做,是因為在 Java 函數的 Hook 中存在一個主動呼叫的範本,這個範本能夠使得後續的主動呼叫發生錯誤時有一個參考答案進行比對。筆者認為主動呼叫時最重要的實際上是參數的建構,而 Hook 中的主動呼叫一定是成功無疑的,因此主動呼叫其實就是建構和 Hook 時使用的參數類型一致的參數。讓我們直接來看程式,其內容如程式清單 3-12 所示,最終 Hook 效果如圖 3-5 所示。

◐ 程式清單 3-12　invoke.js

```
function hook(){
    Java.perform(function(){
```

```
var MainActivity = Java.use("com.example.demoso1.MainActivity");
MainActivity.method01.implementation = function(str){
    var result = this.method01(str); // Hook時的主動呼叫
    console.log("str => ",str);
    console.log("result => ",result);
    return result;
}
})
}
```

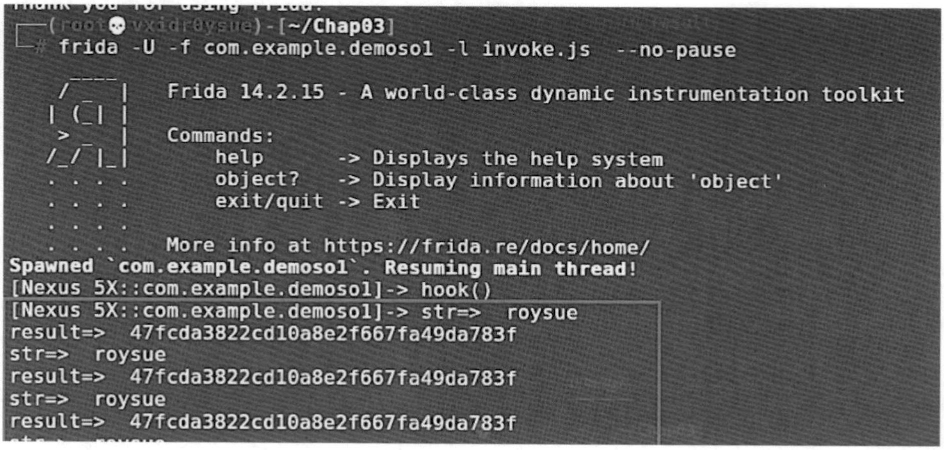

```
──(root💀vxidr0ysue)-[~/Chap03]
─# frida -U -f com.example.demoso1 -l invoke.js  --no-pause

   / _  |      Frida 14.2.15 - A world-class dynamic instrumentation toolkit
  | (_| |
   > _  |      Commands:
  /_/ |_|           help      -> Displays the help system
  . . . .           object?   -> Display information about 'object'
  . . . .           exit/quit -> Exit
  . . . .
  . . . .      More info at https://frida.re/docs/home/
Spawned `com.example.demoso1`. Resuming main thread!
[Nexus 5X::com.example.demoso1]-> hook()
[Nexus 5X::com.example.demoso1]-> str=>   roysue
result=>  47fcda3822cd10a8e2f667fa49da783f
str=>   roysue
result=>  47fcda3822cd10a8e2f667fa49da783f
str=>   roysue
result=>  47fcda3822cd10a8e2f667fa49da783f
```

▲ 圖 3-5 Hook 效果

在參照 Hook 指令稿中的主動呼叫函數後,我們來寫一個簡單的主動呼叫指令稿,呼叫 method01 函數對 r0ysue 字串進行加密,其程式如程式清單 3-13 所示,最終主動呼叫的結果如圖 3-6 所示。

```
"4e8de2f3c674d8157b4862e50954d81c"
[Nexus 5X::com.example.demoso1]-> invokeMethod01()
plaintext =>   r0ysue
result =>   4e8de2f3c674d8157b4862e50954d81c
[Nexus 5X::com.example.demoso1]->
```

▲ 圖 3-6 method01() 函數的主動呼叫結果

⬤ 程式清單 3-13 invoke.js

```
function invokeMethod01(){
    Java.perform(function(){
        var MainActivity = Java.use("com.example.demoso1.MainActivity");
        var javaString = Java.use("java.lang.String")
        var plaintext = "r0ysue"
        var result = MainActivity.method01(javaString.$new(plaintext))
        console.log("plaintext => ",plaintext)
        console.log("result => ",result)

    })
}
```

為了節省篇幅，這裡不再介紹 method02() 函數的 Hook，我們直接介紹 method02() 函數的主動呼叫。

在程式清單 3-13 中，要注意靜態函數的主動呼叫只需要獲取對應類別的控制碼，而如果我們依樣畫葫蘆，同樣寫一個對實例函數 method02() 的呼叫，就會發生錯誤：method02: cannot call instance method without an instance，其大意是指動態的實例函數只能透過對應的實例進行呼叫，而非一個類別控制碼，具體顯示出錯如圖 3-7 所示。

```
result =>   4e8de2f3c674d8157b4862e50954d81c
[Nexus 5X::com.example.demoso1]-> invokeMethod02()
Error: method02: cannot call instance method without an instance
    at value (frida/node_modules/frida-java-bridge/lib/class-factory.js:961)
    at e (frida/node_modules/frida-java-bridge/lib/class-factory.js:547)
    at <anonymous> (/invoke.js:39)
    at <anonymous> (frida/node_modules/frida-java-bridge/lib/vm.js:11)
    at _performPendingVmOps (frida/node_modules/frida-java-bridge/index.js:238)
    at <anonymous> (frida/node_modules/frida-java-bridge/index.js:213)
    at <anonymous> (frida/node_modules/frida-java-bridge/lib/vm.js:11)
    at _performPendingVmOpsWhenReady (frida/node_modules/frida-java-bridge/index.js:232)
    at perform (frida/node_modules/frida-java-bridge/index.js:192)
    at invokeMethod02 (/invoke.js:43)
    at <eval> (<input>:1)
    at eval (native)
    at fridaEvaluate (/invoke.js:57)
    at apply (native)
    at <anonymous> (frida/runtime/message-dispatcher.js:13)
    at c (frida/runtime/message-dispatcher.js:23)[Nexus 5X::com.example.demoso1]->
```

▲ 圖 3-7 method02() 函數主動呼叫發生錯誤

自然而然地，我們會想到在上一章介紹的獲取記憶體中實例的 API:Java.
choose()。因此，最終 method02() 函數一個簡單的主動呼叫應當如程式
清單 3-14 所示，主動呼叫的效果如圖 3-8 所示。

● 程式清單 3-14 invoke.js

```javascript
function invokeMethod02(){
    Java.perform(function(){
        Java.choose("com.example.demoso1.MainActivity",{
            onMatch:function(instance){
            var javaString = Java.use("java.lang.String")
            var ciphertext = "4e8de2f3c674d8157b4862e50954d81c"
            result = instance.method02(javaString.$new(ciphertext))
            console.log("ciphertext => ",ciphertext);
            console.log("result => ",result); // r0ysue
            },onComplete(){}
        })
    })
}
```

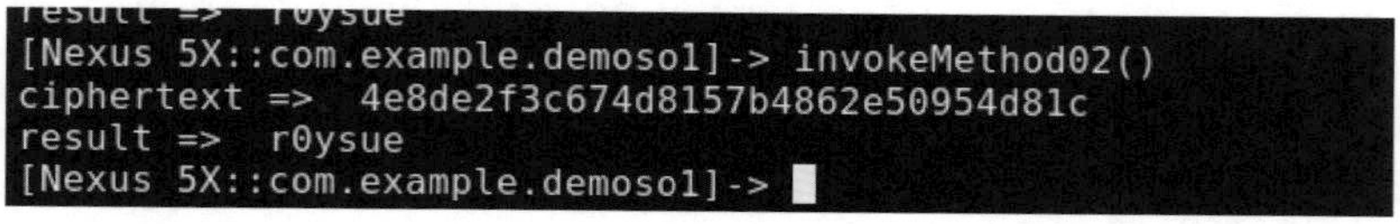

```
result =>   r0ysue
[Nexus 5X::com.example.demoso1]-> invokeMethod02()
ciphertext =>   4e8de2f3c674d8157b4862e50954d81c
result =>   r0ysue
[Nexus 5X::com.example.demoso1]-> █
```

▲ 圖 3-8 method02() 函數的主動呼叫結果

在成功測試主動呼叫後，接下來就是 Frida 三劍客的最後一步：RPC。

這裡為了將主動呼叫提供為外部介面，因此還需要兩步：

步驟 01 將所有的主動呼叫參數設定為 JavaScript 函數的參數並將主動呼
叫的結果返回，以方便外部自訂參數進行主動呼叫，最終程式如程式清
單 3-15 所示。

● 程式清單 3-15　invoke.js

```javascript
function invokeMethod01(plaintext){
    var result;
    Java.perform(function(){
        var MainActivity = Java.use("com.example.demoso1.MainActivity");
        var javaString = Java.use("java.lang.String")
        result = MainActivity.method01(javaString.$new(plaintext))
        console.log("plaintext => ",plaintext)
        console.log("result => ",result)
    })
    return result;
}
function invokeMethod02(ciphertext){
    var result;
    Java.perform(function(){
        Java.choose("com.example.demoso1.MainActivity",{
            onMatch:function(instance){
            var javaString = Java.use("java.lang.String")
            result = instance.method02(javaString.$new(ciphertext))

            },onComplete(){}
        })
    })
    return result
}
```

這裡要注意的是，作為返回值的 result 變數需要定義在 Java.perform() 函數的外部，否則 result 變數在作為 JavaScript 函數返回時會被認為是未定義狀態，如圖 3-9 所示。

▲ 圖 3-9　undefined 錯誤

步驟02 在確認第一步修改的主動呼叫無誤後，再將這兩個主動呼叫的函數匯出，具體程式如程式清單 3-16 所示。

⬥ 程式清單 3-16 invoke.js

```
rpc.exports={
    method01:invokeMethod01,
    method02:invokeMethod02
}
```

至此，RPC 中屬於 JavaScript 的部分已經完成。接下來寫一個 Python 外部呼叫的指令稿，結合在 3.1 節中所學的 RPC 知識，這裡直接舉出 Python 指令稿，程式如程式清單 3-17 所示，最終執行 Python 指令稿的結果如圖 3-10 所示。

⬥ 程式清單 3-17 invoke.py

```
import time
import frida
import json

def my_message_handler(message , payload): #定義錯誤處理
    print(message)
    print(payload)

# 連接Android機上的frida-server
device = frida.get_usb_device()
#device = frida.get_device_manager().add_remote_device("192.168.0.3:8888")

# 啟動demo01這個App
pid = device.spawn(["com.example.demoso1"])
device.resume(pid)
time.sleep(1)
session = device.attach(pid)
# 載入指令稿
```

```
with open("invoke.js") as f:
    script = session.create_script(f.read())
script.on("message" , my_message_handler) #訊息呼叫訊息處理
script.load()
api = script.exports # 獲取匯出函數清單

print('mehtod01 => encode_result: ' + api.method01("roysue")) # 呼叫匯出函數
print('mehtod02 => decode_result: ' + api.method02("47fcda3822cd10a8e2f667f
a49da783f"))

# 指令稿會持續等待輸入
input()
```

▲ 圖 3-10 RPC 呼叫結果

至此，整個 Frida 三劍客的流程已經完全結束，但是單純的 Python 批次
呼叫還不夠簡單，如果可以直接透過瀏覽器存取批次呼叫就更完美了。
因此，還可以透過為 Python 指令稿增加 HTTP 外部呼叫來達到這一目
的，筆者這裡是使用 Flask 協力廠商套件實現這一效果的，最終 Flask 的
相關程式如程式清單 3-18 所示。

◯ 程式清單 3-18 invoke.py

```
from flask import Flask, request
import json
app = Flask(__name__)

@app.route('/encrypt', methods=['POST']) # URL加密
def encrypt_class():
    data = request.get_data()
```

```
    json_data = json.loads(data.decode("utf-8"))
    postdata = json_data.get("data")
    res = script.exports.method01(postdata)
    return res


@app.route('/decrypt', methods=['POST']) # Data解密
def decrypt_class():
    data = request.get_data()
    json_data = json.loads(data.decode("utf-8"))
    postdata = json_data.get("data")
    res = script.exports.method02(postdata)
    return res
if __name__ == '__main__':
    app.run()
```

在確定 Flask 服務成功啟動後，就可以透過存取網址獲取加密結果，這裡為了方便，直接使用 curl 命令完成對加密解密 URL 的存取，獲取加密結果的 curl 命令如下，其結果如圖 3-11 所示。

```
curl -X POST http://127.0.0.1:5000/encrypt -H "{Content-Type: application/
json}" -d '{"data": "rOysue"}'
```

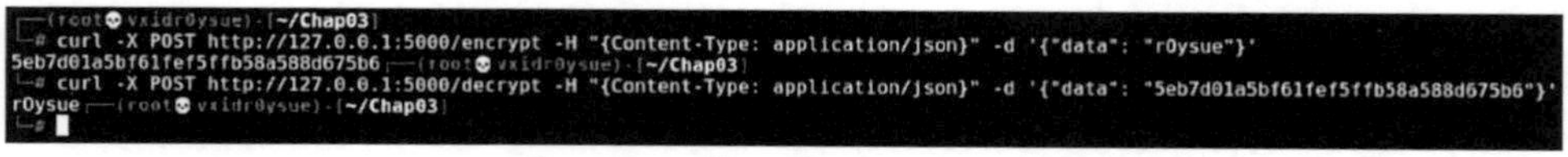

▲ 圖 3-11 curl 的呼叫結果

如果想要進一步將這樣的 RPC 變成批次化的叢集呼叫，可以透過將本地通訊埠透過 FRP/NPS 等內網映射到公網供外部呼叫，或者直接將手機所在通訊埠映射到公網，當使用 Python 進行 RPC 遠端呼叫時，直接選擇連接網路裝置的 API 進行存取即可。

當然，將裝置映射到公網進行 RPC 批次叢集呼叫的前提是這樣的存取性能足夠強勁，而性能則取決於多方面因素，包括手機性能、Frida 版本的穩定性、網路狀況等都需要納入考慮範圍，這裡在使用 Siege 高性能壓力測試工具透過如下命令進行測試時，其結果並不總是盡如人意，當併發數量（-c 參數指定）達到 10，執行測試次數（-r 參數控制）達到 100時，其回應效率僅能達到每秒 100 次左右，並且極其不穩定，其中兩次的測試結果如圖 3-12 所示。當然，這樣的結果不排除是筆者所選的裝置性能不夠卓越以及對應 Frida 版本不太穩定導致的，因此僅供參考。

```
siege -c1 -r1 "http://127.0.0.1:5000/encrypt POST < iloveroyse.json"
```

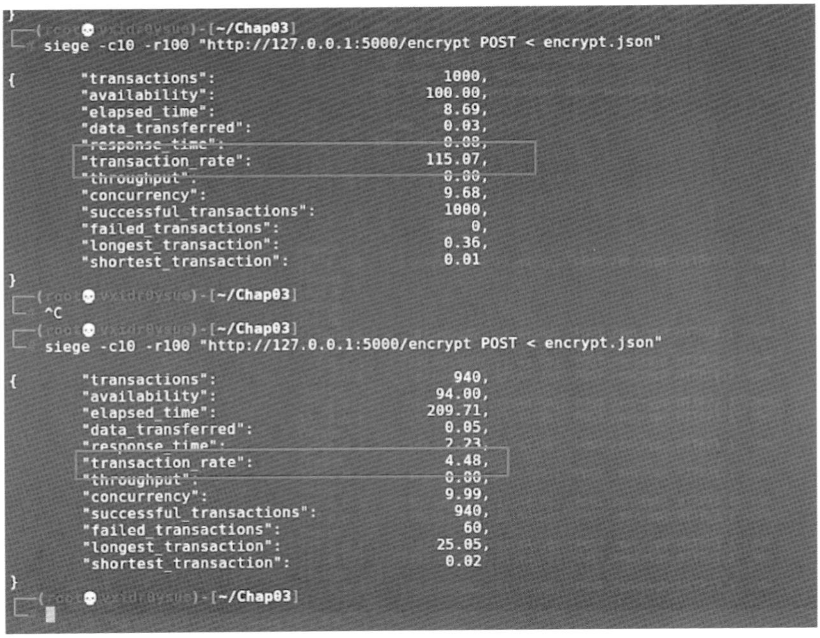

▲ 圖 3-12 Siege 測試結果

另外，在測試過程中發現，相比 method01() 函數的 RPC 呼叫速率，method02() 函數的呼叫速率總是比較慢，這是因為 Java.choose() 這個函數本身非常耗時，每次呼叫函數 method02() 都會在記憶體中重新搜索實例，這樣的操作需要盡可能避免。由於這裡樣本 App 的特殊性，只要 App 不退出，MainActivity 物件就會始終存在於記憶體中，因此這裡將 Java.choose() 搜索實例的部分取出到外部，並將搜索到的實例保存為外部全域變數，以便後續進行 RPC 呼叫，最終修改後的程式如程式清單 3-19 所示。

⬥ 程式清單 3-19 invoke.js

```js
var MainActivityObj = null;
Java.perform(function(){
    Java.choose("com.example.demoso1.MainActivity",{
        onMatch:function(instance){
            MainActivityObj = instance;
        },onComplete(){}
    })
    console.log("MainActivityObj is => ",MainActivityObj)
})
function invokeMethod02(ciphertext){
    var result;
    Java.perform(function(){
        var javaString = Java.use("java.lang.String")
        result = MainActivityObj.method02(javaString.$new(ciphertext))
    })
    return result;
}
```

經過這樣的修改，最終呼叫解密函數的性能獲得了很大的改善。事實上，這裡選取了一種取巧的方法，如果在其他的 App 中，一旦事先保存的物件被系統進行垃圾回收，後續的 RPC 呼叫就完全得不到想要的結果。那麼碰到這種情況怎麼辦呢？遇到這種情況，可使用下一節介紹的另一種主動呼叫方案來解決。

3.3 Frida Native 層函數主動呼叫

在上一節的最後，筆者提出了一種解決由於 Java.choose() 這個 API 導致解密方法 method02 函式呼叫效率過低問題的方案，但是這種解決方案實際上是投機取巧的：測試的樣本 App 中對應物件一定不會被釋放。在其他 App 中，類別物件的回收是非常正常且頻繁的，一旦目標類別物件被堆積進行垃圾回收，那麼對應的動態函數解密方式就會完全故障。

這裡要注意的是，兩個目標函數 method01() 和 method02() 其實都是 native 函數，如果將函數的主動呼叫放到 Native 層呢？事實上，如果是 native 函數的主動呼叫，那麼完全不會存在動態實例釋放的問題。在 Native 層中，無論是動態的實例函數 method02() 還是靜態函數 method01()，其實都會被當成普通函數處理。接下來將介紹 native 函數的主動呼叫。

要完成 native 函數的主動呼叫，筆者同樣堅持 Frida 三劍客的思想。

要 Hook 對應的函數，首先要找到 Java 函數在 Native 層對應的函數符號。在 Native 層要找到對應的函數符號，最終透過函數符號找到對應的函數位址。這時主要存在兩種情況：一種情況是，如果是靜態註冊的 JNI 函數，其對應的 Native 層函數符號只需要在原本的 Java 函數名前加上 Java_< 完整類別名 >_ 即可，比如在樣本 App 中的 stringFromJNI 函數是靜態註冊的，而該函數所在完整類別名為 com.example.demoso1.MainActivity，因此其對應的 Native 層函數簽名即為 Java_com_example_demoso1_MainActivity_stringFromJNI，最終針對該函數在 Native 層的 Hook 程式如程式清單 3-20 所示。

⬤ 程式清單 3-20 Hook 靜態註冊的 JNI 函數

```
function hookmethod(){
    var stringFromJNI= Module.findExportByName('libnative-lib.so',
                        'Java_com_example_demoso1_MainActivity_stringFromJNI')
    Interceptor.attach(stringFromJNI,{
        onEnter:function(args){
            // do something
        },onLeave: function(retval){
            // do something
        }
    })
}
```

當然，如果讀者不確定函數的符號，可以透過 Objection 的如下命令直接
查看匯出的符號清單，最終在列出的函數中根據 stringFromJNI 函數名進
行搜索以得到對應的函數符號，如圖 3-13 所示。

```
# memory list exports <mmodule_name>
```

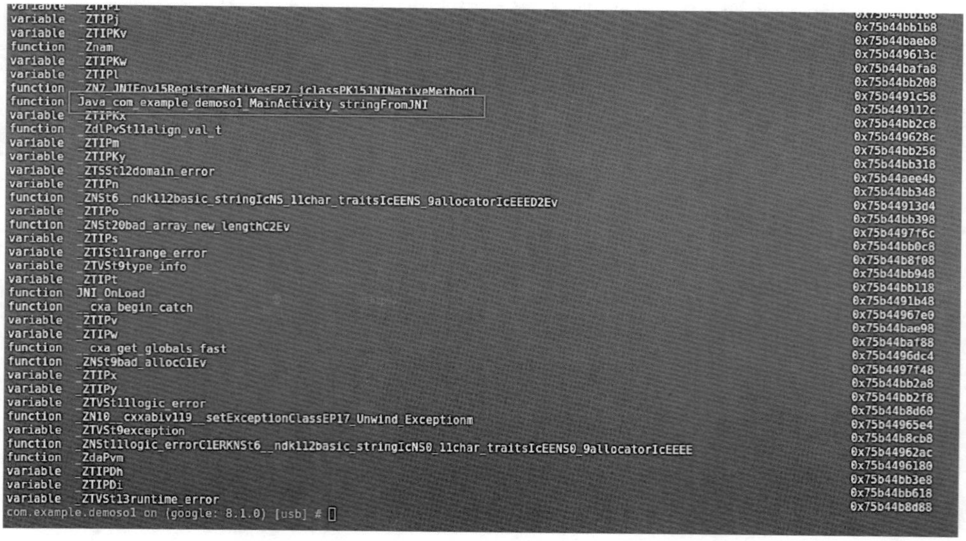

▲ 圖 3-13 Objection 列出的模組匯出函數

另一種情況是，如果函數是動態註冊的，可以使用 frida_hook_libart 專案（對應專案位址為 https://github.com/lasting-yang/frida_hook_libart）中的 hook_RegisterNatives.js 指令稿獲取動態註冊後的函數所在的位址，如圖 3-14 所示。

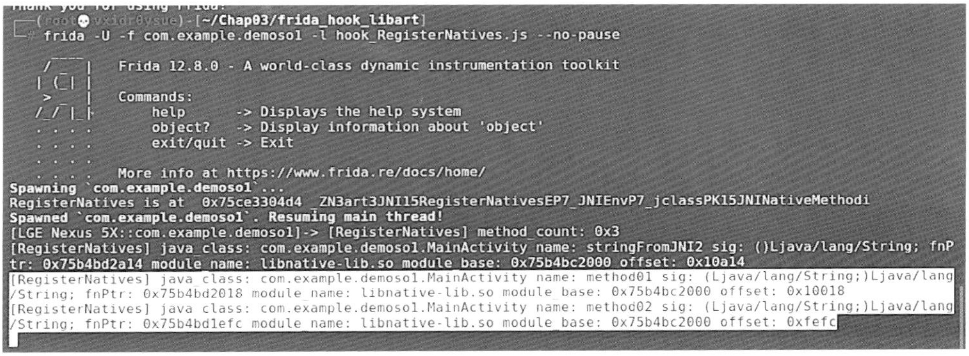

▲ 圖 3-14 hook_RegisterNatives 確定函數偏移

以 method01() 為例，該函數透過該指令稿找到的最終的 native 函數位址偏移分別為 0x10018，因此得到最終的 Hook 指令稿如程式清單 3-21 所示，Hook 的最終結果如圖 3-15 所示。

◎ 程式清單 3-21　Hook 函數 method01()

```
function hook_native_method(addr){
    Interceptor.attach(addr,{
        onEnter:function(args){
            console.log("args[0]=>",args[0]) // JNIEnv*
            console.log("args[1]=>",args[1]) // jclass
            console.log("args[2]=>",
                Java.vm.getEnv().getStringUtfChars(args[2], null)
                              .readCString()) // 呼叫jni函數,參考`frida-
java-bridge`
        },onLeave:function(retval){
```

```
        console.log('result => ',
            Java.vm.getEnv().getStringUtfChars(retval, null)
                            .readCString())
    }
  })
}
function hookmethod01(){
    var base = Module.findBaseAddress('libnative-lib.so')
    var method01_addr  = base.add(0x10018)
    hook_native_method(method01_addr)
}
```

```
Spawned `com.example.demoso1`. Resuming main thread!
[LGE Nexus 5X::com.example.demoso1]-> hookmethod01()
[LGE Nexus 5X::com.example.demoso1]-> args[0]=> 0x75ce6cc1c0
args[1]=> 0x7ff3982384
args[2]=> roysue
result =>   47fcda3822cd10a8e2f667fa49da783f
args[0]=> 0x75ce6cc1c0
args[1]=> 0x7ff3982384
args[2]=> roysue
result =>   47fcda3822cd10a8e2f667fa49da783f
args[0]=> 0x75ce6cc1c0
args[1]=> 0x7ff3982384
args[2]=> roysue
result =>   47fcda3822cd10a8e2f667fa49da783f
args[0]=> 0x75ce6cc1c0
args[1]=> 0x7ff3982384
args[2]=> roysue
result =>   47fcda3822cd10a8e2f667fa49da783f
```

▲ 圖 3-15 Hook method01() 函數

觀察程式清單 3-21 和圖 3-15，會發現這裡並沒有直接列印參數和結果，而是透過 Java.vm.getEnv() 獲取 JNIEnv* env 參數最終呼叫 JNI 函數 GetStringUtfChars()，從而得到儲存對應字串的位址，並透過 readCString() 函數獲取對應的字串，這部分其實是因為 Java 中的 String 參數到 Native 層中變成了 JString 物件，因此要獲取其實際內容，還得透過開發中的方式進行獲取。在 Frida 中，JNI 函數的使用方式參見 https://github.com/frida/frida-java-bridge/blob/master/lib/env.js，這裡不再贅述。

另外，對比程式清單 3-21 和 3-12 中 Hook 函數的方式，細心的讀者一定會發現，在 Native 層中使用 Interceptor.attach 這個 API 的 Hook 方式並沒有涉及函數的主動呼叫，這裡為了與 Java 層中函數 Hook 的方式一致，將 Interceptor.attach 替換為 Interceptor.replace，最終的 Hook 指令稿內容如程式清單 3-22 所示，Hook 結果如圖 3-16 所示。

⬭ **程式清單 3-22 存在主動呼叫的 Hook**

```
function replacehook(addr){
    // 根據位址得到
    var addr_func = new NativeFunction(addr,'pointer',['pointer','pointer',
'pointer']);
    Interceptor.replace(addr,new NativeCallback(function(arg1,arg2,arg3){
        // 確定主動呼叫可以成功，只要參數合法，位址正確
        var result = addr_func(arg1,arg2,arg3) // <== 主動呼叫
        console.log('arg3 =>', Java.vm.getEnv().getStringUtfChars(arg3,
null).readCString() )

        console.log("result is ",Java.vm.getEnv().getStringUtfChars(result,
null).readCString())
        return result;
    },'pointer',['pointer','pointer','pointer']))
}
function hookmethod01(){
    var base = Module.findBaseAddress('libnative-lib.so')
    var method01_addr  = base.add(0x10018)
    replacehook(method01_addr)
}
```

```
[LGE Nexus 5X::com.example.demoso1]->
[LGE Nexus 5X::com.example.demoso1]-> hookmethod01()
[LGE Nexus 5X::com.example.demoso1]-> arg3 => roysue
result is  47fcda3822cd10a8e2f667fa49da783f
arg3 => roysue
result is  47fcda3822cd10a8e2f667fa49da783f
arg3 => roysue
result is  47fcda3822cd10a8e2f667fa49da783f
arg3 => roysue
result is  47fcda3822cd10a8e2f667fa49da783f
arg3 => roysue
result is  47fcda3822cd10a8e2f667fa49da783f
arg3 => roysue
result is  47fcda3822cd10a8e2f667fa49da783f
arg3 => roysue
result is  47fcda3822cd10a8e2f667fa49da783f
arg3 => roysue
result is  47fcda3822cd10a8e2f667fa49da783f
arg3 => roysue
result is  47fcda3822cd10a8e2f667fa49da783f
arg3 => roysue
```

▲ 圖 3-16 Hook method01() 函數

在確定函數能夠被 Hook 後，基於程式清單 3-22 中的主動呼叫部分，最終寫出 method01 函數的主動呼叫方法如程式清單 3-23 所示。

◎ 程式清單 3-23 method01 函數的主動呼叫

```
function invoke_func(addr,contents){
    var result = null;
    var func = new NativeFunction(addr,'pointer',['pointer','pointer','poin
ter']); // new一個native函數
    Java.perform(function(){
        var env = Java.vm.getEnv();
        console.log("contents is ",contents)
        var jstring =  env.newStringUtf(contents)
        result = func(env,ptr(1),jstring)
        // console.log("result is =>",result)
        result = env.getStringUtfChars(result, null)
    })
    return result;
}
function invoke_method01(){
    var base = Module.findBaseAddress('libnative-lib.so')
```

```
    var method01_addr  = base.add(0x10018)
    var result  = invoke_func(method01_addr,"r0ysue")
    console.log("result is ",result.readCString())
}
```

在這個主動呼叫的指令稿中，有以下幾個需要注意的地方：

（1）在 JNI 函數中，第一個參數一定是 JNIEnv 的指標，第二個參數取決於對應 JNI 函數在 Java 層中是靜態還是動態函數，分別對應 jclass 類型和 jobject 類型，用於指示函數在 Java 層中的類別或者實例物件。這兩個參數在主動呼叫時都需要進行建構，其中 JNIEnv 的指標可以透過 Java.vm.getEnv() 進行建構，而第二個參數由於在函數中並沒有使用到，因此可以任意傳遞相同類型的資料，這裡使用 ptr(1) 建構了一個指標。而如果第二個參數在 JNI 函數中被使用，就需要透過 env 物件進行建構，至於如何建構，讀者可以自行研究實現。

（2）由於函數在主動呼叫時使用了 Java.vm.getEnv() 這個 API，因此需要包裹在 Java.perform() 中。最終主動呼叫 method01() 函數的結果如圖 3-17 所示。

▲ 圖 3-17　method01 函數的主動呼叫結果

在確認能夠主動呼叫後，便可以按照 Java 函數主動呼叫的方式對 native 函數進行匯出，並設定最終的 RPC 和批次呼叫，這裡不再贅述。

最後，筆者還要介紹一種脫離特定 APK 載入對應模組並呼叫 native 函數的方式。同樣，以樣本 APK 中的 method01() 函數為例進行介紹。

要做到這一點，首先需要解壓對應 APK 將目標函數所在模組匯出，再透過 ADB 工具推送到 /data/app 目錄下，並以 Root 使用者身份指定對應模組所有權限，最終效果如圖 3-18 所示。

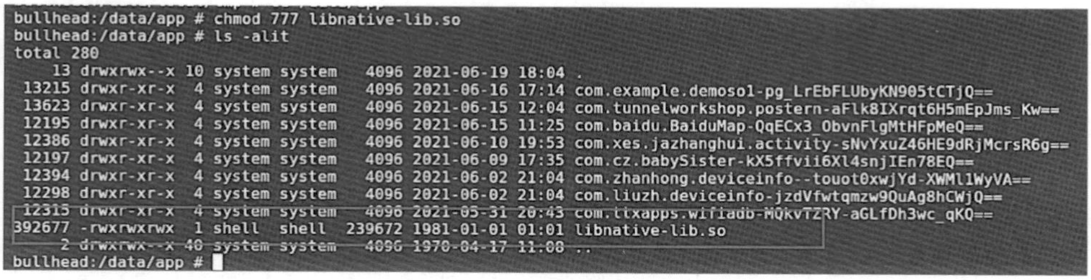

```
bullhead:/data/app # chmod 777 libnative-lib.so
bullhead:/data/app # ls -alit
total 280
      13 drwxrwx--x 10 system system     4096 2021-06-19 18:04 .
   13215 drwxr-xr-x  4 system system     4096 2021-06-16 17:14 com.example.demoso1-pg_LrEbFLUbyKN905tCTjQ==
   13623 drwxr-xr-x  4 system system     4096 2021-06-15 12:04 com.tunnelworkshop.postern-aFlkBIXrqt6H5mEpJms_Kw==
   12195 drwxr-xr-x  4 system system     4096 2021-06-15 11:25 com.baidu.BaiduMap-QqECx3_ObvnFlgMtHFpMeQ==
   12386 drwxr-xr-x  4 system system     4096 2021-06-10 19:53 com.xes.jazhanghui.activity-sNvYxuZ46HE9dRjMcrsR6g==
   12197 drwxr-xr-x  4 system system     4096 2021-06-09 17:35 com.cz.babySister-kX5ffvii6Xl4snjIEn78EQ==
   12394 drwxr-xr-x  4 system system     4096 2021-06-02 21:04 com.zhanhong.deviceinfo--touot0xwjYd-XWMl1WyVA==
   12298 drwxr-xr-x  4 system system     4096 2021-06-02 21:04 com.liuzh.deviceinfo-jzdVfwtqmzw9QuAg8hCWjQ==
   12315 drwxr-xr-x  4 system system     4096 2021-05-31 20:43 com.tixapps.wifiadb-MQkvTZRY-aGLfDh3wc_qKQ==
  392677 -rwxrwxrwx  1 shell  shell   239672 1981-01-01 01:01 libnative-lib.so
       2 drwxrwx-- 40 system system     4096 1970-04-17 11:08 ..
bullhead:/data/app #
```

▲ 圖 3-18 libnative-lib.so 模組許可權

在確認對應模組的許可權後，便可以透過 Frida 所提供的 Module.load() 函數對該模組進行載入並執行其中的函數，最終主動呼叫 method01() 函數的程式如程式清單 3-24 所示。

● 程式清單 3-24 主動載入模組並呼叫其中的函數

```
function invoke_method01_1(){
    var base     = Module.load('/data/app/libnative-lib.so').base
    var method01_addr  = base.add(0x10018)
    var result = invoke_func(method01_addr,"r0ysue")
    console.log("result is ",result.readCString())
}
```

此時無論注入任何應用，method01 函數都可以成功被呼叫，比如這裡注入「設定」應用，主動呼叫 method01() 函數的結果如圖 3-19 所示。

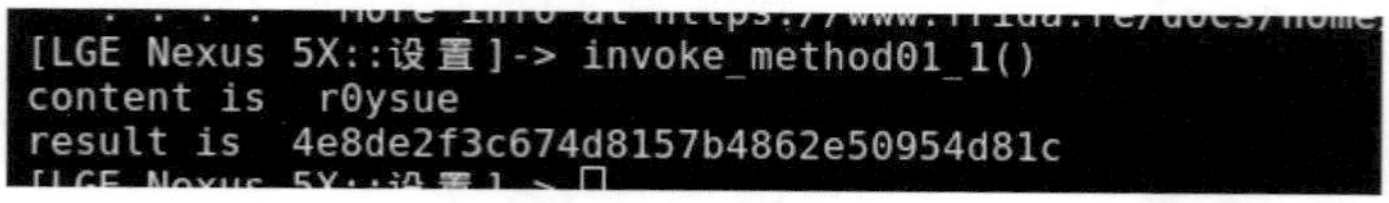

```
[LGE Nexus 5X::设置]-> invoke_method01_1()
content is  r0ysue
result is  4e8de2f3c674d8157b4862e50954d81c
[LGE Nexus 5X::设置]->
```

▲ 圖 3-19 注入「設定」應用主動呼叫 method01() 函數的結果

當然，這裡介紹的是一種簡單的模組脫離具體 App 進行呼叫，真實的情況是在脫離具體 App 進行模組函式呼叫時往往會發生各種問題，比如簽名驗證、Native 層呼叫 Java 函數等，而這些就需要逆向工程師進一步研究和繞過了。

3.4 本章小結

本章主要介紹了利用 Frida 進行函數主動呼叫以及 RPC 的方式，在筆者始終堅持的 Frida 三劍客的理念中，這兩個部分的存在非常重要。Frida 提供的主動呼叫方式給使用者複現固定參數下的函數執行流程提供了一種方式，這為可功耗時漫長的逆向偵錯分析排除了因為不同參數導致的不同執行流程對演算法還原造成的阻礙，而 RPC 和主動呼叫的結合一方面能夠多次呼叫以驗證演算法還原的正確性，另一方面還避免了專案在規定時間內無法完成逆向要求的尷尬。另外，在這一章中還介紹了一些 Frida 的 API 使用方式，相信借助這一章的學習，讀者能夠進一步意識到 Frida 的強大之處。

Frida 逆向之違法 App 協定分析與取證實戰

在之前的章節中，介紹了 Frida 的使用方法以及在實踐中的應用，還介紹了兩種使用 Frida 快速定位關鍵類別的方式。本章將以兩個違法的樣本為例，透過對兩個 App 某些關鍵協定的分析過程，帶領讀者更加深入地了解 Frida 工具的使用與 Hook 的思想。

4.1 保護 App 協定分析

本節將以樣本「行動 TV」為例，透過登入協定的分析介紹分析人員在逆向過程中從最初的資料抓取到最後脫離 App 進行利用的全過程。

4.1.1 抓取封包

在 Android App 的逆向分析中，抓取封包通常是指透過一些手段獲取 App 與伺服器之間傳輸的明文網路資料資訊，這些網路資料資訊往往是分析的切入點，透過抓取封包得到的資訊往往可以快速定位關鍵介面函數的位置，為從浩如煙海的程式中找到關鍵的演算法邏輯提供便利。甚至可以説如果連封包都抓不到，那麼後續的逆向分析也就無從談起了。

一般來説，要抓取 Android App 的資料封包，通常採取中間人抓取封包的形式。分析人員透過在手機上設定代理，將手機上的流量資料轉發到電腦上的代理軟體後再完成上網，這樣就可以實現在電腦上監聽手機上流量資料的效果。因為中間人抓取封包的方式無法應對 App 採用加密協定（比如 HTTPS 等）進行通訊的情況，因此逆向人員還需將代理軟體自身的證書匯入手機系統並加入憑證信任清單中；同時，如果 App 連使用者增加到系統中的證書都不信任，那麼分析人員還需要透過一些手段將增加的證書從使用者信任區移動到系統信任清單中。

通常在使用中間人的方式進行抓取封包時，有些分析人員可能會採用如圖 4-1 所示的在「設定」應用中設定 WiFi 代理的方式完成資料的轉發。但是這種設定代理的方式，一方面無法處理非 HTTP(S) 資料通信（比如 WebSocket 等協定）的資料轉發；另一方面 WiFi 代理方式經常被 App 程式檢測或繞過，比如程式清單 4-1 中這樣的 API 會直接獲取當前網路代理狀況，進而導致最終抓不到資料封包。因此，這裡推薦使用 VPN 代理方式。

⬤ 程式清單 4-1 對抗抓取封包

```
System.getProperty("http.proxyHost");
System.getProperty("http.proxyPort");
```

相對於 WLAN 直接從應用層設定代理的方式，VPN 代理則是透過虛擬出一個新的網路卡並修改手機路由表的方式完成網路通訊的。這樣做不僅可以繞過如程式清單 4-1 這種方式的檢測，而且 VPN 代理的方式能夠抓取的資料封包更加全面和完整。雖然也存在一些對抗 VPN 代理的方式，但是這樣的方式相對較少，也比較容易繞過。

▲ 圖 4-1 WiFi 代理

在選擇 VPN 代理工具時，筆者推薦 Postern，透過 Postern 可以自訂設定伺服器位址與通訊埠，同時還支援選擇代理類型，如圖 4-2 所示。

要注意的是，在選擇圖 4-2 中的代理類型時，建議使用 SOCKS5 而非 HTTP/HTTPS，這是因為相對於使用 HTTP/HTTPS 代理類型，SOCKS5 代理工作於網路 7 層模型的傳輸層，它比工作在應用層的 HTTP/HTTPS 代理能夠觀察到更多的協定資訊。

▲ 圖 4-2 VPN 代理

另外，與之對應的，在電腦上的代理工具這裡選擇的是 Charles 軟體，其代理模式也要選擇為 SOCKS 模式，如圖 4-3 所示。

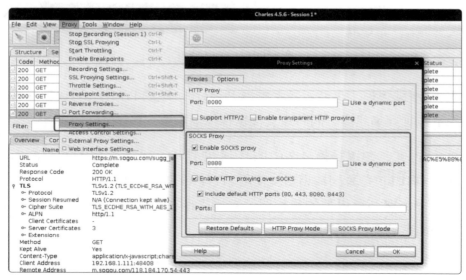

▲ 圖 4-3 Charles 設定為 SOCKS 模式

當手機上的 Postern 代理及抓取封包軟體 Charles 正確設定後，如果手機和執行 Charles 的電腦能夠順利 Ping 通（用於確認手機和電腦能夠相互連通），那麼 Android 手機上的網路通訊資料封包就會成功地被 Charles 攔截。最終設定完成後的樣本「行動 TV」的登入資料封包如圖 4-4 所示。

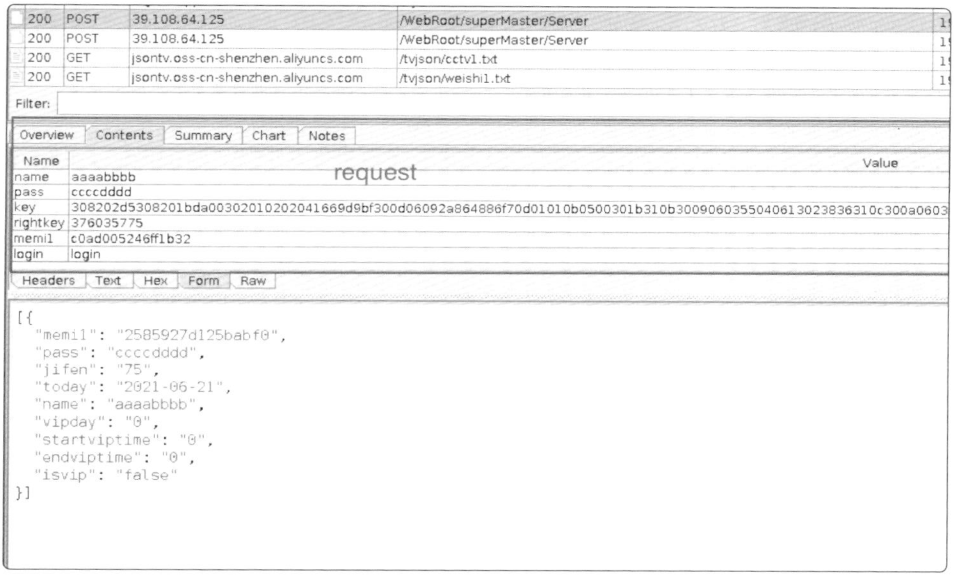

200	POST	39.108.64.125	/WebRoot/superMaster/Server	1
200	POST	39.108.64.125	/WebRoot/superMaster/Server	1
200	GET	jsontv.oss-cn-shenzhen.aliyuncs.com	/tvjson/cctv1.txt	1
200	GET	jsontv.oss-cn-shenzhen.aliyuncs.com	/tvjson/weishi1.txt	1

Filter:

Overview | Contents | Summary | Chart | Notes

Name	request	Value
name	aaaabbbb	
pass	ccccdddd	
key	308202d5308201bda0030201020204b669d9bf300d06092a864886f70d01010b0500301b310b3009060355040613023836310c300a0603	
rightkey	376035775	
memil	c0ad005246ff1b32	
login	login	

Headers | Text | Hex | Form | Raw

```
[{
  "memil": "2585927d125babf0",
  "pass": "ccccdddd",
  "jifen": "75",
  "today": "2021-06-21",
  "name": "aaaabbbb",
  "vipday": "0",
  "startviptime": "0",
  "endviptime": "0",
  "isvip": "false"
}]
```

▲ 圖 4-4 「行動 TV」登入資料封包

本小節並未仔細介紹增加代理軟體證書到系統信任區以及 Postern 和 Charles 的使用方式。

4.1.2 註冊 / 登入協定分析

在多次成功抓取 App 登入資料封包後，為了達到最終離線利用的目的，觀察圖 4-4 的 request 請求資料封包中的參數，會發現其中的 name 和 pass 欄位就是我們在測試時輸入的用戶名和密碼的明文組合，而 login 欄位對應的名稱固定為 login 字串。其他參數的含義如果只是抓取到資料封包,是無法百分百確定的,因此還需要對這些欄位的形成方式進行進一步的分析。

要完成對這些欄位的分析,通常需要先找到欄位形成的地方。如果讀者在學習本章前一直是透過靜態分析工具反編譯 App 檔案進而透過搜索辨識符號字串的方式找到形成對應欄位的程式的,那麼一定會經常被多個搜索結果所干擾。為了避免這種情況,建議讀者使用在第 2 章中介紹的其中一種快速定位關鍵類別的方式一基於記憶體列舉的關鍵類別定位方案以加快逆向速度。這裡基於使用者登入一定要點擊「登入」按鈕這一特性,而按鈕控制項 Button 在 Android 中屬於 View 類別的繼承類別,因此理論上可以透過 Hook View 類別的 onClick 函數快速得到當前控制項 onClick 函數所在類別,程式清單 4-2 hookEvent.js 就是這種理論的一種實現。

● 程式清單 4-2 hookEvent.js

```
var jclazz = null;
var jobj = null;
function getObjClassName(obj) {
    if (!jclazz) {
        var jclazz = Java.use("java.lang.Class");
    }
    if (!jobj) {
        var jobj = Java.use("java.lang.Object");
    }
    return jclazz.getName.call(jobj.getClass.call(obj));
}
function watch(obj, mtdName) {
    var listener_name = getObjClassName(obj);
    var target = Java.use(listener_name);
    if (!target || !mtdName in target) {
```

```
            return;
        }
    target[mtdName].overloads.forEach(function (overload) {
        overload.implementation = function () {
            console.log("[WatchEvent] " + mtdName + ": " +
getObjClassName(this))
            return this[mtdName].apply(this, arguments);
        };
    })
}
function OnClickListener() {
    Java.perform(function () {

        //以spawn啟動處理程序的模式來注入
        Java.use("android.view.View").setOnClickListener.implementation =
function (listener) {
            if (listener != null) {
                watch(listener, 'onClick');
            }
            return this.setOnClickListener(listener);
        };

        //如果Frida以attach的模式進行注入
        Java.choose("android.view.View$ListenerInfo", {
            onMatch: function (instance) {
                instance = instance.mOnClickListener.value;
                if (instance) {
                    console.log("mOnClickListener name is :" +
getObjClassName(instance));
                    watch(instance, 'onClick');
                }
            },
            onComplete: function () {
            }
        })
```

```
    })
}
setImmediate(OnClickListener);
```

在使用 Frida 注入樣本 App 並點擊「登入」按鈕後,最終效果如圖 4-5 所示,從而得到該控制項所在類別為 com.cz.babySister.activity. LoginActivity。

▲ 圖 4-5 點擊「登入」按鈕得到控制項所在類別

別在得到控制項所在的類別後,要繼續得到具體的業務程式,就需要了解靜態分析的流程,透過反編譯結果得到具體的程式細節。但不幸的是,筆者在反編譯時發現這個樣本竟然是保護狀態,慶倖的是,現有的很多脫殼工具可以解決這個問題,比如一代脫殼工具 dexdump,二代取出殼脫殼工具 Frida_fart、FART,等等。

在完成脫殼後,重新使用 Jadx 開啟脫殼後的 DEX 檔案定位到 LoginActivity 類別以及對應的 onClick 函數,程式的大致邏輯如圖 4-6 所示,可以發現輸入的用戶名和密碼都被當作參數傳遞給 b() 函數了。

```
private SharedPreferences t;

public void onClick(View view) {
    if (view.getId() == 2131230865) {
        this.m.setText("");
    }
    if (view.getId() == 2131230868) {
        this.n.setText("");
    } else if (view.getId() == 2131230863) {
        this.k.hideSoftInputFromWindow(view.getWindowToken(), 0);
    } else if (view.getId() == 2131230862) {
        String trim = this.n.getText().toString().trim();
        String trim2 = this.m.getText().toString().trim();
        if ("".equals(trim)) {
            Toast.makeText(this, "用戶名不能为空!", 0).show();
        } else if ("".equals(trim2)) {
            Toast.makeText(this, "密码不能为空!", 0).show();
        } else {
            this.k.hideSoftInputFromWindow(view.getWindowToken(), 0);
            c("");
            b(trim, trim2);
        }
    } else if (view.getId() == 2131230866) {
        this.k.hideSoftInputFromWindow(view.getWindowToken(), 0);
        Intent intent = new Intent();
        intent.setClass(this, RegisterActivity.class);
        startActivity(intent)
    } else if (view.getId() != 2131230860) {
    } else {
        if (this.s) {
            this.s = false;
            this.r.setImageResource(R$mipmap.icon_photo_selected2);
            return;
        }
        this.s = true;
        this.r.setImageResource(R$mipmap.icon_photo_selected1);
    }
}
```

▲ 圖 4-6 onClick 函數邏輯

為了印證靜態分析的正確性，再次使用 Objection 注入應用並使用如下命令針對 b() 函數進行 Hook，最終在再次點擊「登入」按鈕後，Objection 的 Hook 結果如圖 4-7 所示，其中 aaaabbbb 和 ccccdddd 分別是輸入的用戶名和密碼。

```
# android hooking watch class_method com.cz.babySister.activity.Login
Activity.b --dump-args --dump-backtrace --dump-return
```

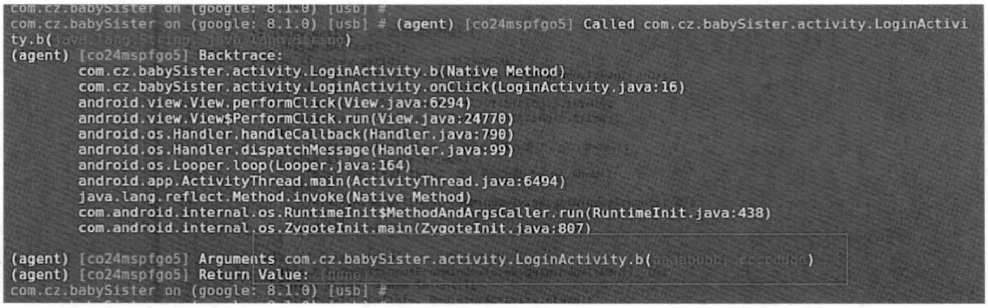

▲ 圖 4-7　b() 函數的 Hook 結果

在確定是 b() 函數傳遞用戶名和密碼後，透過靜態分析追蹤其實現，並透過 Hook 一一驗證，最終確定參數形成的具體位置（如程式清單 4-3 所示）。其中變數 StringResource.URL 為 http://39.108.64.125/WebRoot/superMaster/Server，對比圖 4-4 會發現其實就是登入網址。

⬥ 程式清單 4-3　登入簽名形成關鍵函數

```
class q implements Runnable{
    ...
    public void run(){
        String e = LoginActivity.e(this.c);
        String b2 = this.c.b();
        String a2 = BaseActivity.a((Context) this.c);
        String a3 = a.a(StringResource.URL, "name=" + this.name +
"&pass=" + this.pass + "&key=" + b2 + "&rightkey=" + a2 + "&memi1=" + e +
"&login=login");
        if (a3 == null || "".equals(a3)) {
            LoginActivity.a(this.c, "登入失敗!");
            return;
        }
        ...
    }
}
// b2 key的實現
```

```
public String b() {
    try {
        Signature[] signatureArr = getPackageManager().getPackageInfo(getPa
ckageName(), 64).signatures;
        StringBuilder sb = new StringBuilder();
        for (Signature signature : signatureArr) {
            sb.append(signature.toCharsString());
        }
        return sb.toString();
    } catch (PackageManager.NameNotFoundException e2) {
        e2.printStackTrace();
        return "";
    }
}
// a2 rightkey形成方式
public static String a(Context context) {
    for (PackageInfo packageInfo : context.getPackageManager().
getInstalledPackages(64)) {
        if (packageInfo.packageName.equals(context.getPackageName())) {
            try {
                CertificateFactory instance = CertificateFactory.
getInstance("X.509");
                ByteArrayInputStream byteArrayInputStream = new ByteArrayIn
putStream(packageInfo.signatures[0].toByteArray());
                byteArrayInputStream.close();
                return ((X509Certificate) instance.generateCertificate(byte
ArrayInputStream)).getSerialNumber().toString().trim();
            } catch (Exception e2) {
                e2.printStackTrace();
            }
        }
    }
    return "123";
}
// e memi1形成函數
```

```
public String d() {
    try {
        String string = Settings.Secure.getString(getContentResolver(),
"android_id");
        if (string == null || "".equals(string)) {
            return "0";
        }
        return string;
    } catch (Exception e) {
        e.printStackTrace();
        return "0";
    }
}
```

繼續追蹤程式就會發現，實際上變數 b2 即 key 欄位，對應的是 App 的
簽名，變數 a2 即 rightkey 欄位，對應的是 App 部分簽名資料，而變數 e
即 memi1 欄位，對應的是 android_id。最終發現實際上除了用戶名和密
碼外，其他參數都是固定值，因此如果想要脫離 App 完成使用者的登入
行為十分簡單，只需要輸入正確的用戶名和密碼再傳入固定的 memi1、
rightkey 和 key 參數即可。

同樣，讀者可以按照上述分析的邏輯繼續分析樣本 App 的註冊等業務，
最終筆者撰寫了一段 Python 程式用於離線實現註冊與登入操作，其程式
內容如程式清單 4-4 所示。圖 4-8 是註冊一個新帳號及登入獲取到個人資
訊的結果。

● 程式清單 4-4 invoke.py

```
import base64
import time

import requests
requests.packages.urllib3.disable_warnings()
```

```python
class tv:
    def __init__(self):
        self.root = 'http://39.108.64.125/WebRoot/superMaster/Server'
        self.memi1 = "0ae7635c6a9a0942"
        # APK簽名：可寫可不寫，簽名的頭部都是3082
        self.rightkey = "376035775"

        self.key = "..." # key 太長了，這裡省略，可直接參考附件程式

    def post(self, data=None):
        if data is None:
            data = {}
        return requests.post(url=self.root, data=data)

    def query(self, name, password):
        ret = self.post({'name': name, 'pass': password})
        print("query result is : ")
        print(ret.content.decode('utf-8'))

    def register(self, name, password):
        ret = self.post({'name': name, 'pass': password, 'memi1': self.memi1,
                         'key': self.key, 'rightkey': self.rightkey,
'register': 'register'})
        print("Register response data: ")
        print(ret.content.decode('utf-8'))

    def login(self, name, password):
        ret = self.post({'name': name, 'pass': password, 'memi1': self.memi1,
                         'key': self.key, 'rightkey': self.rightkey,
'login': 'login'})
        print("Login response data: ")
        print(ret.content.decode('utf-8'))

    def updateSocre(self, name, password, jifen):
        t = int(round(time.time() * 1000))
```

```
        sign = base64.b64encode(str(5 * t).encode('utf-8')).decode('utf-8')
        ret = self.post({'name': name, 'pass': password,
                         'jifen': jifen, 'time': t, 'sign': sign})
        print("UpdataScore response data: ")
        print(ret.content.decode('utf-8'))


if __name__ == "__main__":
    tv = tv()

    # 註冊帳號

    print(tv.register("aaaabbbb4", "ccccdddd4"))

    # time.sleep(3)

    # 登入帳號
    print(tv.login("aaaabbbb4", "ccccdddd4"))
```

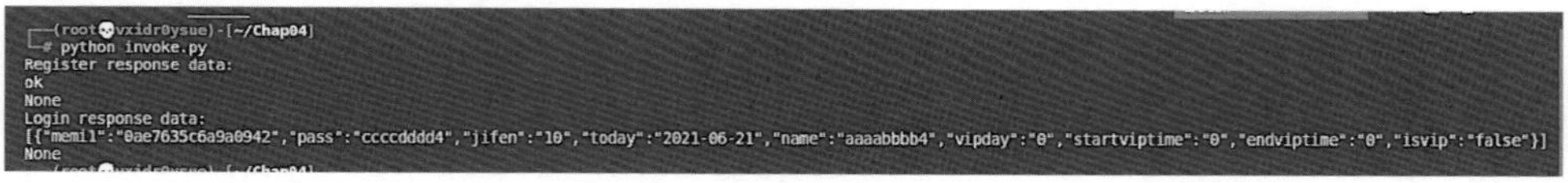

▲ 圖 4-8 離線執行結果

4.2 違法應用取證分析與 VIP 破解

本節將透過介紹另一個違法樣本 App 的逆向分析過程繼續深入 Frida 的學習。

4.2.1 VIP 清晰度破解

如圖 4-9 所示，已知樣本 App 在觀看視訊時如果想切換清晰度，就需要購買 VIP。想讓我們為違法應用付費？這完全就是異想天開，讓我們一步一步來破解這個功能。

▲ 圖 4-9 視訊清晰度切換

面對這種破解性的難題，一般來說可以跳過抓取封包的步驟，直接進入定位關鍵類別的邏輯。那麼如何快速定位清晰度切換的邏輯呢？相信看過上一節的讀者都知道，由於清晰度切換是一個按鈕控制項，因此只需要再次使用程式清單 4-2 中的 watchEvent.js 指令稿即可。最終得到切換清晰度的 View 控制項所在類別名為 com.ilulutv.fulao2.film.l$t，如圖 4-10 所示。

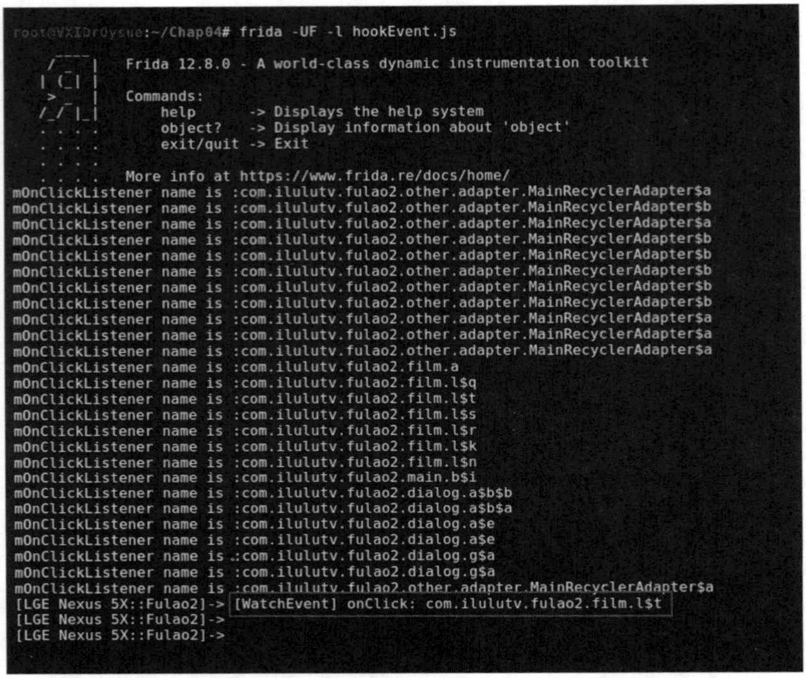

▲ 圖 4-10 Hook 結果

在定位到對應的類別名後，由於這個樣本並未保護處理，因此可以直接使用 Jadx 等靜態分析工具開啟 APK 檔案並搜索定位到類別名。這裡要注意的是，在 Frida 中列印出來的美金符 "$" 代表子類別，而在 Jadx 中子類別的連接方式還是透過 "." 符號，因此搜索時需將 "$" 符號替換為 "." 符號。最終定位到關鍵的類別程式如程式清單 4-5 所示。

🔽 程式清單 4-5 關鍵程式

```
public void i() {
    if (h() != null) {
        androidx.fragment.app.d h2 = h();
        if (h2 != null) {
            ((PlayerActivity) h2).a(true, "playpage_dialog");
```

```
            return;
        }
        throw new TypeCastException("null cannot be cast to non-null type
com.ilulutv.fulao2.film.PlayerActivity");
    }
}
static final class t implements View.OnClickListener {
    /* renamed from: d  reason: collision with root package name */
    final /* synthetic */ l f11236d;

    t(l lVar) {
        this.f11236d = lVar;
    }

    public final void onClick(View view) {
        if (!this.f11236d.q0) {
            this.f11236d.i();
        } else if (!l.a(this.f11236d).d()) {
            this.f11236d.a(true, 8, 0, false, true);
        } else if (this.f11236d.m0) {
            this.f11236d.a(true, 8, 0, false, true);
        } else {
            this.f11236d.a(false, 8, 8, false, false);
        }
    }
}
```

觀察程式清單 4-5 中 onClick 函數會發現有一些判斷敘述。如果第一個 if
敘述中 this.f11236d.q0 變數的值為 false，則會呼叫 i() 函數，而 i() 函數
正是一個彈出視窗的 Dialog，對應圖 4-9 中彈出的「VIP 限定功能」視
窗。

為了印證靜態分析的結果，可以使用 WallBreaker 在記憶體中搜索 l 類別
的實例並列印物件內容。如圖 4-11 所示，最終會發現唯一存在的 l 物件
實例中的 q0 變數的值的確為 false。

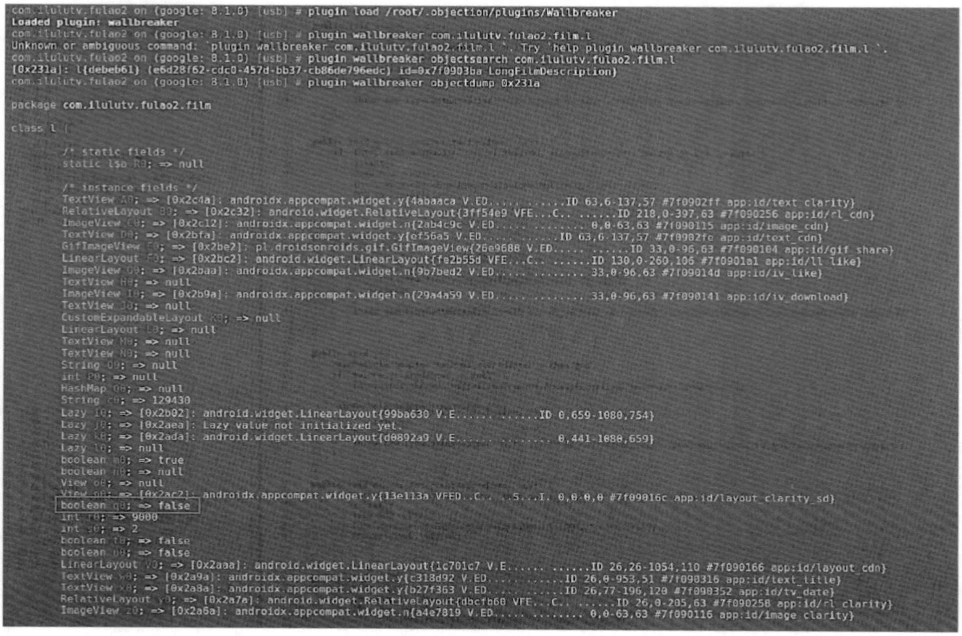

▲ 圖 4-11 使用 WallBreaker 查看實例中的 q0 值

如果這個 q0 的值為 true，那麼會不會繞過清晰度限制呢？可以使用 Frida 指令稿透過記憶體搜索 l 類別的實例並修改其中 q0 變數的值進行驗證。最終 Frida 指令稿的內容如程式清單 4-6 所示。

🔻 程式清單 4-6 hookq0.js

```
function hookq0(){
    Java.perform(function(){
        Java.choose("com.ilulutv.fulao2.film.l",{
            onMatch:function(ins){
                console.log("found ins:=>",ins)
                ins.q0.value = true;

            },onComplete:function(){
```

```
                console.log("search completed!")
            }
        })
    })
}
```

最終，在使用 Frida 重新注入應用並執行 hookq0() 函數後，再次點擊切換畫質的按鈕發現並未快顯視窗並且成功修改了視訊清晰度，因此最終確認程式清單 4-5 中 i() 函數確實是 VIP 限制的彈出視窗。

4.2.2 圖片取證分析

在完成 VIP 許可權的破解後，讓我們來分析樣本 App 的協定內容。正如 4.1 節所介紹的那樣，協定分析的第一步一定是針對樣本流量的抓取與關鍵字段的定位。

本節將介紹另一種抓取封包的方式—Hook 抓取封包。之所以出現 Hook 抓取封包的方式，是因為相比于使用中間人抓取封包的方式，利用 Hook 抓取封包所抓取的流量資料更加「專一」，不會受到手機上其他 App 資料流程量的影響；同時，利用 Hook 抓取封包可以避免 App 本身各種對抗抓取封包的姿勢，比如伺服器驗證用戶端、SSL Pinning 等手段。但相對的，如果 App 有著對抗 Frida 等 Hook 工具的手段，那麼這種方式就需要其他 Bypass Hook 檢測的輔助。

除此之外，Hook 抓取封包的效果取決於找到的 Hook 點，我們知道 Android 上封裝的網路通訊協力廠商函數庫種類豐富，比如 OkHttp3、Retrofit 等，如果選取的 Hook 點只是針對某一個通訊函數庫，那麼在實戰過程中就會面臨無法抓取其他類型的網路通訊資料的尷尬境地；另一方面，由於通訊協定類型的多樣性，僅僅找到 HTTP(S) 協定的 Hook 點是不夠全面的。因此，就有了開發的 Android 應用層抓取封包通殺指令稿：r0capture。該專案貫徹從網路模型下層觀測上層資料的理念，選取系統中在 Socket 層發送和接收資料封包的關鍵函數作為 Hook 點，完美通殺 TCP/IP 四層模型中的最上層應用層中的全部協定，包括 HTTP、WebSocket、FTP、XMPP 等明文協定以及這些明文協定對應的 SSL 通訊加密版本。不僅如此，r0capture 還通殺所有 Android 應用層框架，包括 HttpUrlConnection、OkHttp1/3/4、Retrofit/Volley 等。可以説只要 App 最終透過系統 API 完成通訊資料的發送與接收，就無法逃過 r0capture 的掌心。

在這個樣本中，我們使用 r0capture 指令稿的雛形 hookSocket.js 進行抓取封包工作，並透過 Frida 命令列提供的 -o 參數將抓取到的資料保存到 hookCapture.log 檔案中。在得到資料封包後，筆者發現樣本 App 可以説是武裝到牙齒，甚至每一個圖片的資料封包都是加密狀態的（如圖 4-12 所示，透過將每一個圖片的資料封包起始位元組與標準的 JPEG 格式的圖片檔案表頭的 hex 值相比進行判定，標準 JPEG 起始位元組 hex 為 0a 45 70）。

雖然圖片在傳輸過程中進行加密了，但是最終呈現在使用者眼中的圖片一定是處於解密狀態的，那麼 App 是如何對圖片進行解密的呢？這裡就以圖片的資料解密為切入點對 App 的協定進行分析。

Hook 抓取封包的另一個好處是，可以透過列印呼叫堆疊的方式確定發送封包函數執行前所經過的函數，其中可能就有關鍵的資料加密的部分。但是這裡基於離資料越近就越有效的原則，由於圖片資料在收發送封包函數的時候仍舊處於加密狀態，因此收發送封包函數的地方並不是離真實圖片資料最近的地方，那麼什麼時候離資料最近呢？

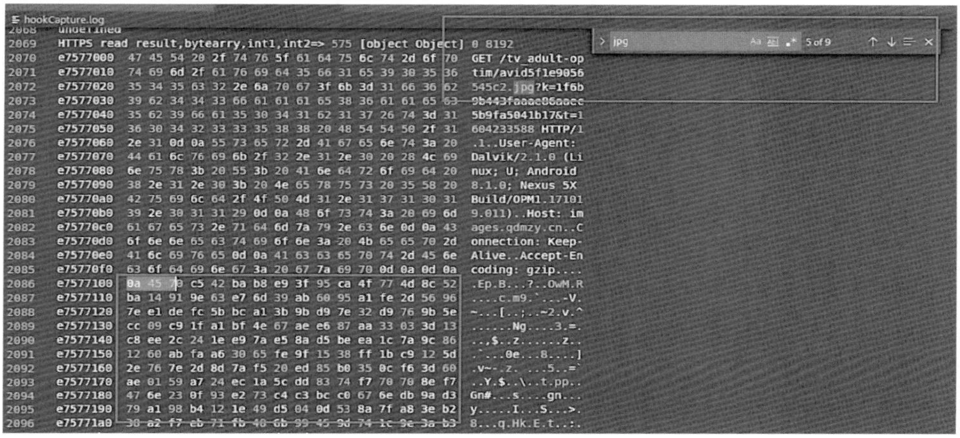

▲ 圖 4-12　圖片資料封包

圖片要載入的時候是離資料最近的時候。

為了進一步了解在 Android 中如何載入圖片，筆者特地去查了開發相關資料發現：在 Android 中，通常使用 BitmapFactory 類別中的函數去載入 Bitmap 物件，最終透過控制項 ImageView 去載入 Bitmap 物件類型的圖片，從而呈現出一個使用者可見的圖片。

透過開發的內容發現，載入圖片的重要的類別是 Bitmap 類別、BitmapFactory 類別以及 ImageView 控制項類別。其中，由於 ImageView 是 View 控制項類別，只是相當於一個放置東西的位置，而這裡更關注填充在位置中的內容，因此這裡更關注 Bitmap 內容本身以及用於建立 Bitmap 內容的 BitmapFactory 類別。

為了印證資料的正確性，這裡首先使用 WallBreaker 在記憶體中搜索 Bitmap 物件，並在手動觸發圖片的載入後再次搜索 Bitmap 物件。如圖 4-13 所示是最終兩次搜索 Bitmap 物件後在記憶體中的物件數量比較，可以確認在案例 App 中是使用 Bitmap 物件來保存圖片的。

▲ 圖 4-13 搜索 Bitmap 物件

在確認 Bitmap 物件是案例 App 所使用的圖片格式後，還需要進一步對圖片建立的方式進行探索。筆者在查詢資料後發現在 Android 開發的過程中，正常使用 BitmapFactory 類別提供的 4 個靜態方法：decodeFile()、decodeResource()、decodeStream() 和 decodeByteArray()，分別用於從檔案系統、資源、輸入流以及位元組陣列中載入出一個 Bitmap 物件。為了確認在這個 App 中使用的具體函數，這裡直接使用如下命令對 BitmapFactory 類別中的所有函數進行 Hook，如圖 4-14 所示是在完成對這個類別中所有函數 Hook 後手動觸發載入圖片邏輯所呼叫的函數清單。

```
# android hooking watch class android.graphics.BitmapFactory
```

觀察圖 4-14，發現實際上在案例 App 中所載入的函數為 decodeByteArray()。查看這個函數的具體使用方式，會發現這個函數的第一個參數就是儲存的原始圖片的位元組資訊。可以肯定的是，App 在這一步已經要載入和呈現圖片，此時 decodeByteArray() 函數的第一個參數所儲存的圖片資訊一定是解密狀態，那麼作為圖片取證的第一步：找到圖片的位元組資訊就已經成功完成。

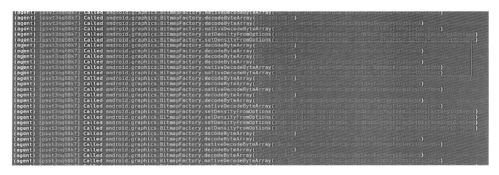

▲ 圖 4-14 BitmapFactory 類別中的函式呼叫

為了進一步確認所獲得的圖片資訊是明文狀態，這裡使用 Frida 指令稿的方式進一步獲取 decodeByteArray() 函數的第一個參數資訊並保存為圖片檔案進行確認，其中 Hook 的目標函數可以是圖 4-14 中出現的任意函數名為 decodeByteArray 的多載函數，這裡選擇 decodeByteArray(byte[] data, int offset, int length, Options opts) 多載。最終 Frida 指令稿內容如程式清單 4-7 所示。

◆ 程式清單 4-7 saveBitmap.js

```
function guid() {
    return 'xxxxxxxx-xxxx-4xxx-yxxx-xxxxxxxxxxxx'.replace(/[xy]/g, function
(c) {
        var r = Math.random() * 16 | 0,
            v = c == 'x' ? r : (r & 0x3 | 0x8);
        return v.toString(16);
    });
}
function saveBitmap_1(){
    Java.perform(function(){
        // public static Bitmap decodeByteArray(byte[] data, int offset,
int length, Options opts)
        Java.use('android.graphics.BitmapFactory').decodeByteArray.
overload('[B', 'int', 'int', 'android.graphics.BitmapFactory$Options').
implementation =  function(data,offset,length,opts){
```

```javascript
        var result = this.decodeByteArray(data,offset,length,opts)
        /*
        var ByteString = Java.use("com.android.okhttp.okio.ByteString");
        console.log("data is =>",ByteString.of(data).hex())
        */
        console.log("data is coming!")

        /*
        File f1 = new File("d:\\ff\\test.txt");
        fos = new FileOutputStream(f1);
        byte bytes[] = new byte[1024];
        fos.write(s.getBytes());
        fos.close();
        */
        var path = '/sdcard/Download/tmp/'+guid()+'.jpg'
        console.log("path is =>",path);
        var f = Java.use("java.io.File").$new(path)
        var fos = Java.use("java.io.FileOutputStream").$new(f)
        fos.write(data);
        fos.close();

        return result
    }
})

}
setImmediate(saveBitmap_1)
```

在程式清單 4-7 中，要注意 guid() 函數是為了生成一個隨機的字串作為保存的圖片名稱，最終圖片保存的目錄是手機的 /sdcard/Download/tmp/ 資料夾下。另外，這裡使用 Java 的 File 類別完成圖片檔案的讀寫（注意，tmp 資料夾需要手動建立，同時要保證 App 具有儲存許可權）。

在使用 Frida 以 attach 模式注入 App 後，手動刷新觸發圖片的載入，最終會發現手機的 /sdcard/Download/tmp/ 目錄下確實出現了很多圖片檔案，從而確定了我們的分析思路無誤。

在完成對關鍵函數的 Hook 後，如果只是簡單地想要對 App 進行圖片取證工作，那麼到這裡就可以考慮最終的 RPC 工作，將圖片資料直接保存到電腦上用於後續取證。程式清單 4-8 和 4-9 分別是最後修改的 JavaScript 指令稿和實現 RPC 呼叫的 Python 指令稿內容。

⬤ 程式清單 4-8　hookBitmap.js

```
function saveBitmap_4(){
    Java.perform(function(){
        // public static Bitmap decodeByteArray(byte[] data, int offset,
int length, Options opts)
        Java.use('android.graphics.BitmapFactory').decodeByteArray.
overload('[B', 'int', 'int', 'android.graphics.BitmapFactory$Options').
implementation = function(data,offset,length,opts){
            var result = this.decodeByteArray(data,offset,length,opts)
            send(data)
            return result
        }
    })

}
```

⬤ 程式清單 4-9　saveBitmap.py

```
import frida
import json
import time
import uuid


def my_message_handler(message, payload):
    if message["type"] == "send":
        image = message["payload"]

        intArr = []
        for m in image:
            ival = int(m)
            if ival < 0:
```

```
            ival += 256
        intArr.append(ival)
     bs = bytes(intArr)

     fileName = "/root/Chap10/tmp/"+str(uuid.uuid1()) + ".jpg"
     print('path is ',fileName)
     f = open(fileName, 'wb')
     f.write(bs)
     f.close()

device = frida.get_usb_device()
target = device.get_frontmost_application()
session = device.attach(target.pid)
# 載入指令稿
with open("hookBitmap.js") as f:
    script = session.create_script(f.read())
script.on("message", my_message_handler)   # 呼叫錯誤處理

script.load()

# 指令稿會持續執行等待輸入
input()
```

修改完畢後，執行 saveBitmap.py 這個 Python 指令檔並再次在手機上對圖片進行刷新，就能在 /root/Chap04/tmp 目錄下看到最終生成的圖片。

但是我們的目的實際上是對圖片解密的協定進行分析。要做到這一點，需要重新回到 hook BitmapFactory 類別的那一步。

為了獲得 App 的業務層相關邏輯，相信讀者一定會想到使用 Objection 去 Hook 在圖 4-14 中出現的 decodeByteArray 函數並列印呼叫堆疊。如圖 4-15 所示，最終在 Hook decodeByteArray 函數並手動觸發載入圖片後，發現 com.ilulutv.fulao2.other.helper.glide.b.a 函數是關鍵的業務層程式（這裡更下層的 com.bumptech.glide 相關函數是 Android 中用於動態載入圖片的協力廠商函數庫）。

▲ 圖 4-15 定位業務層程式

在定位到用於載入圖片的關鍵業務層位置後，使用 Jadx 開啟案例 App 並檢索對應函數，最終得到關鍵的函數內容如程式清單 4-10 所示。

● **程式清單 4-10 業務層關鍵函數**

```java
public v<Bitmap> a(Object obj, int i2, int i3, i iVar) {
    // 加密狀態
    String encodeToString = Base64.encodeToString(com.ilulutv.fulao2.
other.i.b.a((ByteBuffer) obj), 0);
    String decodeImgKey = CipherClient.decodeImgKey();
    Intrinsics.checkExpressionValueIsNotNull(decodeImgKey,
"CipherClient.decodeImgKey()");
    Charset charset = Charsets.UTF_8;
    if (decodeImgKey != null) {
        byte[] bytes = decodeImgKey.getBytes(charset);
        Intrinsics.checkExpressionValueIsNotNull(bytes, "(this as java.
lang.String).getBytes(charset)");
        byte[] decode = Base64.decode(bytes, 0);
        String decodeImgIv = CipherClient.decodeImgIv();
        Intrinsics.checkExpressionValueIsNotNull(decodeImgIv,
"CipherClient.decodeImgIv()");
```

```
         Charset charset2 = Charsets.UTF_8;
         if (decodeImgIv != null) {
             byte[] bytes2 = decodeImgIv.getBytes(charset2);
             Intrinsics.checkExpressionValueIsNotNull(bytes2, "(this as
java.lang.String).getBytes(charset)");
             // 下一行程式執行完後，圖片位元組資訊已經處於解密狀態
             byte[] c2 = com.ilulutv.fulao2.other.i.b.c(decode, Base64.
decode(bytes2, 0), encodeToString);
             if (c2 == null) {
                 Intrinsics.throwNpe();
             }
             // 這裡載入BitmapFactory.decodeByteArray
             return com.bumptech.glide.load.q.d.e.a(BitmapFactory.
decodeByteArray(c2, 0, c2.length), this.f12023a);
         }
         throw new TypeCastException("null cannot be cast to non-null
type java.lang.String");
     }
     throw new TypeCastException("null cannot be cast to non-null type
java.lang.String");
 }
```

通讀程式清單 4-10 中的函數，會發現實際上在 com.ilulutv.fulao2.other.
i.b.c 函數執行完畢後載入的圖片位元組資訊就已經處於明文狀態，而
com.ilulutv.fulao2.other.i.b.c 函數內容如程式清單 4-11 所示。實際上 c
函數是一個 CBC 模式的 AES 解密函數，其中第一個參數是 AES 解密使用
的金鑰 key，第二個參數是 AES 解密使用的向量 IV，而第三個參數就是
圖片的加密位元組陣列進行 Base64 編碼後的字串。

◯ 程式清單 4-11 解密函數

```
public static final byte[] c(byte[] bArr, byte[] bArr2, String
str) throws NoSuchAlgorithmException, NoSuchPaddingException,
IllegalBlockSizeException, BadPaddingException, InvalidAlgorithmParameterEx
```

```
ception, InvalidKeyException {
        Cipher instance = Cipher.getInstance("AES/CBC/PKCS5Padding");
        instance.init(2, new SecretKeySpec(bArr, "AES"), new
IvParameterSpec(bArr2));
        byte[] doFinal = instance.doFinal(Base64.decode(str, 2));
        Intrinsics.checkExpressionValueIsNotNull(doFinal, "cipher.
doFinal(Base64.de…de(text, Base64.NO_WRAP))");
        return doFinal;
}
```

結合對 c 函數的分析，再次回頭看程式清單 4-10 中函數的內容，會發現變數 encodeToString 就是用於儲存圖片加密資料進行 Base64 編碼後的字串，變數 decodeImgKey 就是進行 Base64 編碼後的金鑰 key，變數 decodeImgIv 就是 Base64 編碼後的向量 IV。

為了獲得金鑰和向量的值，這裡可以採取 Hook 的方式進行獲取，但是在觀察程式清單 4-10 中對 decodeImgIv 和 decodeImgKey 變數的獲取方式後，會發現實際上這兩個變數分別是 CipherClient 類別的兩個靜態函數的返回值，因此可以直接透過主動呼叫的方式對 AES 解密的 key 和 IV 進行獲取，最終主動呼叫獲取 key 和 IV 具體程式如程式清單 4-12 所示。

◯ 程式清單 4-12 getKey 函數

```
function getKey(){
    Java.perform(function(){
        var CipherClient = Java.use('net.idik.lib.cipher.so.CipherClient')
        var key = CipherClient.decodeImgKey()
        var iv = CipherClient.decodeImgIv()
        console.log(key,iv)

    })
}
```

最終得到 key 和 IV 進行 base64 編碼後的內容分別為 svOEKGb5WD0ezm
HE4FXCVQ＝＝ 和 4B7eYzHTevzHvgVZfWVNlg＝＝。據此可以根據抓取
封包得到的資料得到最終的明文圖片資料。

考慮到要獲取圖片還需要獲取特定的圖片名稱，這裡還是採取 RPC 的方
式來模擬實現最終的離線抓取圖片資料。根據上述針對程式清單 4-10 業
務層關鍵函數的分析，可以判定在 com.ilulutv.fulao2.other.i.b.a 函數執
行後的返回資料就是在抓取封包時的資料封包內容。最終模擬實現的抓
取封包指令稿如程式清單 4-13 所示，用於解密資料封包的 Python 指令
稿內容如程式清單 4-14 所示。

⦿ 程式清單 4-13 hookBitmap.js

```
function hookEncodedBuffer() {

    Java.perform(function () {
        var base64 = Java.use("android.util.Base64")
        // com.ilulutv.fulao2.other.i.b.a((ByteBuffer) obj)
        Java.use("com.ilulutv.fulao2.other.i.b").a.overload('java.nio.
ByteBuffer').implementation = function (obj) {
            var result = this.a(obj);
            //var ByteString = Java.use("com.android.okhttp.okio.ByteString");
            //console.log("data is =>",ByteString.of(result).hex())
            send(result)
            return result
        }
    })
}
```

⦿ 程式清單 4-14 saveBitmap.py

```
import frida
import json
import time
import uuid
```

```python
import base64
from Crypto.Cipher import AES

def decrypt():
    key = 'svOEKGb5WD0ezmHE4FXCVQ=='
    iv =  '4B7eYzHTevzHvgVZfWVNIg=='

def IMGdecrypt(bytearray):
    imgkey = base64.decodebytes(
        bytes("svOEKGb5WD0ezmHE4FXCVQ==", encoding='utf8'))

    imgiv = base64.decodebytes(
        bytes("4B7eYzHTevzHvgVZfWVNIg==", encoding='utf8'))

    cipher = AES.new(imgkey, AES.MODE_CBC, imgiv)
    # enStr += (len(enStr) % 4)*"="
    # decryptByts = base64.urlsafe_b64decode(enStr)
    msg = cipher.decrypt(bytearray)
    def unpad(s): return s[0:-s[-1]]
    return unpad(msg)

def my_message_handler(message, payload):
    if message["type"] == "send":
        image = message["payload"]

        intArr = []
        for m in image:
            ival = int(m)
            if ival < 0:
                ival += 256
            intArr.append(ival)
        bs = bytes(intArr)

        bs = IMGdecrypt(bs)

        fileName = "/root/Chap04/tmp/"+str(uuid.uuid1()) + ".jpg"
        print('path is ',fileName)
        f = open(fileName, 'wb')
```

```
        f.write(bs)
        f.close()

device = frida.get_usb_device()
target = device.get_frontmost_application()
session = device.attach(target.pid)
# 載入指令稿
with open("hookBitmap.js") as f:
    script = session.create_script(f.read())
script.on("message", my_message_handler)   # 呼叫錯誤處理

script.load()
# 指令稿會持續執行等待輸入
input()
```

在執行 saveBitmap.py 指令稿後,就能夠在電腦中得到明文的圖片資料。

4.3 本章小結

本章透過對兩個樣本協定分析的過程將前兩章介紹的一些關於 Frida 和 Objection 的理論知識應用在實際的逆向分析過程中。可以發現,Frida 在逆向分析中的角色可以認為接近中心位置,而這也正是筆者十分推崇 Frida 的原因。另外,本章還介紹了一些抓取封包的姿勢,作為協定分析 的第一步,它往往是指引我們找到目標,當然本章並未詳細說明其中的 細節,讀者還要注意的是,在破解應用時,只要其中的邏輯是基於本地 判斷的,我們都可以破解;如果目標邏輯是基於伺服器判斷的就很難破 解,此時如果想要實現繞過,就需要尋找其中的業務邏輯漏洞。由於筆 者對這方面的內容不甚了解,在這裡就不再班門弄斧了,讀者如果對這 方面感興趣,可以自行研究。

Xposed Hook 及主動呼叫
與 RPC 實現

前　面的章節從 Hook、主動呼叫以及 RPC 三個方面介紹了 Frida 的
　　使用方式及其在逆向工程中的作用，並透過對 App 的實戰帶領讀
者深入理解了 Frida 三劍客的實際利用價值。本章將介紹另一款 Hook
工具—Xposed。與新興勢力 Frida 相比，Xposed 作為 Android Hook 界
的前輩，雖然在 Android 7.1 後再也沒有新的正式版本發佈，但是其作
為系統框架類型的 Hook 思想還是在 Android 安全界留下了濃墨重彩的
印記，甚至時至今日，仍舊存在著很大一部分 Android 安全研究員使
用 Xposed 作為 Hook 主力進行安全研究。除此之外，EdXposed 等基於
Xposed 後續開發的工具和產品也延續著 Xposed 的生命。因此，作為逆
向研究人員，了解 Xposed 很有必要。本章主要介紹 Xposed 的基本使用
並將其與 Frida 進行對比，以供讀者參考。

5.1 Xposed 應用 Hook

5.1.1 Xposed 安裝與 Hook 外掛程式開發入門

與 Frida 直接將對應版本的 Server push 到手機上的 /data/local/tmp 目錄後以 Root 使用者身份執行即可對目標處理程序進行 Hook 相比，要使用 Xposed 實現 App 的 Hook，首先需要在 Root 環境下透過 XposedInstaller App（Xposed 的外掛程式管理和功能控制 App）安裝對應系統的 Xposed 的框架；同時在 Xposed 框架安裝成功後，還需要安裝對應的 Hook 外掛程式並重新啟動，從而完成對目標處理程序的 Hook。

這裡要注意的是，由於 Xposed 本質上是透過替換 Android 系統中的 zygote 以及 libart.so 庫，從而將 XposedBridge.jar 注入應用中，最終實現針對應用處理程序的 Hook 的，因此 Xposed 框架與系統版本高度相關，但時至 2021 年，Xposed 的正式版最高只支持到 Android 7.1 版本。

要安裝 Xposed 框架，滿足以下兩個條件：

（1）Android 系統版本小於或等於 Android 7.1（雖然 Android 8.1 上仍舊有 Xposed 版本，但是非正式版本）。
（2）系統已 Root。

這裡選用 Android 7.1.2_r8 版本（之所以選擇這個版本，是因為這個版本支持的裝置最多），並透過 TWRP 將 SuperSU 刷入系統進行 Root。具體如何更新韌體與 Root 在第 1 章中已經詳細介紹過了，這裡不再贅述。

在系統 Root 並安裝上 XposedInstaller.apk 後，XposedInstaller 的主介面如圖 5-1 所示。

此時只需透過點擊頁面中的 Version 89 按鈕並在彈出的提示框中點擊 Install 按鈕並授予 Root 許可權，即可在等待 Xposed 框架下載完畢後重新啟動完成安裝，在安裝成功後 XposedInstaller 的主介面如圖 5-2 所示。

▲ 圖 5-1 未安裝 Xposed 框架前 XposedInstaller 的主介面

▲ 圖 5-2 安裝 Xposed 框架後 XposedInstaller 的主介面

在 Xposed 框架安裝完畢後，便可以正式開始學習如何開發一個 Xposed 外掛程式。

事實上，Xposed 外掛程式也是以 App 的形式安裝在系統中的，只是區別于普通 App 的開發，Xposed 外掛程式的開發還需要一些特別的設定。

（1）在 AndroidManifest.xml 中的 application 節點中增加如下 3 個 meta-data 屬性，分別用於表示是不是 Xposed 模組、Xposed 模組的介紹以及支援最低的 Xposed 版本。

```
<meta-data
    android:name="xposedmodule"
    android:value="true" /> <!-- 是不是Xposed模組-->
<meta-data
    android:name="xposeddescription" <!-- Xposed模組的介紹-->
    android:value="這是一個Xposed常式" />
<meta-data
    android:name="xposedminversion"  <!-- 最低的Xposed版本-->
    android:value="53" />
```

（2）在 app/src/main/assets 目錄下新建一個 xposed_init 檔案用於指定 Xposed 模組入口類別的完整類別名，這裡 Xposed 外掛程式的入口類別為 com.roysue.xposed1.HookTest，如圖 5-3 所示。

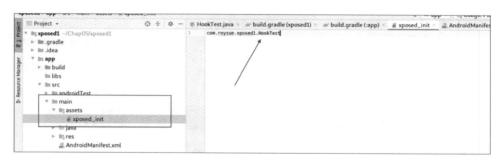

▲ 圖 5-3 xposed_init 檔案的內容

（3）在 App 專案的 app/build.gradle 檔案中的 dependencies 節點中加上以下依賴，並同步用於撰寫 Xposed 相關程式和執行 Hook 操作。

```
dependencies {
    compileOnly 'de.robv.android.xposed:api:82'
    compileOnly 'de.robv.android.xposed:api:82:sources'
    ...
}
```

至此，將一個設定好的 App 安裝到手機上，如圖 5-4 所示，XposedInstaller 即可將對應的 App 辨識為一個 Xposed 模組。此時將圖 5-4 中的核取方塊選取上並重新啟動手機，即可使得 xposed_init 檔案中指定的 Hook 入口類別生效。

當然，由於此時入口類別中無實際程式，因此即使重新啟動後也不會出現任何效果。接下來將正式開始介紹 Xposed 模組開發的一些 API 使用與 Hook 實現。

以 Xposed1 為例，App 的主要業務程式如程式清單 5-1 所示。

▲ 圖 5-4 Xposed 模組

● 程式清單 5-1 MainActivity.java

```java
package com.roysue.xposed1;
...
public class MainActivity extends AppCompatActivity {
    private Button button;
    @Override
    protected void onCreate(Bundle savedInstanceState) {

        super.onCreate(savedInstanceState);
        setContentView(R.layout.activity_main);
        button = findViewById(R.id.button);
        button.setOnClickListener(new View.OnClickListener() {
            public void onClick(View v) {
                Toast.makeText(MainActivity.this, toastMessage("我未被綁架")
                        ,Toast.LENGTH_SHORT).show();
            }
```

```
        });
    }

    public String toastMessage(String message) {
        return message;
    }
}
```

相信有一定開發基礎的讀者都知道如果沒有 Hook 程式的存在,該 Demo 在進入首頁面後,待使用者點擊按鈕,一定會彈出字串「我未被綁架」的 Toast 資訊,如圖 5-5 所示。

本次 Hook 的目標就是將程式清單 5-1 中 toastMessage() 函數的參數列印出來,並且修改該函數的返回值為「你已被綁架」。

要實現這一點,首先要將 xposed_init 檔案中指定的類別 (也就是 HookTest 類別) 實現 IXposedHookLoadPackage 介面,用於引入在安裝 Xposed 框架的系統中,每個 Zygote 孵化出來的 App 處理程序在啟動時都會呼叫函數 handleLoadPackage(),在實現 IXposedHookLoadPackage 介面後,HookTest 類別的程式內容如程式清單 5-2 所示。

▲ 圖 5-5 未 Hook 前

◆ 程式清單 5-2 HookTest 類別

```
package com.roysue.xposed1;

...

import de.robv.android.xposed.IXposedHookLoadPackage;
```

```
import de.robv.android.xposed.callbacks.XC_LoadPackage;

public class HookTest implements IXposedHookLoadPackage {
    public void handleLoadPackage(XC_LoadPackage.LoadPackageParam
loadPackageParam) throws Throwable {
        ...
    }
}
```

由於 handleLoadPackage() 函數在 App 啟動時會被呼叫，此時如果想要 Hook 指定處理程序，就需要透過 handleLoadPackage() 函數的參數 loadPackageParam 進行過濾。loadPackageParam 參數是一個 XC_LoadPackage.LoadPackageParam 類型的參數，它提供了一些有用的成員變數，用於表示應用處理程序的一些資訊，其中主要成員類型資訊如表 5-1 所示。

表 5-1　LoadPackageParam 類別中的成員含義表

編　號	成員變數類型	成員變數名	含　義
1	String	packageName	處理程序套件名
2	String	processName	處理程序名
3	ClassLoader	classLoader	處理程序類別載入器
4	ApplicationInfo	appInfo	應用的更多資訊

在過濾目標處理程序時，由於應用套件名是唯一標識 App 的方式，因此通常是透過 processName 成員進行過濾的。實現過濾後，就可以透過一些真實實現 Hook 的函數對目標處理程序中的函數進行 Hook 實現，而這就涉及 Xposed 中實現 Hook 最關鍵的類別 XposedHelpers。

XposedHelpers 類別中提供了無數關於 Java 類別、類別成員以及函數的介面函數，這裡如果要實現對 toastMessage 函數的 Hook，只需要利用其中一個類別函數 findAndHookMethod() 即可。

顧名思義，findAndHookMethod() 函數，就是用於尋找函數並 Hook 指定函數的函數。在使用該函數時，只需傳入對應函數所在類別的 handle（可以透過表 5-1 中的 classLoader 獲取）、對應函數名和參數清單以及最重要的 Hook 回呼類別 XC_MethodHook 即可。這裡 XC_MethodHook 是一個抽象函數，在具體傳入 findAndHookMethod() 函數作為參數時，需要實現其中兩個抽象回呼函數：beforeHookedMethod() 和 afterHookedMethod()，用於在目標函式呼叫前後進行呼叫。其中 beforeHookedMethod() 函數通常用於獲取和修改目標函數的參數類型，afterHookedMethod() 函數通常用於獲取和修改目標函數的返回值。這裡由於 toatMessage() 函數的參數即返回值，因此若要修改返回值，可採取兩種方式，最終對於 Demo 中該函數的 Hook 實現如程式清單 5-3 所示。

● 程式清單 5-3 Hook 實現

```
if (loadPackageParam.packageName.equals("com.roysue.xposed1")) {
    XposedBridge.log("inner => " + loadPackageParam.processName);
    Class clazz = loadPackageParam.classLoader
                              .loadClass("com.roysue.xposed1.
MainActivity"); // 獲取toastMessage函數所在類別的handle
    XposedHelpers.findAndHookMethod(clazz, "toastMessage", String.class,new
XC_MethodHook() {

        protected void beforeHookedMethod(MethodHookParam param) throws
Throwable {

            String oldText = (String) param.args[0];
            Log.d("din not hijacked=>", oldText);
            param.args[0] = "你已被綁架"; // Hook實現方式1
        }
        protected void afterHookedMethod(MethodHookParam param) throws
Throwable {
            Log.d("getResult is => ",(String) param.getResult());
```

```
        param.setResult("你已被綁架2");   // Hook實現方式2
    }
  });
}
```

在編譯並安裝 Xposed 模組到手機上後，啟動模組並重新啟動系統。此時再次開啟目標 App，會發現 App 的 toast 已成功更改，這也正是 Xposed 模組基礎的開發方式，最終被 Hook 應用彈出的 toast 效果以及對應日誌內容分別如圖 5-6 和圖 5-7 所示。

▲ 圖 5-6　Hook 後的效果

▲ 圖 5-7　Hook 後的日誌

5.1.2 Hook API 詳解

在 5.1.1 節中，透過對 Demo App 的 Hook 講解了透過 Xposed 提供的 API 實現簡單的關於應用函數的 Hook 工作。本節將介紹一個真實的被大量使用的 Xposed 模組—GravityBox，並透過 GravityBox 原始程式對 Xposed 的 Hook 相關 API 做進一步的詳細介紹。

首先，簡單地介紹一下 GravityBox。它可以做很多事情，包括狀態列調整、螢幕鎖定調整、電源調整等，如圖 5-8 所示是 GravityBox 在修改狀態列之後的系統頁面。

▲ 圖 5-8 GravityBox 實現狀態列修改

接下來正式介紹 GravityBox 的程式內容。

在使用 Android Studio 將下載的 Android 7 對應的 GravityBox 原始程式開啟後，按照前面的介紹，首先開啟的檔案是 xposed_init，透過這個檔案可以找到 GravityBox 作為 Hook 模組的入口類別為 com.ceco.nougat.
gravitybox.GravityBox。

在找到入口類別後，除去一些無關 Hook 的程式，首先關注的第一個
關鍵函數為 initZygote，該函數是 IXposedHookZygoteInit 介面中定
義的函數，用於在 Zygote 處理程序啟動時執行，也就是說每次系統開
機時都會執行一次，在實際使用過程中通常用於初始化工具類別，在
GravityBox 中用於初始化一些設定檔（透過 XSharedPreferences 這一包
裝的 SharedPrefrences 類別實現）和在 Xposed 日誌中列印系統關鍵資訊
（透過 XposedBridge.log() 函數實現），具體程式如程式清單 5-4 所示。

◯ 程式清單 5-4 initZygote 函數

```
@Override
// 開機執行
public void initZygote(StartupParam startupParam) throws Throwable {
    MODULE_PATH = startupParam.modulePath;
    if (Utils.USE_DEVICE_PROTECTED_STORAGE) {
        prefs = new XSharedPreferences(prefsFileProt);
        ...
    }
    ...

    ...
    XposedBridge.log("GB:ROM: " + Build.DISPLAY);
    XposedBridge.log("GB:Error logging: " + LOG_ERRORS);

    ...
}
```

在 initZygote 函數之後的是 handleInitPackageResources() 函數，該函數
是 IXposedHookInitPackageResources 介面中定義的函數，是處理程序資
源初始化後呼叫的回呼函數，可以用於替換處理程序資源。

接下來是在實現 Java Hook 中最重要的 handleLoadPackage() 函數，正如 5.1.1 節中所說的，該函數也是處理程序啟動時被呼叫的函數，其時機比 Application.onCreate() 函數還早，通常用於完成 Java 函數的 Hook 工作，而這個函數中的內容正是本節特別注意的部分。

在 handleLoadPackage() 函數中，如程式清單 5-5 所示，仔細閱讀原始程式後，可以發現很多 if 判斷敘述用於區分啟動的 App，並且根據目標處理程序的不同執行不同的 Hook 分支，從而最終實現在一個 Xposed 模組中 Hook 多個應用。

⬤ 程式清單 5-5 handleLoadPackage() 函數中的 Hook 相關分支

```
public void handleLoadPackage(LoadPackageParam lpparam) throws Throwable {
    ...
    if (lpparam.packageName.equals(SystemPropertyProvider.PACKAGE_NAME)) {
        SystemPropertyProvider.init(prefs, qhPrefs, tunerPrefs, lpparam.
classLoader);
    }
    // Common
    if (lpparam.packageName.equals(ModLowBatteryWarning.PACKAGE_NAME)) {
        ModLowBatteryWarning.init(prefs, qhPrefs, lpparam.classLoader);
    }

    if (lpparam.packageName.equals(ModClearAllRecents.PACKAGE_NAME)) {
        ModClearAllRecents.init(prefs, lpparam.classLoader);
    }
    ...
    // anaylsis
    if (lpparam.packageName.equals(ModStatusbarColor.PACKAGE_NAME)) {
        ModStatusbarColor.init(prefs, lpparam.classLoader);
    }
}
```

接下來以控制系統狀態列顏色的 App：ModStatusbarColor 為例進行介紹。

在 追 蹤 ModStatusbarColor.init(prefs, lpparam.classLoader) 函 數 的 實
現後會發現與 5.1.1 節中的程式類似，關鍵是對函數的 Hook 方式實
際 上 都 是 XposedHelpers 類 別 在 進 行 處 理，即 透 過 XposedHelpers
類 別 中 findAndHookMethod() 函 數 對 目 標 函 數 進 行 Hook。 如 程 式
清 單 5-6 所 示，與 5.1.1 節 不 同 的 是，在 init 函 數 中 存 在 著 另 一 個
findAndHookMethod() 函 數 的 多 載，這 個 多 載 的 第 一 個 參 數 不 是 目 標 函
數 的 handle，而是目標函數的類別名。但是相對應的其第二個參數的類
型 也 變 成 了 ClassLoader 類 別 載 入 器。事 實 上，這 兩 個 函 數 的 本 質 內 容 是
一 致 的，其 內 部 實 現 都 是 透 過 指 定 類 別 載 入 器 獲 取 對 應 函 數 的 控 制 碼，
從而完成對目標函數的 Hook 工作。

◉ 程式清單 5-6 ModStatusBarColor.java

```
// 第一種：findAndHookMethod
final Class<?> phoneStatusbarClass = XposedHelpers.findClass(CLASS_PHONE_
STATUSBAR, classLoader);
XposedHelpers.findAndHookMethod(phoneStatusbarClass,
        "makeStatusBarView", new XC_MethodHook(XCallback.PRIORITY_LOWEST)
        //優先順序{
    @Override
    protected void afterHookedMethod(final MethodHookParam param) throws
Throwable {
        mPhoneStatusBar = param.thisObject;
        Context context = (Context) XposedHelpers.getObjectField(param.
thisObject, "mContext"); //獲取物件中實例

        if (SysUiManagers.IconManager != null) {
            SysUiManagers.IconManager.registerListener(mIconManagerListener);
        }
```

```
        Intent i = new Intent(ACTION_PHONE_STATUSBAR_VIEW_MADE);
        context.sendBroadcast(i);
    }
});

// 第二種：findAndHookMethod
private static final String CLASS_SB_ICON_CTRL = "com.android.systemui.
statusbar.phone.StatusBarIconController";
XposedHelpers.findAndHookMethod(CLASS_SB_ICON_CTRL, classLoader,
"applyIconTint", new XC_MethodHook() {
    @Override
    protected void afterHookedMethod(MethodHookParam param) throws
Throwable {
        if (SysUiManagers.IconManager != null) {
            SysUiManagers.IconManager.setIconTint(
                    XposedHelpers.getIntField(param.thisObject,
"mIconTint")); // 獲取成員值
        }
    }
});
```

事實上，在程式清單 5-6 中，除了外面包裝的兩種 findAndHookMethod() 函數的實現外，還有以下幾點需要注意的地方：

（1）在第一種實現函數 Hook 的方法中，最後一個 XC_MethodHook 介面類別的實現方式不同：存在一個參數 XCallback.PRIORITY_LOWEST。事實上，在查詢官方 API 介紹後，筆者發現 XCallback 這個類別中存在著 3 個變數：PRIORITY_LOWEST、PRIORITY_DEFAULT 和 PRIORITY_HIGHEST，它們都用於表示變數 Hook 的優先順序，其中 PRIORITY_LOWEST 宣告的函數 Hook 執行最晚，PRIORITY_HIGHEST 宣告的函數 Hook 執行優先順序最高，PRIORITY_DEFAULT 宣告的函數 Hook 是預設的優先順序順序。

（2）在 Xposed 中獲取實例物件的方式十分簡單，透過 param.thisObject 即可拿到對應 Hook 的函數所在實例的物件。

（3）在 Frida 中，如果要獲取實例物件中的成員值，只需要透過實例物件加上成員名稱再加上 .value 即可，Xposed 同樣透過 XposedHelpers 類別中的函數 get＜type＞Field 獲取（這裡 type 可替換為基礎類型或者 Object），比如在程式清單 5-6 中獲取 context 物件，就是透過 XposedHelpers.getObjectField() 這個 API 實現的。事實上，XposedHelpers 類別中還會有著與 get＜type＞Field 方式對應的 set 類型方法，用於設定物件中的成員值。

（4）由程式清單 5-6 中獲取 context 資料可以發現，因為 Xposed 開發本身就是基於 Java 函數的，在獲取到特定類型的資料後，只需要透過 Java 轉換類型的方式進行強行轉換即可，而 Frida 必須透過 Java.cast 完成類型的轉換。

多看 GravityBox 中其他函數的 Hook 會發現，Xposed 還提供了 findAndHookConstructor() 函數用於 Hook 類別的建構函數。區別於 findAndHookMethod() 函數，Hook 類別的建構函數參數中無須傳遞函數名稱，具體展示如程式清單 5-7 所示。

◯ 程式清單 5-7 ModAudio.java

```
...
XposedHelpers.findAndHookConstructor("android.media.AudioManager",
classLoader, Context.class,
        new XC_MethodHook() {
    @Override
    protected void afterHookedMethod(MethodHookParam param) throws
Throwable {
        Object objService = XposedHelpers.callMethod(param.thisObject,
```

```
"getService");
        Context mApplicationContext = (Context) XposedHelpers.
getObjectField(param.thisObject,
                "mApplicationContext");
        if (objService != null && mApplicationContext != null) {
            XposedHelpers.callMethod(param.thisObject,
"disableSafeMediaVolume");
        }
    }
});
...
```

從上述程式的分析中可以發現，幾乎所有針對 App 中的類別、函數、變數的處理都是透過 XposedHelpers 類別中提供的函數完成的，而進一步觀察其中函數的實現會發現內部都是利用 Java 本身提供的反射相關 API 實現的，程式清單 5-8 是 getBooleanField() 函數的具體實現。

⭕ 程式清單 5-8 XposedHelpers.java 中 getBooleanField() 函數實現

```
public static boolean getBooleanField(Object obj, String fieldName) {
    try {
        return findField(obj.getClass(), fieldName).getBoolean(obj);// 反射
    } catch (IllegalAccessException e) {
        // should not happen
        XposedBridge.log(e);
        throw new IllegalAccessError(e.getMessage());
    } catch (IllegalArgumentException e) {
        throw e;
    }
}

public static Field findField(Class<?> clazz, String fieldName) {
  ...
  try {
    Field field = findFieldRecursiveImpl(clazz, fieldName);
```

```
    field.setAccessible(true);
    fieldCache.put(fullFieldName, field);
    return field;
  } catch (NoSuchFieldException e) {
    fieldCache.put(fullFieldName, null);
    throw new NoSuchFieldError(fullFieldName);
  }
}
```

關於 Xposed 中實現函數 Hook 相關的 API 暫時就介紹到這裡。可以發現
在 Hook 方面，Frida 與 Xposed 相比，Xposed 在函數 Hook 上的優勢在
於在一個函數中完成針對所有處理程序的 Hook，在 Zygote 啟動後即可
生效；而 Xposed 則是單處理程序等級的 Hook 框架。Xposed 可以視為
系統框架，作為系統本身來考慮；Frida 則更加類似於偵錯器，只能在被
Hook 的處理程序內生效。另外，Frida 是一個類似於熱載入的框架，能
夠隨時附加到處理程序中進行 Hook；而 Xposed 在每次 Hook 生效前都
需要重新啟動。

5.1.3 Xposed Hook 保護應用

相信用過 Xposed 的讀者都會發現一個問題，在對保護應用進行 Hook
時，如果直接對應用中的函數進行 Hook，則會提示如圖 5-9 所示的
ClassNotFoundException 錯誤。而在使用 Frida 時，以 spawn 模式對保
護 App 進行 Hook 時也會提示同樣的錯誤。那麼為什麼在對未保護應用
進行 Hook 時不會出現這樣的問題呢？或者説為什麼 Frida 在以 attach 模
式對保護 App 進行 Hook 時不會出現找不到類別的情況呢？

沒錯，就是時機問題。以更加專業、準確的方式來表述，其實就是類別載
入器 ClassLoader 在保護應用啟動時切換導致的問題。比較熟悉 Android

虛擬機器機制或者熟悉 JVM 的讀者應該知道，App 中的所有類別其實都是由對應的 ClassLoader 載入到 ART 虛擬機器中的。如果 ClassLoader 不正確，那麼一定無法找到對應的類別，最終造成圖 5-9 中的錯誤。之所以當使用 Frida 以 attach 模式對保護應用進行 Hook 時，進行函數 Hook 不會出現圖 5-9 這種情況，是因為以 attach 模式注入應用時，App 的當前 ClassLoader 已經被切換到載入對應類別的 loader，這也就是所謂的時機問題。

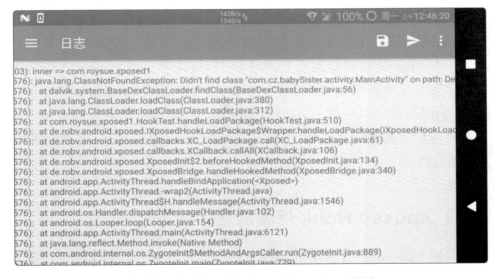

▲ 圖 5-9 ClassNotFoundException 錯誤

Frida 可以透過切換為 attach 模式的方式解決保護應用 Hook 的問題，那麼 Xposed 能夠實現 attach 模式注入處理程序嗎？

答案是不可以，由於 Xposed 本身可以近似於系統框架類型的性質，Xposed 注入處理程序的時機不可更改，而 Xposed 注入處理程序時 App 的 Application 類別並未完成載入，這也就導致真實用於載入 App 業務

相關類別的 ClassLoader 並未出現，導致最終無法實現 App 業務函數的
Hook 工作。

那麼有辦法解決這樣的問題嗎？

答案是有。我們可以透過手動切換 ClassLoader 的方式完成對應用函
數的 Hook。具體來說，當我們靜態分析保護 App 時會發現，殼程式
總是透過在應用處理程序中最先獲得執行許可權的 Application 類別
中的 attachBaseContext 和 onCreate 函數完成對真實 DEX 的釋放以及
ClassLoader 的切換，因此就有研究人員提出透過 Hook 對應殼程式的
Application 類別中的 attachBaseContext 或者 onCreate 函數得到真實
App 的上下文，再透過上下文獲取真實程式釋放後的 ClassLoader，用
於後續的函數 Hook。比如針對某保護 App，要完成對 App 業務函數的
Hook 的方式如程式清單 5-9 所示。

◒ 程式清單 5-9 某保護 App Hook 方式

```
XposedHelpers.findAndHookMethod("com.xxx.StubApp", loadPackageParam.
classLoader, "attachBaseContext", Context.class, new XC_MethodHook() {
    @Override
    protected void afterHookedMethod(MethodHookParam param) throws
Throwable {
        super.afterHookedMethod(param);
        Context context = (Context) param.args[0];
        // 獲取真實業務程式的classLoader
        ClassLoader finalClassLoader =context.getClassLoader();
        //下面就是強classloader修改成360的classloader就可以成功的hook了
        XposedHelpers.findAndHookMethod(clzz, finalClassLoader, "method",
..., new XC_MethodHook() {
            ...
        });
    }
});
```

雖然這樣的方式可以解決特定保護 App 的真實業務函數的 Hook 工作，但是這樣的方式不具有通用性，一旦保護廠商改變對應繼承 Application 類別的類別名，這樣的方式就故障了。那麼有沒有一種通用的可以解決任意保護應用函數 Hook 問題的方式呢？

答案是肯定的，這也正是這裡想要介紹的方式。

如果讀者對 App 啟動的流程十分了解，那麼一定知道 App 被 Zygote 處理程序孵化後，會透過 ActivityThread.main() 函數進入 App 的世界，在該函數中建立一個 ActivityThread 實例以及 App 處理程序的一些初始化工作。

事實上，ActivityThread 這個類別在應用中非常重要，它根據 ActivityManager 發送的請求對 activities、broadcast Receviers 等操作進行排程和執行。其中 performLaunchActivity() 函數用於回應 Activity 相關的操作。另外，ActivityThread 類別中還會有著一個 Application 類型的 mInitialApplication 成員，應用程式中有且僅有一個 Application 元件，而 Application 物件就儲存著應用當前的 ClassLoader，考慮到在應用響應 Activity 訊息活動訊息時，真實 App 的程式已經被釋放到記憶體中，如果此時透過 mInitialApplication 成員獲取應用當前 ClassLoader，即可完成真實 App 業務程式的 Hook 工作。這裡以行動 TV 樣本的 MainActivity 為例，具體實現 Hook 的程式如程式清單 5-10 所示，模組生效後，最終 Hook 效果如圖 5-10 所示。

◯ 程式清單 5-10 Hook 程式

```
if (loadPackageParam.packageName.equals("com.cz.babySister")) {

    XposedBridge.log(" has Hooked!");
    XposedBridge.log("inner  => " + loadPackageParam.processName);
```

```java
    Class ActivityThread = XposedHelpers.findClass("android.app.
ActivityThread", loadPackageParam.classLoader);
    XposedBridge.hookAllMethods(ActivityThread, "performLaunchActivity",
new XC_MethodHook() {
        @Override
        protected void afterHookedMethod(MethodHookParam param) throws
Throwable {
            super.afterHookedMethod(param);
            Application mInitialApplication = (Application) XposedHelpers.
getObjectField(param.thisObject, "mInitialApplication");

            ClassLoader finalLoader = mInitialApplication.getClassLoader();
            XposedBridge.log("found classloader is => " + finalLoader.
toString());

            Class BabyMain = finalLoader.loadClass("com.cz.babySister.
activity.MainActivity");
            XposedBridge.hookAllMethods(BabyMain,"onCreate", new XC_
MethodHook() {
                @Override
                protected void beforeHookedMethod(MethodHookParam param)
throws Throwable {
                    super.beforeHookedMethod(param);

                    XposedBridge.log("MainActivity onCreate called");
                }
            });
        }
    });
}
```

▲ 圖 5-10 Hook 效果

至此，無論任何保護類型的 App，只要其存在一個 Activity，按照程式清單 5-10 的方式進行 Xposed Hook 理論上都可以完美解決。在真實世界中，不存在 Activity 活動的使用者安裝的應用幾乎是不可能的，因此可以說這裡的 Xposed Hook 方式幾近通殺了。

5.1.4 使用 Frida 一探 Xposed Hook

之前介紹了關於 Xposed Hook 的一些 API 以及對應的使用方式，本小節將透過 Frida Hook 的方式來介紹 Xposed Hook 的一些技術細節，從而幫助讀者進一步理解 Xposed Hook 的原理。

這裡以在本章中撰寫的簡單的 Xposed Demo 為例，首先使用 Objection 注入目標處理程序，搜索 xposed_init 檔案中宣告的入口類別 com.

roysue.xposed1.HookTest，會發現出現兩個搜索結果（見圖 5-11），
但是進一步想要使用 list class_methods 命令列出類別中的函數或使用
watch class 命令對類別中的函數進行 Hook 時，就會發現始終會顯示出
錯；ClassNotFoundException，如圖 5-12 所示。

▲ 圖 5-11　HookTest 相關類別

▲ 圖 5-12　找不到 HookTest 類別中的函數顯示出錯

既然 Xposed Hook 的效果已經實現，那麼按道理實現 Hook 的類別和函
數就應該已經載入到記憶體中了，但是為什麼 Objection 在 Hook 時會顯
示出錯呢？

實際上按照上一小節的介紹，與 Xposed 無法直接應對保護 App 的
Hook 問題類似，這裡之所以無法 Hook，是因為 ClassLoader 不對，
而 Objection 工具並未結合 Frida 中切換 ClassLoader 的功能，因此這
裡如果想要針對這些實現 Hook 的類別和函數進行 Hook，則需要自
己撰寫指令稿完成這部分內容，以實現 ClassLoader 切換的功能。這
裡最終實現 ClassLoader 的遍歷和切換是利用 Frida 提供的 enumerate

ClassLoader API 函數完成的，關鍵程式如程式清單 5-11 所示。

◯ 程式清單 5-11 traceXposed.js

```
Java.enumerateClassLoaders({
    onMatch: function (loader) {
        try {
            if(loader.findClass("com.roysue.xposed1.HookTest")){
                console.log("Successfully found loader")
                console.log(loader);
                // 切換classLoader
                Java.classFactory.loader = loader;
            }
        }
        catch(error){
            // console.log("find error:" + error)
        }
    },
    onComplete: function () {
        console.log("end")
    }
})
```

最終在重新使用 Frida 將切換 ClassLoader 的程式注入處理程序中時，發現其所在類別載入器為處理程序的 PathClassLoader 中，如圖 5-13 所示。

▲ 圖 5-13　目標類別載入器

在完成類別載入器的切換後，再次對目標類別中的函數進行列印或者 Hook 會發現此時不會再顯示出錯，且 Hook 列印一切正常，這裡針對 HookTest 類別中的 PrintStack 函數進行 Hook，其結果如圖 5-14 所示，對應的 Hook 指令稿如程式清單 5-12 所示。

◆ 程式清單 5-12 traceXposed.js

```
Java.enumerateClassLoaders({
    ...
})
// 注意 Hook指令稿一定是在切換classloader工作完成後執行的
Java.use("com.roysue.xposed1.HookTest").PrintStack.implementation =
function () {
    console.log("entering PrintStack!")
    return true
}
```

▲ 圖 5-14 切換類別載入器後的 Hook 結果

筆者在後續陸續研究的過程中發現，如果在切換完 ClassLoader 後直接 Hook Xposed 框 架 提 供 的 API 函 數， 比 如 XposedBridge.log()、XposedHelpers.findAndHookMethod() 等，同樣會出現無法找到類別 ClassNotFoundException 的顯示出錯，修改切換 ClassLoader 程式後發現，實際上這部分程式在另一個 ClassLoader 中（XposedBridge.jar 檔案的 ClassLoader 中），其搜索結果如圖 5-15 所示。

▲ 圖 5-15 Xposed 框架中的函數所在 loader

那麼為什麼同樣是基於 Xposed 框架的類別，對應的 ClassLoader 還不同呢？

如圖 5-16 所示，對比 HookTest 這裡自實現的類別和 Xposed 框架本身程式時會發現，HookTest 類別實際上實現的是 Xposed 框架中的 IXposedHookLoadPackage 介面，而 XposedBridge 等框架類別則是直接呼叫 XposedBridge.jar 檔案中的附帶函數，也正因此，兩個類別所在的 loader 不同。

```java
public class HookTest implements IXposedHookLoadPackage {
    public void PrintStack(){

        XposedBridge.log( text: "Dump Stack: "+ "---------------start----------------");
        Throwable ex = new Throwable();
        StackTraceElement[] stackElements = ex.getStackTrace();
        if (stackElements != null) {
            for (int i = 0; i < stackElements.length; i++) {

                XposedBridge.log( text: "Dump Stack"+i+": "+ stackElements[i].getClassName()
                        +"----"+stackElements[i].getFileName()
                        +"----" + stackElements[i].getLineNumber()
                        +"----" +stackElements[i].getMethodName());

            }
        }
        XposedBridge.log( text: "Dump Stack: "+ "---------------over---------------");


        RuntimeException e = new RuntimeException("<Start dump Stack !>");
        e.fillInStackTrace();
        Log.i( tag: "<Dump Stack>:", msg: "++++++++++++", e);

    }

    //...


    public void handleLoadPackage(XC_LoadPackage.LoadPackageParam loadPackageParam) throws Throwable {
        //XposedBridge.log(loadPackageParam.processName);
        if (loadPackageParam.packageName.equals("com.roysue.xposed1")) {
            XposedBridge.log( text: " has Hooked!");
            XposedBridge.log( text: "inner"+loadPackageParam.processName);
            Class clazz = loadPackageParam.classLoader.loadClass( name: "com.roysue.xposed1.MainActivity");
            XposedHelpers.findAndHookMethod(clazz, methodName: "toastMessage", String.class,new XC_MethodHook() {

                protected void beforeHookedMethod(MethodHookParam param) throws Throwable {

                    String oldText = (String) param.args[0];
                    Log.d( tag: "din not hijacked=>", oldText);

                    //param.args[0] = "test";

                    param.args[0] = "你已被劫持";
                    PrintStack();
```

▲ 圖 5-16 程式對比

在弄明白上述問題後，我們再針對真實執行 Hook 的 Xposed 外掛程式程式：beforeHookedMethod 函數和 afterHookedMethod 函數進行 Frida Hook 工作。

事實上，beforeHookedMethod 函數和 afterHookedMethod 函數由圖 5-16 可以發現，其所在類別其實是在 HookTest 類別中透過 new XC_MethodHook() 方式建構的匿名內部類別 HookTest$1 中，因此其最終的 Hook 程式需要先將 ClassLoader 切換為 HookTest$1 所在的 loader，才可以執行對目標函數的 Hook，最終針對兩個函數的 Hook 結果如圖 5-17 所示。

▲ 圖 5-17 Hook 結果

在完成針對簡單 Xposed Hook 外掛程式的研究工作後，我們繼續來研究在一個成熟的 Hook 外掛程式中想要對所有 beforeHookedMethod 函數和 afterHookedMethod 函數進行 Hook 的方法。

首先，這裡以 GravityBox 修改系統狀態列顏色的類別 com.ceco.nougat.gravitybox.ModStatusbarColor 為例，測試上述針對 Demo 研究的成果驗證，在成功注入 GravityBox 處理程序後，筆者發現無論如何切換 ModStatusbarColor 類別中關於 beforeHookedMethod 函數或者 afterHookedMethod 函數所在匿名內部類別的 ClassLoader，始終無法找到完整的匿名內部類別列表，能夠找到的內部類別結果如圖 5-18 所示，而這明顯是不對的。這是為什麼呢？

```
-1/base.apk"],nativeLibraryDirectories=[/data/app/com.ceco.nougat.gravi
 /system/lib64, /vendor/lib64]]]
end1
found inner class =>   com.ceco.nougat.gravitybox.ModStatusbarColor
found inner class =>   com.ceco.nougat.gravitybox.ModStatusbarColor$5
found inner class =>   com.ceco.nougat.gravitybox.ModStatusbarColor
search completed!
```

▲ 圖 5-18 搜索匿名內部類別

別筆者在後續的研究過程中發現，由於 Frida 是處理程序等級的，而 Xposed 本身在將這部分程式對處理程序進行注入時做了處理程序的判斷，其對應程式如程式清單 5-13 所示，因此是不是只有針對 com. android.systemui 處理程序進行 Hook 才能順利找到所有目標類別呢？答案是「是的」。

◉ 程式清單 5-13 GravityBox 針對狀態列顏色修改的入口程式

```java
// GravityBox.java
if (lpparam.packageName.equals(ModStatusbarColor.PACKAGE_NAME)) {
    ModStatusbarColor.init(prefs, lpparam.classLoader);
}
// ModStatusbarColor.java
public class ModStatusbarColor {
    private static final String TAG = "GB:ModStatusbarColor";
    public static final String PACKAGE_NAME = "com.android.systemui";
    // ...
}
```

筆者在使用 Frida 再次對 systemui 處理程序進行注入後，搜索到的類別結果如圖 5-19 所示，最終發現這部分執行 Hook 的程式確實總是在 Hook 的目標處理程序中生效，而之所以在測試 Demo 時並未出現這樣的問題，是因為被 Hook 的處理程序和 Xposed 外掛程式處理程序兩者是同一個處理程序─xposed1 中。

```
end1
found inner class => com.ceco.nougat.gravitybox.ModStatusbarColor
found inner class => com.ceco.nougat.gravitybox.ModStatusbarColor$1
found inner class => com.ceco.nougat.gravitybox.ModStatusbarColor$2
found inner class => com.ceco.nougat.gravitybox.ModStatusbarColor$3
found inner class => com.ceco.nougat.gravitybox.ModStatusbarColor$4
found inner class => com.ceco.nougat.gravitybox.ModStatusbarColor$5
search completed!
end2
```

▲ 圖 5-19 注入 systemui 處理程序後搜索匿名內部類別

別此時再次對搜索到的目標類別進行 Hook，最終成功地對所有目標類別的所有函數進行了 Trace，其 Hook 成功的截圖如圖 5-20 所示，其中對類別中的所有函數進行 Hook 的部分程式如程式清單 5-14 所示（事實上這部分程式摘自 ZenTracer 中的程式）。

```
e/SystemUIGoogle.apk"],nativeLibraryDirectories=[/system/priv-app/SystemUIGoogle/lib/arm
64, /system/lib64, /vendor/lib64, /system/lib64, /vendor/lib64]]
end1
found inner class => com.ceco.nougat.gravitybox.ModStatusbarColor$1
Tracing com.ceco.nougat.gravitybox.ModStatusbarColor$1.afterHookedMethod [1 overload(s)]
found inner class => com.ceco.nougat.gravitybox.ModStatusbarColor$2
Tracing com.ceco.nougat.gravitybox.ModStatusbarColor$2.afterHookedMethod [1 overload(s)]
found inner class => com.ceco.nougat.gravitybox.ModStatusbarColor$3
Tracing com.ceco.nougat.gravitybox.ModStatusbarColor$3.afterHookedMethod [1 overload(s)]
found inner class => com.ceco.nougat.gravitybox.ModStatusbarColor$4
Tracing com.ceco.nougat.gravitybox.ModStatusbarColor$4.afterHookedMethod [1 overload(s)]
found inner class => com.ceco.nougat.gravitybox.ModStatusbarColor$5
Tracing com.ceco.nougat.gravitybox.ModStatusbarColor$5.onIconManagerStatusChanged [1 over
load(s)]
search completed!
end2
[Pixel::com.android.systemui]->
[Pixel::com.android.systemui]->
[Pixel::com.android.systemui]->
```

▲ 圖 5-20 增加 trace 函數內容

◯ 程式清單 5-14 traceClass 函數的主要內容

```
// trace a specific Java Method
function traceMethod(targetClassMethod) {
    var delim = targetClassMethod.lastIndexOf(".");
    if (delim === -1) return;

    var targetClass = targetClassMethod.slice(0, delim)
    var targetMethod = targetClassMethod.slice(delim + 1,
targetClassMethod.length)
```

```
    var hook = Java.use(targetClass);
    var overloadCount = hook[targetMethod].overloads.length;

    console.log("Tracing " + targetClassMethod + " [" + overloadCount + "
overload(s)]");
    for (var i = 0; i < overloadCount; i++) {

        hook[targetMethod].overloads[i].implementation = function () {
            console.warn("\n*** entered " + targetClassMethod);
            // print args
            if (arguments.length) console.log();
            for (var j = 0; j < arguments.length; j++) {
                console.log("arg[" + j + "]: " + arguments[j]);
            }
            // print retval
            var retval = this[targetMethod].apply(this, arguments); // rare
crash (Frida bug?)
            console.log("\nretval: " + retval);
            console.log(Java.use("android.util.Log").
getStackTraceString(Java.use("java.lang.Throwable").$new()));
            console.warn("\n*** exiting " + targetClassMethod);
            return retval;
        }
    }

}
function traceClass(targetClass) {
    //Java.use是新建的一個物件
    var hook = Java.use(targetClass);
    //利用反射的方式拿到當前類別的所有方法
    var methods = hook.class.getDeclaredMethods();
    //建完物件之後，記得將物件釋放掉
    hook.$dispose;
    //將方法名保存到陣列中
```

```
    var parsedMethods = [];
    methods.forEach(function (method) {
        parsedMethods.push(method.toString().replace(targetClass + ".",
"TOKEN").match(/\sTOKEN(.*)\(/)[1]);
    });
    //去掉一些重複的值
    var targets = uniqBy(parsedMethods, JSON.stringify);
    //對陣列中所有的方法進行Hook，traceMethod函數為對所有函數多載進行Hook的指
令稿
    targets.forEach(function (targetMethod) {
        traceMethod(targetClass + "." + targetMethod);
    });
}
```

上述 Trace 過程是基於已知實現 Xposed Hook 實現類別的類別名的
前提下完成的，但是如果想要在不知道類別名的前提下完成對所有
beforeHookedMethod 函數和 afterHookedMethod 函數的 Trace 工作，
應該如何實現呢？

觀察圖 5-16 中 Xposed Hook Demo 中的 beforeHookedMethod 函數所
在類別在 Java 程式中的實現，此時會發現是透過 new XC_MethodHook
類別的形式完成的，追蹤該類別的宣告會發現，該類別是一個 abstract 抽
象類別，那麼對應在記憶體中所觀察到的 HookTest$1 匿名內部類別就是
該抽象類別的繼承類別。因此，如果對記憶體中的 HookTest$1 類別透
過反射進而透過 getSuperClass() 函數獲取對應父類別，就應該能夠得到
XC_MethodHook 類別。因此，最終獲取所有父類別為 XC_MethodHook
類別的程式如程式清單 5-15 所示，在被 GravityBox Hook 的 systemui 處
理程序中針對所有父類別為 XC_MethodHook 類別的類別名進行搜索，
其結果如圖 5-21 所示。

⬇ 程式清單 5-15 獲取父類別為 XC_MethodHook 類別的關鍵程式

```
Java.enumerateLoadedClasses({
    onMatch: function (className) {
        // console.log("found => ", className)
        // 如果存在父類別
        if (Java.use(className).class.getSuperclass()) {
            // 獲取父類別名
            var superClass = Java.use(className).class.getSuperclass().
getName();
            // console.log("superClass is => ",superClass);
            // 父類別包含XC_MethodHook關鍵字
            if (superClass.indexOf("XC_MethodHook") > 0) {
                console.log("found XC_methodHook Child => ", className.
toString())
            }
        }
    },
    onComplete: function () {
        console.log("search completed!")
    }
})
```

```
found XC_methodhook Child =>    [Lcom.ceco.nougat.gravitybox.pie.PieController$Position;
found XC_methodhook Child =>    com.ceco.nougat.gravitybox.ProgressBarController
found XC_methodhook Child =>    com.ceco.nougat.gravitybox.ModHwKeys$HwKeyTrigger
found XC_methodhook Child =>    com.ceco.nougat.gravitybox.SystemIconController$4
found XC_methodhook Child =>    com.ceco.nougat.gravitybox.TrafficMeterAbstract$1
found XC_methodhook Child =>    com.ceco.nougat.gravitybox.managers.GpsStatusMonitor
found XC_methodhook Child =>    com.ceco.nougat.gravitybox.TrafficMeterAbstract$2
found XC_methodhook Child =>    [Lcom.ceco.nougat.gravitybox.CmCircleBattery$Style;
found XC_methodhook Child =>    com.ceco.nougat.gravitybox.pie.PieSliceContainer
found XC_methodhook Child =>    com.ceco.nougat.gravitybox.managers.BatteryInfoManager
found XC_methodhook Child =>    com.ceco.nougat.gravitybox.ModLockscreen$25
found XC_methodhook Child =>    com.ceco.nougat.gravitybox.ModStatusbarColor$5
found XC_methodhook Child =>    com.ceco.nougat.gravitybox.SystemIconController$BtMode
found XC_methodhook Child =>    com.ceco.nougat.gravitybox.TrafficMeterOmni$1
found XC_methodhook Child =>    com.ceco.nougat.gravitybox.ModPieControls$PieSettingsObserv
er
found XC_methodhook Child =>    [Lcom.ceco.nougat.gravitybox.managers.AppLauncher$DialogThe
me;
found XC_methodhook Child =>    com.ceco.nougat.gravitybox.ModLockscreen$26
```

▲ 圖 5-21 搜索所有繼承 XC_MethodHook 類別的子類別

別最終增加 trace 函數的部分 Hook 結果如圖 5-22 所示。

```
*** entered com.ceco.nougat.gravitybox.ModStatusBar$14.beforeHookedMethod
arg[0]: de.robv.android.xposed.XC_MethodHook$MethodHookParam@2e2d1d5
retval: undefined
*** exiting com.ceco.nougat.gravitybox.ModStatusBar$14.beforeHookedMethod
*** entered com.ceco.nougat.gravitybox.ModStatusBar$14.beforeHookedMethod
arg[0]: de.robv.android.xposed.XC_MethodHook$MethodHookParam@6406dea
retval: undefined
*** exiting com.ceco.nougat.gravitybox.ModStatusBar$14.beforeHookedMethod
*** entered com.ceco.nougat.gravitybox.ModStatusBar$14.beforeHookedMethod
arg[0]: de.robv.android.xposed.XC_MethodHook$MethodHookParam@7e848db
retval: undefined
*** exiting com.ceco.nougat.gravitybox.ModStatusBar$14.beforeHookedMethod
*** entered com.ceco.nougat.gravitybox.ModStatusBar$14.beforeHookedMethod
arg[0]: de.robv.android.xposed.XC_MethodHook$MethodHookParam@87bef78
retval: undefined
*** exiting com.ceco.nougat.gravitybox.ModStatusBar$14.beforeHookedMethod
```

▲ 圖 5-22　增加 trace 函數的部分 Hook 結果

關於 Frida 對 Xposed Hook 的分析到此為止。縱觀上述分析過程會發現 Xposed Hook 過程中，實際上是透過新建一個專門用於實現 Hook 的 PathClassLoader 注入目標處理程序中完成對目標函數的插樁工作，該 PathClassLoader 所載入的類別清單中包含著對目標處理程序 Hook 的程式邏輯，而原生的 Xposed API 則處於另一個指向 XposedBridge.jar 檔案的 ClassLoader。這實際上與 App 保護原理的本質類似，希望讀者能夠透過這一小節對 Xposed Hook 的細節有進一步的理解。

5.2 Xposed 主動呼叫與 RPC 實現

與 Frida 主動呼叫類似，Xposed 框架同樣支援針對函數的主動呼叫及 RPC 實現，本節將介紹其實現方法。

5.2.1 Xposed 主動呼叫函數

本小節將透過一個樣本案例介紹 Xposed 主動呼叫的一些細節。

如圖 5-23 所示，樣本 App 的功能十分簡單，只是一個 Crackme 程式，我們的目標非常明確：獲取正確的 PIN 值。

▲ 圖 5-23 樣本 App 介面

如圖 5-24 所示，為了快速定位驗證 PIN 函數的關鍵程式，樣本 App 的邏輯十分簡單：只有一個 MainActivity 類別。因此這裡直接使用

Objection Hook MainActivity 類別中的所有函數，從而幫助快速定位
到驗證函數入口：org.teamsik.ahe17.qualification.MainActivity.verify
PasswordClick 函數。

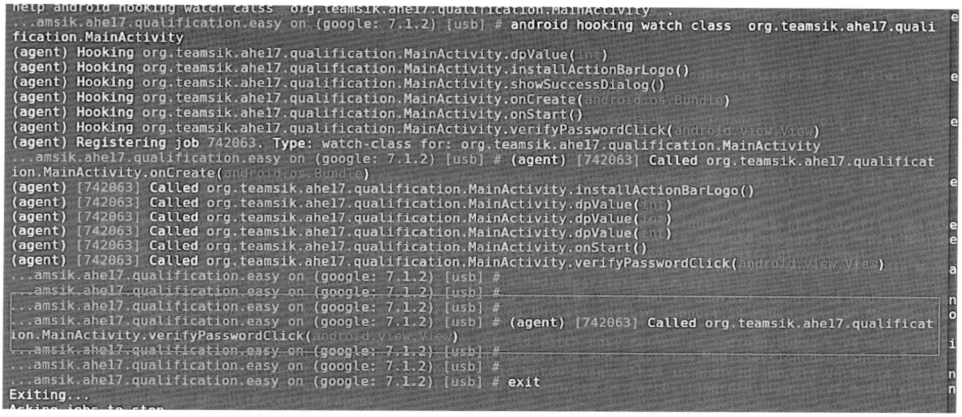

▲ 圖 5-24　定位關鍵函數

在 Jadx 的靜態分析輔助下，如程式清單 5-16 所示，最終定位到真實
用於驗證 PIN 碼正確與否的函數實際上是 Verifier 類別的靜態函數
verifyPassword，其中函數的第二個參數為使用者的輸入。根據函數中的
內容可以得到兩個關鍵性條件：第一，使用者輸入的 PIN 碼長度為 4；第
二，忽略函數的多層呼叫邏輯，實際上驗證方式是將使用者的輸入進行
標準的 SHA-1[1] 加密後與指定加密對比，進而判定 PIN 碼是否正確。

1　SHA-1（Secure Hash Algorithm，安全雜湊演算法）是一種密碼雜湊函數，由美國國家安
　全局設計，並由美國國家標準技術研究所（NIST）發佈為聯邦資料處理標準（FIPS）。
　SHA-1 訊息可以生成一個被稱為訊息摘要的 160 位元（20 位元組）雜湊值，雜湊值通常
　的呈現形式為 40 個十六進位數。

● **程式清單 5-16 驗證 PIN 碼正確與否的邏輯**

```
public void verifyPasswordClick(View view) {
    if(!Verifier.verifyPassword(this,this.txPassword.getText().toString()))
{
        Toast.makeText(this, (int) R.string.dialog_failure, 1).show();
    } else {
        showSuccessDialog();
    }
}
public static boolean verifyPassword(Context context, String input) {
    if (input.length() != 4) {
        return false;
    }
    byte[] v = encodePassword(input);
    byte[] p = "09042ec2c2c08c4cbece042681caf1d13984f24a".getBytes();
    if (v.length != p.length) {
        return false;
    }
    for (int i = 0; i < v.length; i++) {
        if (v[i] != p[i]) {
            return false;
        }
    }
    return true;
}
private static byte[] encodePassword(String input) {
    byte[] SALT = {95, 35, 83, 73, 75, 35, 95};
    try {
        ...
        return SHA1(sb.toString()).getBytes("iso-8859-1"); // sha1加密
    } catch (UnsupportedEncodingException e) {
        e.printStackTrace();
        return null;
    }
}
```

```
private static String SHA1(String text) {
    try {
        MessageDigest md = MessageDigest.getInstance("SHA-1");
        byte[] bArr = new byte[40];
        md.update(text.getBytes("iso-8859-1"), 0, text.length());
        return convertToHex(md.digest());
    } catch (NoSuchAlgorithmException e) {
        e.printStackTrace();
    } catch (UnsupportedEncodingException e2) {
        e2.printStackTrace();
    }
    return null;
}
```

結合以上兩筆關鍵資訊得出：由於 SHA-1 是一個不可逆的 Hash 函數，要確定正確的明文，一般情況下是無法完成的，但是由於在 verifyPassword 函數中對輸入的長度進行了長度為 4 的限定，因此最佳的解決方案是透過暴力窮舉得到明文。這裡可以使用演算法還原的方式在電腦上爆破出正確的 PIN 碼，同時提供另一種解決方案—透過 Xposed 主動呼叫的方式完成 PIN 碼的計算。

作為一個優秀的 Hook 框架，Xposed 提供了關於主動呼叫的 API，其對應函數簽名分別為：

- callMethod(Object obj, String methodName, Object... args)。
- callStaticMethod(Class<?> clazz, String methodName, Object... args)。

其中 callMethod 函數用於供實例物件呼叫對應實例所在類別中的動靜態函數，而 callStaticMethod 函數則用於呼叫指定類別的靜態函數。因此，這兩個函數的第一個參數分別為指定物件或者對應類別的 class 物件，這裡如果要主動呼叫 encodePassword(String input) 這個靜態函數，可以透過 callStaticMethod 函數實現，最終主動呼叫的關鍵函數程式如程式清單 5-17 所示。

◯ 程式清單 5-17 Xposed API 主動呼叫

```
// 靜態函數直接Hook
if (loadPackageParam.packageName.equals("org.teamsik.ahe17.qualification.
easy")) {

    XposedBridge.log("inner: "+loadPackageParam.processName);
    // 獲取類別物件
    Class clazz = XposedHelpers.findClass("org.teamsik.ahe17.qualification.
Verifier",loadPackageParam.classLoader);
    // 透過迴圈暴力窮舉
    for(int i = 999;i<10000;i++){
        // 主動呼叫目標函數
        if((boolean) XposedHelpers.callStaticMethod(clazz,"encodePassword",
String.valueOf(i))){
            XposedBridge.log("1). Current i is => "+ String.valueOf(i));
        }
    }
}
```

事實上，在 Java 中本身就存在函數的主動呼叫方式—透過 invoke 函數反射呼叫，在追蹤 Xposed 提供的兩個主動呼叫相關的 API 函數具體實現時，會發現該函數不過是封裝了 Java 的反射呼叫程式。如果想要使用 Java 反射的方式進行目標函數的主動呼叫，則關於 encodePassword (String input) 函數的主動呼叫關鍵程式如程式清單 5-18 所示。

◯ 程式清單 5-18 Java 反射的主動呼叫

```
if (loadPackageParam.packageName.equals("org.teamsik.ahe17.qualification.
easy")) {
```

```java
    XposedBridge.log("inner: "+loadPackageParam.processName);
    // 反射獲取對應類別的Class物件
    Class clazz = loadPackageParam.classLoader.loadClass("org.teamsik.
ahe17.qualification.Verifier");
    // 反射獲取Method物件
    Method encodePassword = clazz.getDeclaredMethod("encodePassword",
String.class);
    //  允許函數透過外部反射呼叫，這裡主要是為了避免目標函數是private私有函數
    encodePassword.setAccessible(true);
    byte[] p = "09042ec2c2c08c4cbece042681caf1d13984f24a".getBytes();
    String pStr = new String(p);
    // 迴圈方式暴力窮舉
    for(int i = 999;i<10000;i++){
        // invoke函數反射呼叫
        byte[] v = (byte[]) encodePassword.invoke(null,String.valueOf(i));
        if (v.length != p.length) {
            break;
        }
        String vStr = new String(v);
        if( vStr == pStr ){
            XposedBridge.log("2). Current i is => "+ String.valueOf(i));
        }
    }
}
```

另外，與 Frida 主動呼叫類似，Xposed 的主動呼叫主要存在兩個問題：
第一，參數建構問題；第二，如果目標函數是動態函數，如何獲取對應
的物件實例。

首先來看參數建構的問題。與 Frida 相同，如果想要建構一個與目標函
數的參數相同類型的資料，存在兩種方式：第一，Hook 獲取一個相同
類型的資料，這裡暫不介紹；第二，主動建構一個實例。但相比於 Frida
而言，由於 Xposed 是使用原生的 Java 進行開發的，因此 Xposed 在參

數的建構問題上更有優勢。比如這裡想要使用 Xposed 主動呼叫程式清
單 5-16 中的 verifyPassword(Context,String) 函數，如果是 Frida，想要建
構 Context 類型的資料，其對應的程式如程式清單 5-19 所示，透過各種
Java.use() 函數封裝才能最終獲取 Context 物件；而在 Xposed 中，則只
需透過 AndroidAppHelper.currentApplication() 即可獲取對應 Context 物
件內容，最終 Xposed 中對應的主動呼叫函數程式如程式清單 5-20 所示。

⊙ 程式清單 5-19 Frida 得到 Context 參數

```
var ActivityThread = Java.use('android.app.ActivityThread');
var Context = Java.use('android.content.Context');
var ctx = Java.cast(ActivityThread.currentApplication().getApplicationConte
xt(),Context);
```

⊙ 程式清單 5-20 複雜參數的函數主動呼叫

```
if (loadPackageParam.packageName.equals("org.teamsik.ahe17.qualification.
easy")) {

  XposedBridge.log("inner: "+loadPackageParam.processName);
  Class clazz = loadPackageParam.classLoader.loadClass("org.teamsik.ahe17.
qualification.Verifier");
  Method verifyPassword = clazz.getMethod("verifyPassword", Context.class,
String.class);

  // Context獲取的兩種情況：
  // - Hook獲取一個
  // - 自己建構一個（假的），這裡是第二種方式
  Context context = AndroidAppHelper.currentApplication();
  for(int i = 999;i<10000;i++){
    if((boolean)verifyPassword.invoke(null,context,String.valueOf(i))){
      XposedBridge.log("3). Current i is => "+ String.valueOf(i));
    }
  }
}
```

同樣，在主動呼叫中，另一個問題：獲取物件實例的解決方法也有兩種：
第一，主動建構一個實例物件；第二，Hook 獲取一個實例物件。

這裡要注意的是，與主動呼叫函數相同，在主動建構一個實例物件的方
法時，Java 原生函數本身就支援透過對應類別的 Constructor 物件呼叫
newInstance() 函數的方式建構一個物件，Java 原生函數實現本案例中函
數的主動呼叫方式如程式清單 5-21 所示。而 Xposed 更是在 Java 原生函
數的基礎上進一步進行封裝，從而提供了 XposedHelpers.newInstance()
函數用於建構實例物件，對應的主動呼叫方法如程式清單 5-22 所示。

◉ 程式清單 5-21 Java newInstance()

```
if (loadPackageParam.packageName.equals("org.teamsik.ahe17.qualification.
easy")) {
    XposedBridge.log("inner" + loadPackageParam.processName);

    Constructor cons = XposedHelpers.findConstructorExact("org.teamsik.
ahe17.qualification.Verifier",loadPackageParam.classLoader);
    Object Verifier = cons.newInstance();

    Context context = AndroidAppHelper.currentApplication();

    for (int i = 999; i < 10000; i++) {
        if ((boolean) XposedHelpers.callMethod(Verifier, "verifyPassword",
context, String.valueOf(i))) {
            XposedBridge.log("5). Current i is => " + String.valueOf(i));
        }
    }
}
```

◉ 程式清單 5-22 Xposed newInstance() 建構物件

```
if (loadPackageParam.packageName.equals("org.teamsik.ahe17.qualification.
easy")) {
```

```
XposedBridge.log("inner" + loadPackageParam.processName);
Class clazz = XposedHelpers.findClass("org.teamsik.ahe17.qualification.
Verifier", loadPackageParam.classLoader);

Object Verifier = XposedHelpers.newInstance(clazz); // 獲取物件

Context context = AndroidAppHelper.currentApplication();

for (int i = 999; i < 10000; i++) {
    if ((boolean) XposedHelpers.callMethod(Verifier, "verifyPassword",
context, String.valueOf(i))) {
        XposedBridge.log("6). Current i is => " + String.valueOf(i));
    }
}
}
```

最終分別使用以上幾種主動呼叫函數的方式對答案進行暴力窮舉，得到
正確的 PIN 碼為 9083，對應的 Xposed 日誌列印如圖 5-25 所示。

▲ 圖 5-25 PIN 碼日誌

仔細閱讀的讀者可能會發現，筆者並未介紹透過 Hook 方式得到實例物件的方法，這是因為在這個範例 App 中，Verifier 這個類別中的函數均為靜態函數，導致該類別始終並未初始化，因此要透過 Hook 得到對應的實例物件是不可能實現的。要介紹透過 Hook 方式得到對應實例的方法，只能退而求其次，透過其他實例方法的主動呼叫介紹。

這裡以輸入正確的 PIN 碼後展示成功彈出視窗的 showSuccessDialog 函數為例進行介紹。由於 showSuccessDialog 函數是 MainActivity 類別中的一個動態方法，因此要獲取對應的實例物件，首先該物件需要被成功建立。這裡選擇 Hook MainActivity 類別的 onCreate() 函數作為得到對應實例物件的跳板函數，最終在 afterHookedMethod(param) 回呼函數被呼叫後，透過 param.thisObject 得到對應物件實例，從而進一步完成 showSuccessDialog 這個實例函數的呼叫，最終主動呼叫 showSuccessDialog 函數的程式如程式清單 5-23 所示。

● 程式清單 5-23 Hook 方式得到實例

```
// 測試一個新的彈出成功彈出視窗
if (loadPackageParam.packageName.equals("org.teamsik.ahe17.qualification.
easy")) {

    XposedBridge.log("inner" + loadPackageParam.processName);

    Class clazz = loadPackageParam.classLoader.loadClass("org.teamsik.
ahe17.qualification.MainActivity");

    XposedBridge.hookAllMethods(clazz, "onCreate",new XC_MethodHook() {
        @Override
        protected void afterHookedMethod(MethodHookParam param) throws
Throwable {
            super.afterHookedMethod(param);
            // 獲取MainActivity物件
```

```
        Object mMainAciticity = param.thisObject;
        // 主動呼叫showSuccessDialog函數
        XposedHelpers.callMethod(mMainAciticity,"showSuccessDialog");

    }
  });
}
```

在將模組重新編譯安裝啟動並重新啟動裝置後，重新開啟樣本 App，就會發現頁面會立即彈出 Congratulations 成功的提示，最終效果如圖 5-26 所示。

▲ 圖 5-26 Congratulations 成功提示

至此，關於 Xposed 主動呼叫相關的方法就暫時告一段落。

5.2.2 Xposed 結合 NanoHTTPD 實現 RPC 呼叫

在前面的章節中,我們學習了 Frida 中 RPC 的方式,那麼同樣作為 Hook 框架的 Xposed 是否支持 RPC 呢?如果不支持,有什麼方法能夠使得它支持嗎?

事實上,Xposed 框架本身並未提供 RPC 呼叫方式的支援,但是基於 Xposed 模組在成為 Hook 外掛程式前首先是一個普通的應用 App,因此在其他普通 Android App 上的開發方式可以完美地移植到 Xposed 模組中,這大大拓展了 Xposed 模組的可能性。基於此,這裡提供一種 Xposed RPC 的解決方案—結合 NanoHTTPD 將主動呼叫匯出為 Web 服務實現。

NanoHttpD 是一個免費的、羽量級的(甚至只有一個 Java 檔案)HTTP 伺服器,可以極佳地嵌入 Java 程式中。NanoHttpD 支持 GET、POST、PUT、HEAD 和 DELETE 等多種 HTTP 請求方式,除此之外,它還支持檔案上傳實現,在使用時佔用的記憶體很小。基於以上優點,NanoHttpD 正好符合這裡做 RPC 呼叫的方案。

和第 4 章一樣,同樣以 demoso1 為例介紹如何使用 NanoHTTPD 實現對 Hook 函數的遠端程序呼叫。

與 Frida 實現 RPC 呼叫的過程類似,首先需要測試該函數的主動呼叫無誤,這裡使用上述介紹得到物件實例的兩種方法,最終分別實現了 Xposed 主動呼叫的 method01() 函數和 method02() 函數,其具體程式如程式清單 5-24 所示。

● 程式清單 5-24 主動呼叫測試

```
if (loadPackageParam.packageName.equals("com.example.demoso1")) {
    final Class clazz = loadPackageParam.classLoader.loadClass("com.
example.demoso1.MainActivity");

    //得到物件：Hook(想透過Hook方式得到一個obj的話，得Hook一個實例方法)
    XposedBridge.hookAllMethods(clazz, "onCreate", new XC_MethodHook() {
        @Override
        protected void beforeHookedMethod(MethodHookParam param) throws
Throwable {
            super.beforeHookedMethod(param);
            Object mMainAciticity = param.thisObject;
            String cipherText = (String) XposedHelpers.
callMethod(mMainAciticity, "method01", "roysue");
            String clearText = (String) XposedHelpers.
callMethod(mMainAciticity, "method02", "47fcda3822cd10a8e2f667fa49da783f");
            XposedBridge.log("1). Cipher text is => " + cipherText);
            XposedBridge.log("1). Clear text is => " + clearText);
        }
    });

    //xposed.newInstance獲取物件 active call
    Object newMainActivity = XposedHelpers.newInstance(clazz);
    String cipherText = (String) XposedHelpers.callMethod(newMainActivity,
"method01", "roysue");
    String clearText = (String) XposedHelpers.callMethod(newMainActivity,
"method02", "47fcda3822cd10a8e2f667fa49da783f");
    XposedBridge.log("2). Cipher text is => " + cipherText);
    XposedBridge.log("2). Clear text  is => " + clearText);
}
```

要注意的是，筆者這裡 Hook onCreate() 函數的方式並未使用 XposedHelpers.findAndHookMethod() 函數，而是使用 XposedBridge. hookAllMethods() 函數，使用後者的優勢在於能夠一次性 Hook 指定函

數名的所有多載函數，這樣就無須考慮函
數參數的問題，最終主動呼叫的結果如圖
5-27 所示。

在主動呼叫成功後，我們正式開始進入
Xposed-RPC 的開發工作，實際上就是
NanoHTTPD 的開發。

在使用 NanoHTTPD 之前，首先需要匯
入 NanoHTTPD 的依賴套件，要完成這
一點，只需在專案的 app/build.gradle 檔
案的 dependencies 層級下增加以下一行
程式並進行同步，以支持 NanoHTTPD
的 API 的使用。

▲ 圖 5-27　主動呼叫的結果

```
implementation 'org.nanohttpd:nanohttpd:2.3.1'
```

NanoHTTPD 非常簡單，只存在一個類別 NanoHTTPD，且該類別是一個
抽象類別，主要存在 3 個重要函數：start、stop 和 serve。其中 start 和
stop 函數用於啟動和停止 Web 伺服器，serve 函數是一個收到 Web 請
求後的回呼函數，瀏覽器得到的頁面資料都是透過這個函數返回的，其
唯一的 IHTTPSession 類型參數可以用於判斷瀏覽器的請求內容，包括請
求方法、參數、URL 等。另外，透過 NanoHTTPD 的建構函數可以指定
Web 服務監聽的通訊埠。

在同步完成後，只需要繼承 NanoHTTPD 類別並呼叫和實現 start 函數和
serve 函數即可為 App 啟動一個 Web 服務，這裡簡單地實現一個 hello
world 頁面，其具體程式如程式清單 5-25 所示。

● 程式清單 5-25 NanoHTTPD 服務

```
class App extends NanoHTTPD {

    public App() throws IOException {
        // 指定監聽通訊埠
        super(8899);
        // start函數啟動HTTP服務
        start(NanoHTTPD.SOCKET_READ_TIMEOUT, true);
        XposedBridge.log("\nRunning! Point your browsers to http://
localhost:8899/ \n");
    }

    @Override
    // 處理HTTP請求的回呼函數
    public NanoHTTPD.Response serve(IHTTPSession session) {

        // 獲取HTTP存取方法 POST、GET等
        Method method = session.getMethod();
        // 獲取URI
        String uri = session.getUri();
        // 獲取存取者的IP位址
        String RemoteIP = session.getRemoteIpAddress();
        // 獲取存取者的HostName
        String RemoteHostName = session.getRemoteHostName();
        Log.i("r0ysue nanohttpd ","Method => " + method + " ;Url => " + uri
+ "' ");
        Log.i("r0ysue nanohttpd ","Remote IP  => " + RemoteIP + "
;RemoteHostName => " + RemoteHostName + "' ");
        // 頁面返回值
        String msg = "<html><body><h1>Hello NanoHttpd</h1>\n";
        return newFixedLengthResponse(Response.Status.OK, NanoHTTPD.MIME_
PLAINTEXT, msg);
    }
}
```

```
// 呼叫建構函數啟動 NanoHTTPD 服務
new App();
```

最終在外掛程式更新後，對應的 log 日誌的列印結果以及使用瀏覽器存取的結果如圖 5-28 所示。

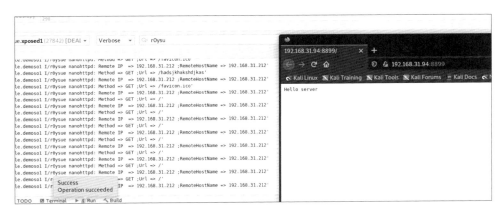

▲ 圖 5-28　NanoHTTPD 服務啟動成功

在測試 NanoHTTPD 服務啟動無誤後，就可以向 serve 函數增加對函數的主動呼叫並返回呼叫結果的內容，最終 RPC 的具體程式如程式清單 5-26 所示。

◎ 程式清單 5-26　serve() 函數

```
@Override
public NanoHTTPD.Response serve(IHTTPSession session) {
    String uri = session.getUri(); //獲取uri

    // 解析POST方法存取時傳遞的參數內容
    String paramBody = "";
    Map<String, String> params = new HashMap<>();
    try {
        session.parseBody(params);
```

```
        paramBody = session.getQueryParameterString();
    } catch (IOException e) {
        e.printStackTrace();
    } catch (ResponseException e) {
        e.printStackTrace();
    }

    String result = "";
    // 如果請求URL中包含encrypt關鍵字，則呼叫method01加密函數
    if(uri.contains("encrypt")){
        result = (String) XposedHelpers.callMethod(getActivity(),
"method01", paramBody);
    // 如果請求URL中包含decrypt關鍵字，則呼叫method02解密函數
    }else if (uri.contains("decrypt")){
        result = (String) XposedHelpers.callMethod(getActivity(),
"method02", paramBody);
    }else{
        result = paramBody;
    }
    // 返回函數執行結果
    return newFixedLengthResponse(Response.Status.OK, NanoHTTPD.MIME_
PLAINTEXT, result + '\n');
}
```

最終透過 curl 命令列測試加密 RPC 的效果如圖 5-29 所示。

▲ 圖 5-29 RPC 效果

如果想要透過 Python 實現 RPC 呼叫，則需要呼叫 Python requests 包中的函數完成資料的收發送封包，最終 Python 方式實現的 RPC 呼叫程式如程式清單 5-27 所示，呼叫結果如圖 5-30 所示。

❍ 程式清單 5-27　Python 實現 RPC

```python
import requests

def encrypt(enParam):
    url = "http://192.168.31.94:8899/encrypt"
    param = enParam
    headers = {"Content-Type":"application/x-www-form-urlencoded"}
    r = requests.post(url = url ,data=param,headers = headers)
    print(r.content)

def decrypt(enParam):
    url = "http://192.168.31.94:8899/decrypt"
    param = enParam
    headers = {"Content-Type":"application/x-www-form-urlencoded"}
    r = requests.post(url = url ,data=param,headers = headers)
    print(r.content)

if __name__ == '__main__' :
    encrypt("roysue")
    decrypt("47fcda3822cd10a8e2f667fa49da783f")
```

▲ 圖 5-30　Python RPC 效果

當然，與 Frida 類似，讀者如果感興趣，還可以與第 3 章一樣，使用 Siege 對 Xposed 的 RPC 進行壓力測試，也可以將裝置通訊埠映射到公網中進行呼叫，這裡限於篇幅，就不再展開説明了。

5.3 本章小結

本章按照 Frida 模式介紹了在 Xposed 中對應的技術方案，並透過幾個案例的介紹讓 Xposed 的學習顯得不那麼枯燥乏味。事實上，Xposed 和 Frida 在 Hook 功能上不分伯仲，基本上 Frida 在 Java Hook 上的功能，Xposed 都會有對應的實現，即使沒有對應的實現，也可以充分利用其本身就是基於 Java 和 Andorid 平台的優勢結合其他方式實作。

但是，Xposed 和 Frida 還是存在一些差異性的。從微觀上來看，Xposed 支持透過 setAdditionalInstanceField()、setAdditionalStaticField() 等函數為實例物件增加動靜態成員，Frida 支援透過 Java.choose() 函數從處理程序堆積中搜索目標物件。從宏觀上來看，Frida 表現得更加靈活多變，支持熱多載，其作用物件是特定處理程序，Hook 原理更類似於偵錯器；而 Xposed 則更加穩重老成，Hook 效果針對系統全部處理程序都會生效，類似於系統框架級服務。

當然，本章只是對 Xposed 框架進行了簡介，具體 Xposed 和 Frida 兩者誰更適合，還需要讀者自己在實踐中測試。

06

Android 原始程式編譯
與 Xposed 魔改

市面上大多數保護廠商或者大型 App 都會或多或少地對 Xposed 框架執行基於特徵的檢測，而突破這些檢測的基本思路就是找到檢測的地方，無論是在 Java 層還是 Native 層，然後透過 Hook 的方式修改返回結果，或者以強制寫入、直接置零返回等方式來繞過檢測邏輯。但是，不論直接修改二進位能不能透過完整性驗證，大多數逆向工程師或許連 App 具體在哪裡完成的 Xposed 檢測邏輯都很難找到，更不用提保護廠商再使用 Ollvm、VMP 這樣的工具對程式邏輯混淆和保護，進一步隱藏了檢測程式的實現方式。

其實，作為個人，與做保護的團隊甚至廠商在特徵檢測層面鬥智鬥勇是很不明智的。一方面，敵在暗，我在明，尋找檢測點宛如大海撈針；另一方面，團隊的努力總是賽過個人的。因此，倒不如從源頭消滅特徵，任你萬般檢測，我自笑傲江湖。

本章將帶領讀者從 Android 原始程式的編譯開始一步一步介紹如何編譯和魔改 Xposed，最終實現對一個開放原始碼的 Xposed 檢測工具一Xposed Checker 的繞過效果。

6.1 Android 原始程式環境架設

如果讀者想要自己編譯 Xposed，那麼 Android 系統原始程式的編譯工作定然是首當其衝。那麼為什麼需要編譯 Android 原始程式呢？如果讀者研究過 Xposed 原始程式，就會發現 Xposed 原始程式的編譯過程對 Android 系統原始程式環境的依賴甚大。因此，本節將首先從零到一介紹 Android 原始程式的下載編譯過程，為後續的 Xposed 原始程式編譯打下基礎。

6.1.1 編譯環境準備

據 Android 官網介紹，Android 原始程式的編譯極其消耗記憶體與空間，因此這裡選擇開啟一個新的虛擬機器環境進行後續操作。

在第一次使用虛擬機器軟體開啟虛擬機器前，由於預設硬碟空間分配為 80GB，這樣的空間對於可能多達 100 多 GB 的 Android 系統原始程式，甚至在編譯完成後幾百 GB 的需求來說肯定是不夠的，因此在第一次開啟虛擬機器系統前，還需要透過「編輯虛擬機器設定」對虛擬機器硬碟空間進行擴充。如圖 6-1 所示，這裡直接分配了 450GB 大小的硬碟空間。

除此之外，由於編譯 Android 系統原始程式十分消耗記憶體，因此在分配記憶體時需要盡可能多分配一點，否則在編譯時可能會報 out of memory（記憶體溢位）等錯誤，這裡直接設定為 12GB。

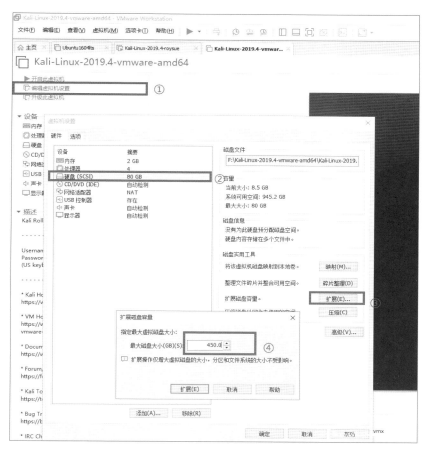

▲ 圖 6-1 硬碟空間擴充

當然，在擴大硬碟空間和增加虛擬機器記憶體時，要注意物理機本身的硬碟空間和記憶體大小，注意量力而行，這裡之所以肆無忌憚地舉出如此多的硬碟空間和記憶體，是因為筆者物理機本身就有 48GB 的記憶體，同時外接有 1TB 的 SSD 硬碟空間。

在重新設定好硬碟空間和記憶體空間後，開啟虛擬機器系統，還需要將擴充的硬碟空間使用 Gparted 等磁碟分割工具進行格式化和分配並再次重新啟動，才能使用後來增加的硬碟空間，在一切準備就緒後，就可以開始下載 Android 系統原始程式以進行後續的編譯工作了。

事實上，Google 官方提供了每個版本的原生系統原始程式供開發者自取使用，其官方網址為 https://android.googlesource.com/，由於 Android 原始程式引用了很多外部的開放原始碼工具，比如 OpenSSL，其每一個子專案都是一個 Git 倉庫，為了更方便地管理這些 Git 倉庫，Android 官方推出了另一個程式版本管理工具—Repo。Repo 封裝了一系列的 Git 命令，可用於方便地對多個 Git 倉庫進行管理。因此，要下載 Android 系統原始程式，首先必須下載 Repo 工具，非官方來源的官網推薦使用如下方式下載安裝 Repo 工具。

```
# mkdir ~/bin
# PATH=~/bin:$PATH
# curl -sSL  'https://gerrit-googlesource.proxy.ustclug.org/git-repo/+/
master/repo?format=TEXT' | base64 -d > ~/bin/repo
# chmod a+x ~/bin/repo
```

在下載並設定 Repo 工具後，還需要進行 Git 的一些相關設定，包括用戶名和電子郵件的設定，對應的命令如下：

```
# git config --global user.name "Your Name"
# git config --global user.email "you@example.com"
```

在設定完成後，便可以使用 Repo 工具同步 Android 原始程式到本地。需要注意的是，儘管提供了透過程式清單 6-1 的方式直接同步原始程式，但是由於這裡我們並不需要最新的 Android 原始程式，因此建議選擇程式清單 6-2 中同步特定原始程式的方式。其中 repo init 命令中 -b 參數後的 android-7.1.2_r8 是指具體的系統版本編號，Android 系統版本編號、對應 Build ID 和支援裝置關係見 https://source.android.com/setup/start/build-numbers#source-code-tags-and-builds，其最終效果如圖 6-2 所示。

🔽 程式清單 6-1 同步原始程式方式

```
# repo init -u git://mirrors.ustc.edu.cn/aosp/platform/manifest
# repo sync
```

🔽 程式清單 6-2 推薦同步原始程式方式

```
# mkdir aosp712_r8 && cd aosp712_r8
# repo init -u git://mirrors.ustc.edu.cn/aosp/platform/manifest -b
android-7.1.2_r8
# repo sync
```

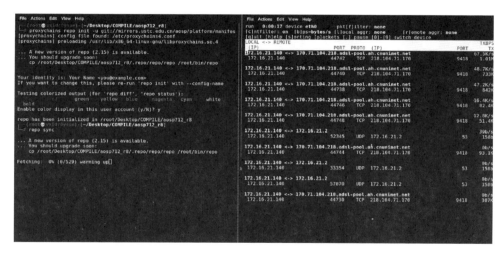

▲ 圖 6-2 同步原始程式

在開始同步後，透過 Jnettop 等查看系統網路連接的工具即可發現存在多個網路連接同步下載 Android 原始程式，此時需要做的是讓虛擬機器處於一個好的網路環境下靜待，最終在等待數小時後 Android 原始程式即可下載完畢。

請注意，透過 repo sync 命令同步到本地的程式只是包含了系統執行必不可少的程式，此時如果編譯，只能編譯出執行 Android Emulator 的虛擬機器系統，如果想將自編譯的系統執行到特定裝置中，還需要完成一個必不可缺的步驟—下載對應裝置的驅動，裝置驅動的作用在於在一個執行在物理裝置的系統上造成協調上層系統與底層硬體的通訊與互動的作用。幸運的是，Android 官網同樣提供了 Pixel 和 Nexus 系列裝置對應對應官方原始程式的驅動二進位檔案。

要下載正確的系統驅動檔案，只需要記住兩個要點：第一，要與目標裝置的型號相對應，比如這裡用於更新軔體的 Pixel 裝置，再比如在第 1 章中用於刷入 Kali NetHunter 的裝置 Nexus 5X，都是裝置型號；第二，所下載原始程式對應的 Build ID 要一致，比如這裡下載的系統原始程式版本是 android-7.1.2_r8，其對應的 Build ID 為 N2G47O，那麼在找對應的裝置驅動時，就需要 Build ID 為 N2G47O。具體系統版本和 Build ID 的對應關係可參考官方提供的 Source code tags and builds 對照表（對應網址：https://source.android.com/setup/start/build-numbers#source-code-tags-and-builds）。在確定裝置型號以及所下載原始程式的對應 Build ID 後，便可以在 Android 官方提供的 Driver Binaries for Nexus and Pixel Devices 網頁 https://developers.google.com/android/drivers）找到對應驅動的二進位檔案，最終這裡所需的適用於 Pixel 裝置的 android-7.1.2_r8 版本的驅動二進位檔案如圖 6-3 所示。

Pixel binaries for Android 7.1.2 (N2G47O)			
Hardware Component	Company	Download	SHA-256 Checksum
Vendor image	Google	Link	4dacefdd2d13a9b4ea28d3356d71fa34c42942a47f3d21a0fa4cd40919dbb945
GPS, Audio, Camera, Gestures, Graphics, DRM, Video, Sensors	Qualcomm	Link	c27798a7d5e796d055bf819fe026bd21a7e641475cb9878137978475e431e313

▲ 圖 6-3 Pixel 裝置的 android-7.1.2_r8 版本對應的驅動二進位檔案

透過圖 6-3 中的連結所下載的兩個驅動二進位檔案是打包好的壓縮檔，下載並解壓完畢後，會發現實際上是兩個 Shell 指令檔。為了安裝裝置驅動到編譯系統中，要將這兩個 Shell 腳本檔案移動到上一步下載的原始程式根目錄中，並透過終端執行兩個指令稿。透過圖 6-4 發現，在執行後會出現一個類似於使用者協定的宣告，此時需要一直按 Enter 鍵，直到出現需要使用者輸入 "I ACCEPT" 的頁面來釋放二進位驅動檔案，在輸入 "I ACCEPT" 後，最終裝置驅動相關檔案就會下載到 vendor 目錄中。

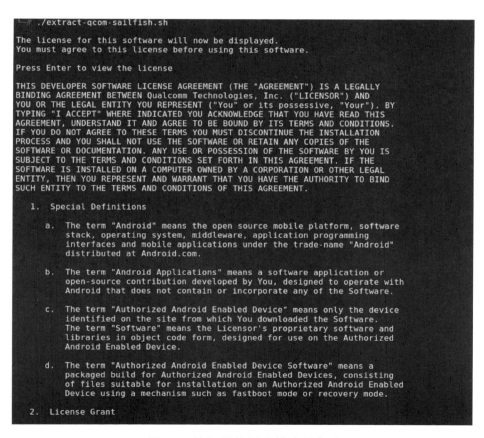

▲ 圖 6-4 執行釋放驅動檔案的指令稿

最後還要提醒的是，裝置驅動檔案一定要按照前面介紹的關鍵點來選擇
對應版本的驅動，否則如果選擇了不相匹配的驅動，在編譯完成後，一
旦將鏡像刷入裝置中，就可能會造成系統無法啟動、開機動畫頁面循環
等異常情況，甚至最終使得手機變「磚」，請讀者務必小心。

6.1.2 原始程式編譯

在完成原始程式與驅動檔案的下載工作之後，在開始 Android 系統的
編譯與更新韌體之前，還需要為用於編譯原始程式的系統安裝一些依賴
庫，以保障編譯過程不會因為缺少依賴檔案而失敗。這裡為 Kali Linux
2021.1 安裝的依賴函數庫檔案以及對應操作如程式清單 6-3 所示。

◉ 程式清單 6-3 安裝依賴函數庫

```
# apt update
# apt install bison tree
# dpkg --add-architecture i386
# apt update
# apt install libc6:i386 libncurses5:i386 libstdc++6:i386 libxml2-utils
```

在安裝完畢上述基礎的依賴函數庫後，由於 Android 原始程式的編譯依
賴於 Java 環境，因此還需要為系統安裝原始程式所需的 JDK 環境，而
Android 7 以上所有的 Android 系統所依賴的 JDK 版本都為 JDK 8，這裡
直接以如下命令安裝 JDK 8 時會發現報如圖 6-5 所示的 "Unable to locate
package" 錯誤，實際上這是因為 Kali Linux 2021 的系統來源中已經不保
留 JDK 8 的函數庫檔案，甚至系統附帶的 JDK 版本已經更新到 11 了。

```
# apt install openjdk-8-jdk
```

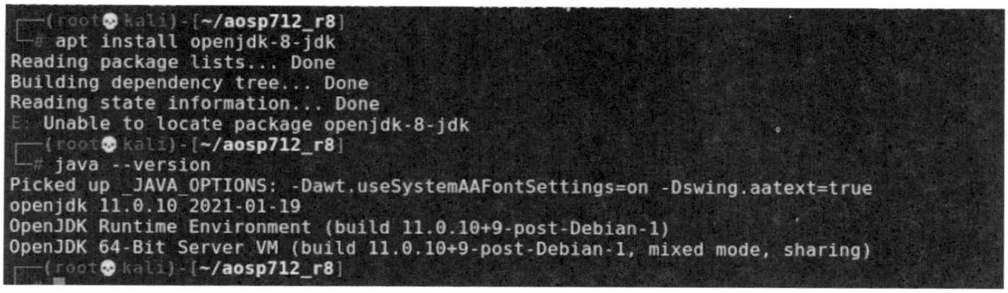

▲ 圖 6-5 apt 命令安裝 JDK 8 顯示出錯

幸運的是，在下載的 Android 系統原始程式中實際上已經內建了 JDK 8 的支援，其相對於所下載原始程式的目錄為 prebuilts/jdk/jdk8，因此原本需要安裝 JDK 8 才能解決的問題變成了只需將 Android 系統原始程式內建的 JDK 8 替換為系統預設的 Java 環境即可。具體替換的操作步驟如程式清單 6-4 所示。

◯ 程式清單 6-4 替換系統 OpenJDK 的預設版本

```
# cd  ~/aosp712_r8 // 切換到所下載的Android原始程式根目錄
# apt install openjdk-11-jdk // 幫助Kali補全Java環境，防止後續步驟找不到javac
命令
# update-alternatives --install /usr/bin/java java $(pwd)/prebuilts/jdk/
jdk8/linux-x86/bin/java 1
# update-alternatives --install /usr/bin/javac javac $(pwd)/prebuilts/jdk/
jdk8/linux-x86/bin/javac 1
# update-alternatives --set java $(pwd)/prebuilts/jdk/jdk8/linux-x86/bin/java
# update-alternatives --set javac $(pwd)/prebuilts/jdk/jdk8/linux-x86/bin/javac
# echo "export JAVA_HOME=$(pwd)/prebuilts/jdk/jdk8/linux-x86/" >> ~/.bashrc
# source ~/.bashrc
# echo "export PATH=$PATH:$HOME/bin:$JAVA_HOME/bin" >> ~/.bashrc
# source ~/.bashrc
```

在執行完成程式清單 6-4 中的命令後，觀察圖 6-6，可以發現最後系統預設的 Java 環境變成了 Android 編譯的 JDK 8。

```
┌──(root💀kali)-[~/aosp712_r8]
└─ # update-alternatives --install /usr/bin/java java $(pwd)/prebuilts/jdk/jdk8/linux-x86/bin/java 1
┌──(root💀kali)-[~/aosp712_r8]
└─ # update-alternatives --install /usr/bin/javac javac $(pwd)/prebuilts/jdk/jdk8/linux-x86/bin/javac 1
┌──(root💀kali)-[~/aosp712_r8]
└─ # update-alternatives --set java $(pwd)/prebuilts/jdk/jdk8/linux-x86/bin/java
┌──(root💀kali)-[~/aosp712_r8]
└─ # update-alternatives --set javac $(pwd)/prebuilts/jdk/jdk8/linux-x86/bin/javac
update-alternatives: using /root/aosp712_r8/prebuilts/jdk/jdk8/linux-x86/bin/javac to provide /usr/bin/javac (javac) in manual mode
┌──(root💀kali)-[~/aosp712_r8]
└─ # echo "export JAVA_HOME=$(pwd)/prebuilts/jdk/jdk8/linux-x86/" >> ~/.bashrc
┌──(root💀kali)-[~/aosp712_r8]
└─ # source ~/.bashrc
┌──(root💀kali)-[~/aosp712_r8]
└─ # echo "export PATH=$PATH:$HOME/bin:$JAVA_HOME/bin" >> ~/.bashrc
┌──(root💀kali)-[~/aosp712_r8]
└─ # source ~/.bashrc
┌──(root💀kali)-[~/aosp712_r8]
└─ # java -version
Picked up _JAVA_OPTIONS: -Dawt.useSystemAAFontSettings=on -Dswing.aatext=true
openjdk version "1.8.0_152-android"
OpenJDK Runtime Environment (build 1.8.0_152-android-4163371-1)
OpenJDK 64-Bit Server VM (build 25.152-b1, mixed mode)
┌──(root💀kali)-[~/aosp712_r8]
└─ █
```

▲ 圖 6-6 替換系統 JDK 為 JDK 8

在準備好上述依賴環境後,即可正式開始進行原始程式的編譯。

首先,如圖 6-7 所示,切換到系統原始程式根目錄並透過 Android 原始程式中附帶的指令稿設定編譯所需的環境變數,具體操作如下:

```
# cd ~/aosp712_r8 && source build/envsetup.sh
```

```
┌──(root💀kali)-[~/aosp712_r8]
└─ # source build/envsetup.sh
including device/asus/fugu/vendorsetup.sh
including device/generic/car/vendorsetup.sh
including device/generic/mini-emulator-arm64/vendorsetup.sh
including device/generic/mini-emulator-armv7-a-neon/vendorsetup.sh
including device/generic/mini-emulator-mips64/vendorsetup.sh
including device/generic/mini-emulator-mips/vendorsetup.sh
including device/generic/mini-emulator-x86_64/vendorsetup.sh
including device/generic/mini-emulator-x86/vendorsetup.sh
including device/generic/uml/vendorsetup.sh
including device/google/dragon/vendorsetup.sh
including device/google/marlin/vendorsetup.sh
including device/google/muskie/vendorsetup.sh
including device/google/taimen/vendorsetup.sh
including device/huawei/angler/vendorsetup.sh
including device/lge/bullhead/vendorsetup.sh
including device/linaro/hikey/vendorsetup.sh
including sdk/bash_completion/adb.bash
┌──(root💀kali)-[~/aosp712_r8]
```

▲ 圖 6-7 設定環境變數

然後，如圖 6-8 所示，設定最終編譯出來的系統版本。這裡要注意的是，由於在下載驅動時選擇的是 Pixel 裝置，因此處理模擬器相關編譯目標只能選擇 Pixel 對應裝置的代號 sailfish 對應的選項；另外，在進行這一步時，注意系統的終端必須是 Bash 環境，不可以是 Kali Linux 2021 原生的 Zsh 環境或者其他終端環境，否則可能會出現明明設定的編譯目標是 sailfish，但是真正最終編譯出來的裝置卻是 bullhead 等其他裝置鏡像，或者出現其他不可知的異常。具體選擇編譯目標的命令如下：

```
# lunch （這裡選擇24，表示編譯目標為sailfish可偵錯版本）
```

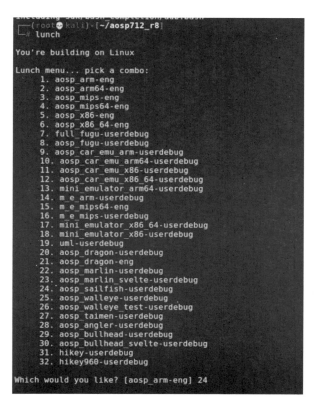

▲ 圖 6-8 選擇編譯目標

最終,在一切執行無誤的情況下,只需輸入如下命令即可開始編譯系統
的處理程序。

```
# make
```

如圖 6-9 所示,當輸入 make 命令後,終端開始出現百分比的進度指示
器時,基本上編譯過程就進入正軌,此時一般不會顯示出錯,只需靜待
即可。當然,編譯過程是非常耗費系統記憶體的,一旦記憶體分配的不
夠,就可能會導致出現 out of memory 等記憶體耗盡錯誤。如果出現這樣
的情況,只要給虛擬機器加記憶體就行,這裡一開始設定的是 12GB 記
憶體,實測過程中發現基本可以滿足原始程式編譯需求。

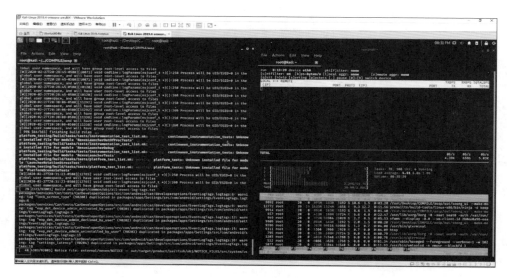

▲ 圖 6-9 編譯過程

不出意外，最終在等待數小時後，如出現圖 6-10 中提示的 build completed successfully 即表示系統原始程式編譯成功。

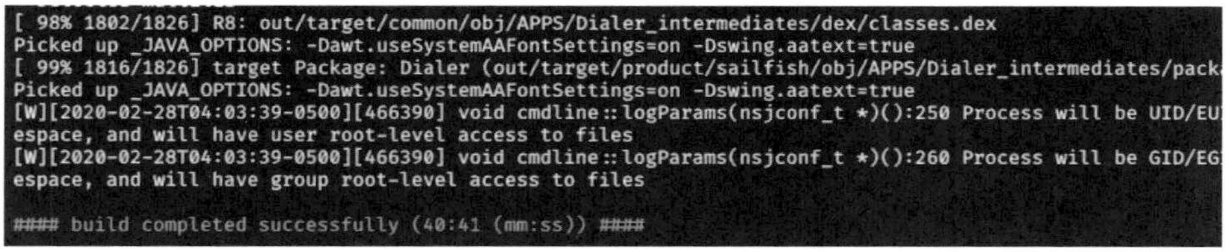

```
[ 98% 1802/1826] R8: out/target/common/obj/APPS/Dialer_intermediates/dex/classes.dex
Picked up _JAVA_OPTIONS: -Dawt.useSystemAAFontSettings=on -Dswing.aatext=true
[ 99% 1816/1826] target Package: Dialer (out/target/product/sailfish/obj/APPS/Dialer_intermediates/pack
Picked up _JAVA_OPTIONS: -Dawt.useSystemAAFontSettings=on -Dswing.aatext=true
[W][2020-02-28T04:03:39-0500][466390] void cmdline::logParams(nsjconf_t *)():250 Process will be UID/EU
espace, and will have user root-level access to files
[W][2020-02-28T04:03:39-0500][466390] void cmdline::logParams(nsjconf_t *)():260 Process will be GID/EG
espace, and will have group root-level access to files

#### build completed successfully (40:41 (mm:ss)) ####
```

▲ 圖 6-10 編譯成功

6.1.3 自編譯系統更新韌體

上一小節編譯得到的系統鏡像預設保存在原始程式根目錄的 out/target/ product/< 裝置代號 >/ 目錄下（這裡裝置代號即為 Pixel 對應的 sailfish 代號），所有編譯得到的鏡像 IMG 檔案如圖 6-11 所示。

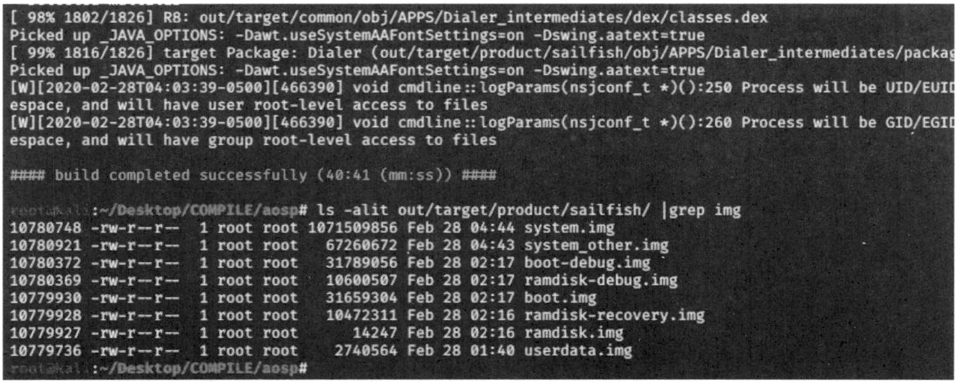

```
[ 98% 1802/1826] R8: out/target/common/obj/APPS/Dialer_intermediates/dex/classes.dex
Picked up _JAVA_OPTIONS: -Dawt.useSystemAAFontSettings=on -Dswing.aatext=true
[ 99% 1816/1826] target Package: Dialer (out/target/product/sailfish/obj/APPS/Dialer_intermediates/package
Picked up _JAVA_OPTIONS: -Dawt.useSystemAAFontSettings=on -Dswing.aatext=true
[W][2020-02-28T04:03:39-0500][466390] void cmdline::logParams(nsjconf_t *)():250 Process will be UID/EUID
espace, and will have user root-level access to files
[W][2020-02-28T04:03:39-0500][466390] void cmdline::logParams(nsjconf_t *)():260 Process will be GID/EGID
espace, and will have group root-level access to files

#### build completed successfully (40:41 (mm:ss)) ####

root@kali:~/Desktop/COMPILE/aosp# ls -alit out/target/product/sailfish/ |grep img
10780748 -rw-r--r--  1 root root 1071509856 Feb 28 04:44 system.img
10780921 -rw-r--r--  1 root root   67260672 Feb 28 04:43 system_other.img
10780372 -rw-r--r--  1 root root   31789056 Feb 28 02:17 boot-debug.img
10780369 -rw-r--r--  1 root root   10600507 Feb 28 02:17 ramdisk-debug.img
10779930 -rw-r--r--  1 root root   31659304 Feb 28 02:17 boot.img
10779928 -rw-r--r--  1 root root   10472311 Feb 28 02:16 ramdisk-recovery.img
10779927 -rw-r--r--  1 root root      14247 Feb 28 02:16 ramdisk.img
10779736 -rw-r--r--  1 root root    2740564 Feb 28 01:40 userdata.img
root@kali:~/Desktop/COMPILE/aosp#
```

▲ 圖 6-11 編譯結果

此時，如果想要將編譯出來的鏡像刷入裝置，還需要從 Android 鏡像官網（https://developers.google.com/android/images）下載如圖 6-12 所示的對應 Pixel 裝置 N2G47O 版本的官方鏡像。

▲ 圖 6-12 系統官方鏡像

之所以要下載官方鏡像，是因為對比圖 6-10 中的 IMG 鏡像檔案和解壓後的官方鏡像套件（見圖 6-13），會發現實際上編譯出來的鏡像只是官方鏡像套件內部壓縮檔的部分鏡像檔案，關鍵的 BootLoader 鏡像檔案以及其他（比如 odem）鏡像檔案在自編譯的系統鏡像中都未出現。

▲ 圖 6-13 系統官方鏡像

因此，如果想刷入自編譯的系統，可以先將編譯出來的鏡像檔案替換到官方鏡像套件中的對應檔案並重新壓縮，然後將手機以 BootLoader 模式啟動，並在保證 Android 裝置透過 USB 資料線與虛擬機器系統相連的基礎上，在虛擬機器中執行 flash-all.sh 檔案一鍵更新韌體，最終刷入自編譯的系統效果如圖 6-14 所示。

▲ 圖 6-14　自編譯系統版本

至此，一個自訂的 Android 系統就成功編譯並更新韌體成功了。接下來正式開始 Xposed 的編譯與訂製工作。

6.2 Xposed 訂製

6.2.1 Xposed 原始程式編譯

在開始 Xposed 原始程式的編譯之前,我們先來了解一下 Xposed 框架的原始程式結構。如圖 6-15 所示,查看其官方倉庫(https://github.com/rovo89),會發現存在 5 個和 Xposed 原始程式相關的專案,其中每個專案與對應功能的對照關係如表 6-1 所示。

▲ 圖 6-15 Xposed 框架的相關專案

表 6-1 Xposed 專案功能對照表

模 組	功 能
XposedInstaller	用於安裝到手機上下載和安裝 Xposed.zip 更新韌體套件,下載安裝和管理模組。注意,這個 App 要正常無誤地執行,裝置必須拿到 Root 許可權
XposedBridge	位於 Java 層的 API 提供者,外掛程式模組呼叫 Xposed 相關 API 時,首先呼叫 XposedBridge 中的函數,然後進一步「轉發」到 Native 方法

模 組	功 能
Xposed	位於 Native 層的 Xposed 實際實現，實現了「方法替換」等功能，本質上是對 Zygote 的延伸開發
android_art	在原版 art 上進行的延伸開發，目錄及檔案基本上與原版 art 相同，稍加修改以提供對 Xposed 的支援
XposedTools	在編譯過程中使用，負責編譯和打包更新韌體用的 ZIP 壓縮檔

事實上，以上 5 個專案在 Xposed 原始程式編譯時都有其對應的作用，簡單來說安裝 Xposed 框架的主要邏輯是，在將 XposedInstaller 應用安裝到裝置上後，XposedInstaller 會下載由 XposedTools 打包的含有 XposedBridge、Xposed 和 android_art 檔案的 ZIP 套件，並將該 ZIP 套件刷入系統，而這裡所謂的「刷入」實際上就是利用 Root 許可權放置和替換對應的系統檔案。

接下來分別對這些模組進行編譯。

1. 編譯模組管理器（XposedInstaller）

首先，透過 git 命令將系統原始程式下載到本地：

```
# git clone https://github.com/rovo89/XposedBridge.git
```

在下載成功後，使用 Android Studio 開啟 XposedInstaller 專案，會發現報一個錯誤：Failed to find target with hash string 'android-27' in: /root/Android/Sdk，如圖 6-16 所示。這是因為缺少 android-27 這個版本的 SDK 套件，此時只需點擊圖 6-16 中的提示 "Install missing platform(s) and sync project" 對對應 SDK 進行安裝即可。當然，在解決這個錯誤後可能還會出現如圖 6-17 所示的錯誤，這時只需繼續點擊提示連結進行安裝即可。

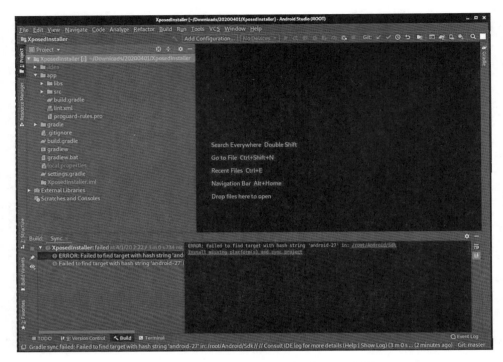

▲ 圖 6-16 XposedInstaller 同步顯示出錯 1

```
ERROR: Failed to find Build Tools revision 26.0.2
Install Build Tools 26.0.2 and sync project
Upgrade plugin to version 3.6.2 and sync project
```

▲ 圖 6-17 XposedInstaller 同步顯示出錯 2

在確認同步完成且無誤後，就可以將這個 App 安裝在使用 SuperSU 進行 Root 後的 android-7.1.2_r8 官方系統上（之所以還是刷入官方提供的系統中，是因為按照上述步驟編譯的系統雖然存在 Root 許可權，但是並沒有提供對 App 獲取 Root 許可權的支援，且上一節中編譯更新韌體

的系統只是用作驗證編譯過程是否有誤,具體進行更新軔體和 Root 的
操作在第 1 章中已經介紹過了,這裡不再贅述),最終手動編譯安裝的
XposedInstaller 功能如圖 6-18 所示。

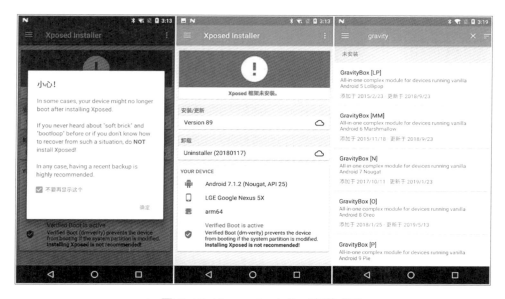

▲ 圖 6-18 XposedInstaller 安裝成功

2. 編譯執行時期支援函式庫(XposedBridge)

與 XposedInstaller 編譯安裝過程一致,首先下載專案原始程式:

```
# git clone https://github.com/rovo89/XposedBridge.git
```

這裡要注意的是,雖然這個專案最終編譯的目的檔案是一個 JAR 檔案,
但是同樣可以使用 Android Studio 開啟專案。當然,編譯過程中同樣可
能會出現錯誤,這裡在開啟專案後,經過漫長時間的同步後發現提示如
下錯誤:

```
ERROR: assert sdkSources.exists()
         |              |
         |             false
       /root/Android/Sdk/sources/android-23
```

事實上，這個錯誤與在編譯 XposedInstaller 時出現的缺少特定 SDK 的錯誤是相同的，只是這裡缺失的是 android-23，也就是 Android 6.0 對應的 SDK 套件。此時只需依次點擊 Android 頁面左上角的 File → Settings → Android SDK，選取 Android 6.0(Marshmallow)，再依次點擊 Apply → OK 按鈕即可開始下載對應的 SDK 套件，最終在下載完畢後再次同步專案直到同步成功。

同步成功後，再點擊 Android Studio 中的編譯按鈕，對專案進行編譯之後，又會發現出現如圖 6-19 所示的錯誤：while loading shared libraries: libz.so.1。

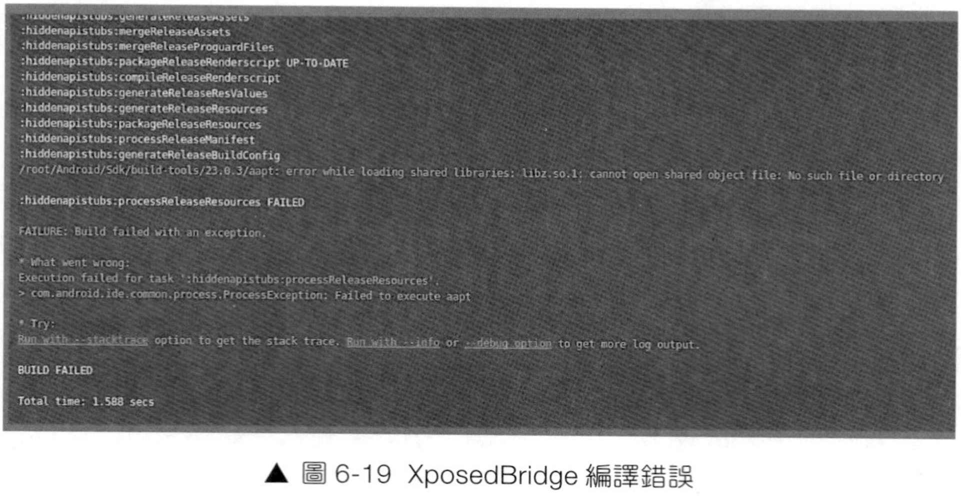

▲ 圖 6-19 XposedBridge 編譯錯誤

經過筆者研究發現，此時只需使用 apt 命令為系統安裝 lib32z1 庫即可解決問題。

如圖 6-20 所示，在解決上述依賴函數庫錯誤後，最終編譯出來的檔案預設保存在 app/build/outputs/apk/release 目錄下。

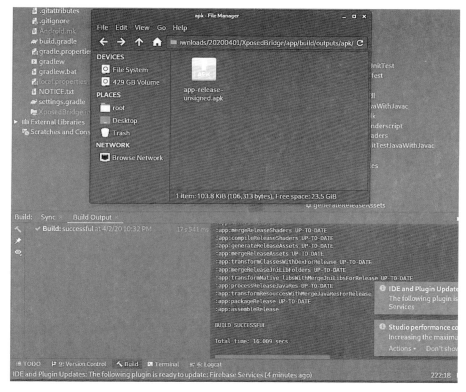

▲ 圖 6-20　XposedBridge 編譯結果

當然，此時生成的檔案實際上是一個 APK 檔案，但是實際上這裡的 JAR 檔案和 APK 檔案只是尾碼不同而已，因此只需將生成的 APK 檔案重新命名為 XposedBridge.jar 並放置於 Android 原始程式目錄中的 out/java 目錄下供後續使用即可（java 資料夾若不存在，則可自行新建）。

另外，這裡值得一提的是，我們在上一章中介紹的關於 Xposed 模組的編譯方式是透過設定 build.gradle 檔案完成的，但實際上也可以手動匯入對應 JAR 開發套件進行 Xposed 模組的開發，而對應的 api.jar 開發套件檔案實際上也是在這個專案中透過 gradle 命令編譯生成的，具體生成方式為：先點擊 Android Studio 右側的 Gradle 展開 Gradle 面板，然後依次展開 XposedBridge → app → Tasks → Other，按兩下 jarStubs 就會自動開始編譯，最終生成開發外掛程式模組用的 JAR 套件，如圖 6-21 所示。

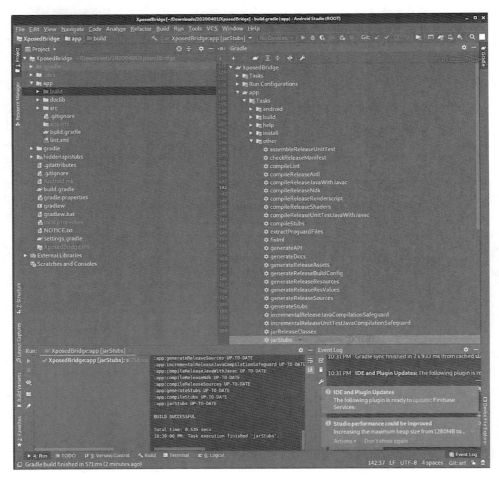

▲ 圖 6-21　編譯 api.jar

另外，jarStubs 下方還會有一個 jarStubsSource，它用於編譯 api-sources.jar 檔案。最終在 app/build/api/ 目錄下會發現生成的 api.jar 和 api-sources.jar。至此，我們寫 Xposed 模組專案時匯入的專案套件和原始程式套件都已生成，這兩個 JAR 檔案實際上用於自訂 Xposed 後的外掛程式開發工作，這裡暫且不討論。

3. 編譯定製版 art 解譯器（android_art）

編譯 android_art 相比以上步驟比較簡單，可在 Android 系統原始程式的環境下編譯。具體編譯步驟如下：

首先，要保證系統原始程式編譯成功過一次並且刷入手機後系統能夠正常開機使用，這也正是 6.1 節所做的工作。

然後，按照如下命令移動原版 art 目錄到系統原始程式外部（實際上這一步只是用於備份原版 art 原始程式），這裡保存在系統桌面上。

```
# mv art/ ~/Desktop/art_backup/
```

再次，如程式清單 6-5 所示，在 Android 系統原始程式目錄下下載 android_art 專案到本地並重新命名為 art，然後重新編譯系統。這個過程中不會出現任何問題，如圖 6-22 所示，在等待幾十分鐘後即可編譯成功。

◉ 程式清單 6-5 編譯 android_art

```
# git clone https://github.com/rovo89/android_art.git  art
# source build/envsetup.sh
# lunch 24 // sailfish-userdebug對應數字
# make
```

```
root@roysue: ...2r8/aosp712r8
Jack server already installed in '/root/.jack-server'
Server is already running
Bad request, see Jack server log
[  6% 160/2298] host Java: ahat (out/host/common/obj/JAVA_LIBRARIES/ahat_intermediates/classes)
Picked up _JAVA_OPTIONS: -Dawt.useSystemAAFontSettings=on -Dswing.aatext=true
Note: Some input files use unchecked or unsafe operations.
Note: Recompile with -Xlint:unchecked for details.
Picked up _JAVA_OPTIONS: -Dawt.useSystemAAFontSettings=on -Dswing.aatext=true
Picked up _JAVA_OPTIONS: -Dawt.useSystemAAFontSettings=on -Dswing.aatext=true
[ 94% 1059/1123] target R.java/Manifest.java: LiveTv (out/target/common/obj/APPS/LiveTv_intermediates/src/R.stamp)
warning: string 'title_br_tv_10' has no default translation.
warning: string 'title_br_tv_12' has no default translation.
warning: string 'title_br_tv_14' has no default translation.
warning: string 'title_br_tv_16' has no default translation.
warning: string 'title_br_tv_18' has no default translation.
warning: string 'title_br_tv_l' has no default translation.
warning: string 'title_kr_tv_12' has no default translation.
warning: string 'title_kr_tv_15' has no default translation.
warning: string 'title_kr_tv_19' has no default translation.
warning: string 'title_kr_tv_7' has no default translation.
warning: string 'title_kr_tv_all' has no default translation.
Warning: AndroidManifest.xml already defines minSdkVersion (in http://schemas.android.com/apk/res/android); using existing va
in manifest.
Warning: AndroidManifest.xml already defines targetSdkVersion (in http://schemas.android.com/apk/res/android); using existing
ue in manifest.
[ 96% 1079/1123] host Java: ahat-tests (out/host/common/obj/JAVA_LIBRARIES/ahat-tests_intermediates/classes)
Picked up _JAVA_OPTIONS: -Dawt.useSystemAAFontSettings=on -Dswing.aatext=true
Note: art/tools/ahat/test/SortTest.java uses unchecked or unsafe operations.
Note: Recompile with -Xlint:unchecked for details.
Picked up _JAVA_OPTIONS: -Dawt.useSystemAAFontSettings=on -Dswing.aatext=true
[ 99% 1104/1123] build out/target/product/sailfish/obj/NOTICE.html
Combining NOTICE files into HTML
Combining NOTICE files into text
[ 99% 1112/1113] Target system fs image: out/target/product/sailfish/obj/PACKAGING/systemimage_intermediates/system.img
Running: mkuserimg.sh -s /tmp/tmpsdYYzs out/target/product/sailfish/obj/PACKAGING/systemimage_intermediates/system.img ext4
13941504 -D out/target/product/sailfish/system -L / out/target/product/sailfish/root/file_contexts.bin
make_ext4fs -s -T -1 -S out/target/product/sailfish/root/file_contexts.bin -L / -l 2113941504 -a / out/target/product/sailfis
j/PACKAGING/systemimage_intermediates/system.img /tmp/tmpsdYYzs out/target/product/sailfish/system
Creating filesystem with parameters:
    Size: 2113941504
    Block size: 4096
    Blocks per group: 32768
    Inodes per group: 8064
    Inode size: 256
    Journal blocks: 8064
    Label: /
    Blocks: 516099
    Block groups: 16
    Reserved block group size: 127
Created filesystem with 2359/129024 inodes and 213371/516099 blocks
build_verity_tree -A aee087a5be3b982978c923f566a94613496b417f2af592639bc80d141e34dfe7 out/target/product/sailfish/obj/PACKAGI
ystemimage_intermediates/system.img /tmp/tmpb1HG7F_verity_images/verity.img
system/extras/verity/build_verity_metadata.py build 2113941504 /tmp/tmpb1HG7F_verity_images/verity_metadata.img 64dd075c54653
da5db970d0313aaa41238456d3dea260f6cd83549175e3a aee087a5be3b982978c923f566a94613496b417f2af592639bc80d141e34dfe7 /dev/block/b
evice/by-name/system verity_signer build/target/product/security/verity.pk8
cat /tmp/tmpb1HG7F_verity_images/verity_metadata.img >> /tmp/tmpb1HG7F_verity_images/verity.img
fec -e -p 0 out/target/product/sailfish/obj/PACKAGING/systemimage_intermediates/system.img /tmp/tmpb1HG7F_verity_images/verit
g /tmp/tmpb1HG7F_verity_images/verity_fec.img
cat /tmp/tmpb1HG7F_verity_images/verity_fec.img >> /tmp/tmpb1HG7F_verity_images/verity.img
append2simg out/target/product/sailfish/obj/PACKAGING/systemimage_intermediates/system.img /tmp/tmpb1HG7F_verity_images/verit
g
[100% 1113/1113] Install system fs image: out/target/product/sailfish/system.img
out/target/product/sailfish/system.img+ maxsize=2192424960 blocksize=135168 total=873995832 reserve=22167552

#### make completed successfully (16:38 (mm:ss)) ####

root@roysue:~/Desktop/COMPILE/aosp712r8/aosp712r8# ls
```

▲ 圖 6-22　android_art 編譯成功

4. 編譯本體（Xposed）

Xposed 專案的編譯同樣非常簡單，只需要將其放在原始程式的指定目錄 frameworks/base/cmds 中即可，後續直接透過 XposedTools 自動尋找並 進行編譯工作，具體執行命令如程式清單 6-6 所示。

🔽 程式清單 6-6 下載編譯 Xposed

```
# cd ~/aosp712_r8  // 切換到Android系統原始程式根目錄
# cd frameworks/base/cmds
# git clone https://github.com/rovo89/Xposed xposed
```

5. 編譯更新軔體套件（XposedTools）

最後一步是最難、最複雜的，就是使用 XposedTools 編譯出可以刷入手機的 xposed-v89-sdk25-arm64.zip 更新軔體套件，供 XposedInstaller 下載且刷入至手機。

首先，在下載對應 XposedTools 專案原始程式後，按照如下命令將資料夾內的編譯設定範本複製一份做備份用：

```
# cd XposedTools && cp build.conf.sample build.conf
```

複製完成後，如程式清單 6-7 所示，對 build.conf 設定檔進行自訂修改，其中 outdir 對應系統原始程式輸出目錄，javadir 對應 XposedBridge.jar 所在目錄，version 對應最終生成的版本編號，AospDir 屬性設定為系統原始程式對應的 SDK numbers（7.1 對應的 SDK number 為 25）以及對應的原始程式目錄，最後一個屬性 BusyBox 中對應的數字也要修改為原始程式對應的 SDK number。

🔽 程式清單 6-7 build.conf

```
[General]
outdir = /root/aosp712_r8/out # 原始程式輸出目錄
javadir = /root/aosp712_r8/out/java  # XposedBridge.jar放置目錄

[Build]
# Please keep the base version number and add your custom suffix
version = 89 (custom build by r0ysue / %s)
# makeflags = -j4
```

```
[GPG]
sign = release
user = 852109AA!

# Root directories of the AOSP source tree per SDK version
[AospDir]
25 = /root/aosp712_r8 # 25代表android 7.1

# SDKs to be used for compiling BusyBox
# Needs https://github.com/rovo89/android_external_busybox
[BusyBox]
arm = 25
x86 = 25
armv5 = 25
```

在設定完 XposedTools 的編譯選項後還不能夠直接編譯。這是因為
XposedTools 的編譯依賴於 Perl 語言的開發環境，而 Kali 系統本身並沒
有這樣的環境，所以還需要輸入如下命令以安裝一系列的 Perl 環境及協
力廠商套件。

```
# apt install libconfig-inifiles-perl libauthen-ntlm-perl libclass-load-perl\
  libcrypt-ssleay-perl libdata-uniqid-perl libdigest-hmac-perl \
  libdist-checkconflicts-perl libfile-copy-recursive-perl libfile-tail-perl
```

在安裝完 Perl 的開發環境後，還需要為 XposedTools 安裝一系列 Perl 的
協力廠商工具套件，這裡推薦使用 cpan 命令進行安裝，具體命令如下：

```
# cpan install Config::IniFiles File::Tail File::ReadBackwards Archive::Zip
```

在所有 Perl 的協力廠商工具函數庫都安裝完畢後，就可以按照如下命令
開始 XposedTools 模組的安裝。這裡 arm64 為目標裝置 sailfish 的裝置架
構，25 代表 Android 7.1 對應的 SDK number。

```
# ./build.pl -t arm64:25
```

最終編譯完成的成品如圖 6-23 所示。

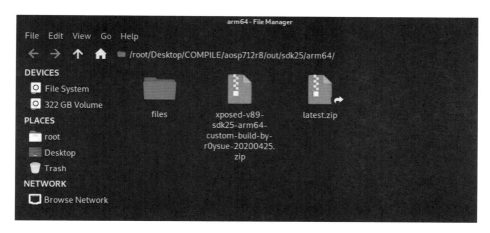

▲ 圖 6-23 XposedTools 編譯成功

6. 驗證

上一步最終編譯出來的更新韌體套件名為 xposed-v89-sdk25-arm64-custom-build-by-r0ysue-20200425.zip，為了方便後續刷入，這裡將其重新命名為與官網相同的 xposed-v89-sdk25-arm64.zip。

在透過 XposedInstaller 進行更新韌體套件的刷入步驟中，採用了本地架設簡單伺服器並替換 XposedInstaller 專案中設定遠端下載網址函數的方式，具體修改的是 XposedInstaller 專案中 DownloadsUtil.java 檔案的 setUrl 函數，這裡最終修改後的 DownloadsUtil.java 檔案內容如程式清單 6-8 所示。

⬥ 程式清單 6-8 DownloadsUtil.java

```
package de.robv.android.xposed.installer.util;

import android.app.DownloadManager;
...
```

```java
import de.robv.android.xposed.installer.repo.ReleaseType;

public class DownloadsUtil {
    public static final String MIME_TYPE_APK = "application/vnd.android.
package-archive";
    public static final String MIME_TYPE_ZIP = "application/zip";
    private static final Map<String, DownloadFinishedCallback> mCallbacks =
new HashMap<>();
    private static final XposedApp mApp = XposedApp.getInstance();
    private static final SharedPreferences mPref = mApp
            .getSharedPreferences("download_cache", Context.MODE_PRIVATE);

    public static class Builder {
        private final Context mContext;
        private String mTitle = null;
        private String mUrl = null;
        private DownloadFinishedCallback mCallback = null;
        private MIME_TYPES mMimeType = MIME_TYPES.APK;
        private File mDestination = null;
        private boolean mDialog = false;

        public Builder(Context context) {
            mContext = context;
        }

        public Builder setTitle(String title) {
            mTitle = title;
            return this;
        }

        public Builder setUrl(String url) {
            //mUrl = url;
            //將這裡改成指向本機伺服器的更新韌體類別檔案放置於本地架設的HTTP伺
服器根目錄下，具體本機伺服器架設方式可自行研究
            mUrl = "http://192.168.0.9/xposed-v89-sdk25-arm64.zip";
```

```
            return this;
        }
        ...
    }
}
```

在修改完 XposedInsaller 專案原始程式後,將其重新編譯安裝到測試手機上,此時按照正常下載安裝 Xposed 的步驟即可完成自編譯 Xposed 的刷入,最終在啟動並重新啟動裝置後,XposedInstaller 應用的介面如圖 6-24 所示。可以發現筆者自編譯的 Xposed 框架 89(custom build by r0ysue)已成功刷入系統且啟動,此時如果安裝協力廠商外掛程式程式,就會發現 Hook 效果同樣表現正常(事實上這裡已經成功安裝了 GravityBox 外掛程式並修改了狀態列顏色,但限於印刷效果,可能無法觀察到)。

▲ 圖 6-24 自編譯版本 Xposed 啟動成功

6.2.2 Xposed 魔改繞過 XposedChecker 檢測

在正式開始進行 Xposed 的魔改前,我們使用 XposedChecker 這個檢測 Xposed 特徵的開放原始碼工具對自編譯的 Xposed 框架進行檢測,最終檢測結果如圖 6-25 所示。

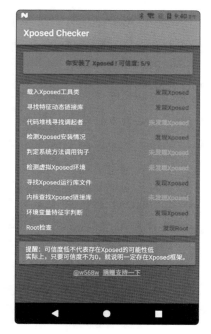

▲ 圖 6-25 Xposed 框架檢測結果

觀察圖 6-25 中的檢測結果可以發現,共存在 5 個被檢測出來的 Xposed 特徵。當然,如 XposedChecker 宣告的那樣,雖然某些檢測專案顯示「未發現 Xposed」,但並非說明不存在 Xposed,只是可能因為該檢測函數並未被 Xposed Hook 或其他原因導致無法滿足檢測特徵。這裡為了方便後續有目的性地修改 Xposed 原始程式,我們先來了解一下 XposedChecker 對 Xposed 的檢測方法。

在將 XposedChecker 專案原始程式從 GitHub 下載到本地並載入 Android Studio 後，會發現所有的檢測邏輯都集中在 app\src\main\java\ml\w568w\checkxposed\ui\MainActivity.java 檔案中，具體的檢測項如程式清單 6-9 所示。

◯ 程式清單 6-9 XposedChecker 檢測項

```
private static final String[] CHECK_ITEM = {
    "載入Xposed工具類別",
    "尋找特徵動態連結程式庫",
    "程式堆疊尋找調起者",
    "檢測Xposed安裝情況",
    "判定系統方法呼叫鉤子",
    "檢測虛擬Xposed環境",
    "尋找Xposed執行函數庫檔案",
    "核心查詢Xposed程式庫",
    "環境變數辨識符號判斷",
};
```

這裡介紹幾個具體的檢測項對應的實現。

（1）如程式清單 6-10 所示，「載入 Xposed 工具類別」檢測項是透過呼叫系統類別載入器嘗試載入 XposedHelpers 類別，並透過 try-catch 異常處理機制確認是否存在 XposedHelpers 類別，進而判斷裝置系統中是否存在 Xposed 框架的。

◯ 程式清單 6-10「載入 Xposed 工具類別」檢測項

```
private boolean testClassLoader() {
    try {
        ClassLoader.getSystemClassLoader()
            .loadClass("de.robv.android.xposed.XposedHelpers");

        return true;
```

```
    }
    catch (ClassNotFoundException e) {
        e.printStackTrace();
    }
    return false;
}
```

（2）在「尋找特徵動態連結程式庫」檢測項對應的實現如程式清單 6-11 所示，可以發現對應檢測方法是透過查看 /proc/self/maps 檔案，從而判斷處理程序自身記憶體模組名中是否包含 XposedBridge 字串的模組的方式進行的。

● 程式清單 6-11「尋找特徵動態連結程式庫」檢測項

```
private int check2() {
    return checkContains("XposedBridge") ? 1 : 0;
}
public static boolean checkContains(String paramString) {
    try {
        HashSet<String> localObject = new HashSet<>();
        // 讀取maps檔案資訊
        BufferedReader localBufferedReader =
    new BufferedReader(new FileReader("/proc/" + Process.myPid() + "/
maps"));
        while (true) {
            String str = localBufferedReader.readLine();
            if (str == null) {
                break;
            }
            localObject.add(str.substring(str.lastIndexOf(" ") + 1));
        }
        //應用程式的程式庫不可能是空，除非高於7.0
        if (localObject.isEmpty() && Build.VERSION.SDK_INT <= Build.
VERSION_CODES.M) {
            return true;
```

```
        }
        localBufferedReader.close();
        for (String aLocalObject : localObject) {
            if (aLocalObject.contains(paramString)) {
                return true;
            }
        }
    }
}
catch (Throwable ignored) {
    }
    return false;
}
```

（3）如程式清單 6-12 所示，在「檢測 Xposed 安裝情況」檢測項中，實際上就是透過獲取系統已安裝 App 清單的方式判斷其中是否包含特定 Xposed 相關 App 的存在，這裡僅關注 XposedInstaller 對應的套件名 de.robv.android.xposed.installer。

❍ 程式清單 6-12「檢測 Xposed 安裝情況」檢測項

```
private static final String[] XPOSED_APPS_LIST = new String[]{"de.robv.
android.xposed.installer", "io.va.exposed", "org.meowcat.edxposed.manager",
"com.topjohnwu.magisk", "com.doubee.ig", "com.soft.apk008v", "com.soft.
controllers", "biz.bokhorst.xprivacy"};
private int check4() {
        StringBuilder builder = new StringBuilder(String.
format(getString(R.string.item_4_1), 0));
        try {
            List<PackageInfo> list = getPackageManager().
getInstalledPackages(0);
            builder = new StringBuilder(String.format(getString(R.string.
item_4_1), list.size()));
            for (PackageInfo info : list) {
                for (String pkg : XPOSED_APPS_LIST) {
                    if (pkg.equals(info.packageName)) {
```

```
                                builder.append(getString(R.string.item_4_2)).
append(pkg).append("\n");
                                techDetails.add(builder.toString());
                                return 1;
                            }
                        }

                }
        } catch (Throwable ignored) {
        }
        builder.append("[").append(toStatus(false)).append("]");
        techDetails.add(builder.toString());
        return 0;
    }
```

（4）關於「環境變數辨識符號判斷」的檢測方式，如程式清單 6-13 所示，其對應實現是透過獲取系統中 CLASSPATH 環境變數的方式進行檢測的。同樣，這種檢測方式也是判斷字串中是否包含 XposedBridge 這個固定檔案名稱。

● 程式清單 6-13　環境變數檢查

```
private int check9() {
    return System.getenv("CLASSPATH").contains("XposedBridge") ? 1 : 0;
}
```

對比上述介紹的檢測方式以及其他未介紹的檢測方式，會發現除了第 5 項「判定系統方法呼叫鉤子」檢測方式外，其他針對 Xposed 的檢測方式都高度依賴字串的匹配，而這樣的檢測方式正是筆者進行魔改的突破點。簡單來說，就是將全域包含 Xposed 字串可供檢測的點修改為其他字串，這裡的魔改目標就是全域修改 Xposed 字串為 Xppsed。

接下來讓我們正式開始「魔改」的流程。

首先，修改 XposedInstaller 這個 App 的 Xposed 字串特徵，具體測試後，筆者發現實際上要修改的部分非常少：只要修改整體的套件名以及 prop 設定檔名相關的字串即可。這兩項的修改方式在 Android Studio 的幫助下異常簡單，具體步驟如下：

步驟 01 按照如圖 6-26 所示的步驟選擇 Project 檔案樹並取消 Android Studio 目錄折疊的預設設定：取消選取 Compact Middle Packages 選項。

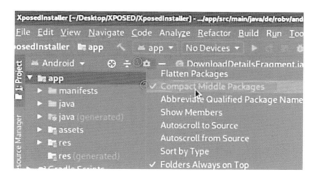

▲ 圖 6-26 取消目錄折疊

步驟 02 按照如圖 6-27 所示的方式按滑鼠右鍵選中 xposed 套件名並重新命名為 xppsed。

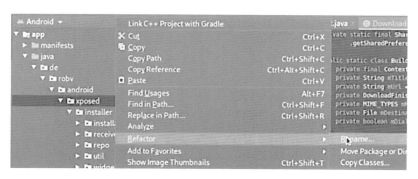

▲ 圖 6-27 重新命名套件名

當然，這裡 Android Studio 會提示搜索專案中所有出現 xposed 套件名的
地方，如圖 6-28 所示，只需在出現的 Preview 視窗點擊 Do Refactor 按
鈕執行所有修改即可。

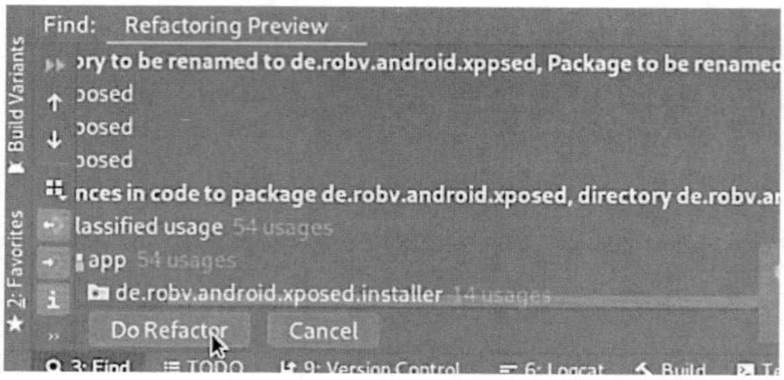

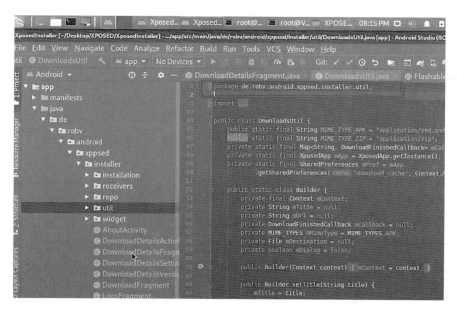

▲ 圖 6-28　確認修改

步驟 03　確認修改成功。在執行完 Do Refactor 後，如圖 6-29 所示，任意
開啟 Java 專案檔案，會發現其套件名中的 xposed 都變成了 xppsed。

▲ 圖 6-29　確認修改成功

步驟 04 到上一步為止，只是將所有套件名中包含 xposed 的程式檔案中相關的字串替換為 xppsed，但是實際上在一個 Android 專案中，除了程式檔案外，還會有很多設定檔和字串強制寫入，比如 AndroidManifest.xml 檔案，這是 Android Studio 智慧修改無法完成的事情。因此，這種可能包含 xposed 的相關套件名記錄的情況，還需要透過如圖 6-30 所示的方式按滑鼠右鍵選中 app 目錄，在專案中全域搜索，替換 de.robv.android.xposed.installer 字串為 de.robv.android.xppsed.installer（注意這裡不能直接全域搜索 xposed）。

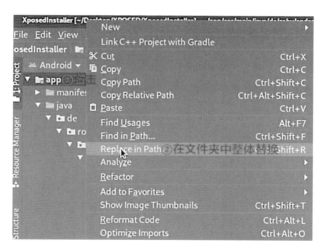

▲ 圖 6-30 全域搜索 de.robv.android.xposed.installer 字串

步驟 05 在完成修改整體的套件名的步驟後，還需要修改 XposedInstaller 專案中強制寫入 prop 設定檔名的位置。對應的設定檔名是在 XposedApp.java 中定義的，圖 6-31 所示是修改後的 prop 設定檔名。

在完成以上所有修改後，再次將應用編譯安裝到手機上，以驗證以上修改是否成功，這裡在點擊 Build → Clean 後，再次編譯安裝到手機上的結果如圖 6-32 所示。到這裡，XposedInstaller 的魔改也就順利完成了。

```
nloadsUtil.java    DownloadFragment.java    XposedApp.java    Flas
package de.robv.android.xppsed.installer;

import ...

public class XposedApp extends Application implements ActivityLifecycleCallbac
    public static final String TAG = "XposedInstaller";

    @SuppressLint("SdCardPath")
    private static final String BASE_DIR_LEGACY = "/data/data/de.robv.android.

    public static final String BASE_DIR = Build.VERSION.SDK_INT >= 24
            ? "/data/user_de/0/de.robv.android.xppsed.installer/" : BASE_DIR_L

    public static final String ENABLED_MODULES_LIST_FILE = XposedApp.BASE_DIR

    private static final String[] XPOSED_PROP_FILES = new String[]{
            "/su/xposed/xppsed.prop", // official systemless
            "/system/xppsed.prop",    // classical
    };

    public static int WRITE_EXTERNAL_PERMISSION = 69;
```

▲ 圖 6-31 修改 prop 設定檔名

▲ 圖 6-32 XposedInstaller 安裝啟動正常

XposedBridge 專案的修改點同樣也是套件名，只需要按照針對XposedInstaller 專案的修改方式依樣畫葫蘆即可完成修改，在修改完成並重新編譯後，無須安裝到手機上，而是將編譯出來的檔案複製並命名為XppsedBridge.jar 即可。

另外，為了後續的 Xposed 模組撰寫工作，還需要按照 6.2 節中編譯 api.jar 和 api-source.jar 檔案的方式再次編譯出一份適用於自訂 Xposed 框架的開發套件待用。

在自訂 Xposed 框架的過程中，實際上最複雜的是針對 Xposed 專案原始程式的修改。如圖 6-33 所示，雖然這部分的程式不多，在 6.2 節中對其編譯的難度也最低，但是這部分在自訂的過程中修改的地方卻是最多的。

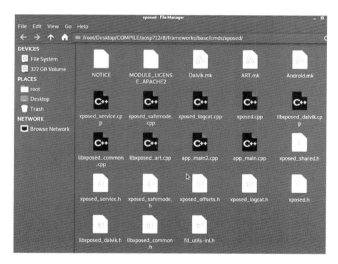

▲ 圖 6-33 Xposed 框架原始程式檔案

其中具體要修改的部分如下：

（1）替換 libxposed_common.h 檔案中包含 xposed 套件名的位置為xppsed，具體修改內容如程式清單 6-14 所示。

● 程式清單 6-14 libxposed_common.h 修改內容

```
// 修改前
#define CLASS_XPOSED_BRIDGE "de/robv/android/xposed/XposedBridge"
#define CLASS_ZYGOTE_SERVICE "de/robv/android/xposed/services/
ZygoteService"
#define CLASS_FILE_RESULT "de/robv/android/xposed/services/FileResult"
// 修改後
#define CLASS_XPOSED_BRIDGE "de/robv/android/xppsed/XposedBridge"
#define CLASS_ZYGOTE_SERVICE "de/robv/android/xppsed/services/
ZygoteService"
#define CLASS_FILE_RESULT "de/robv/android/xppsed/services/FileResult"
```

（2）如程式清單 6-15 所示，修改 Xposed.h 檔案中包含 xposed 子字串的
字串為 xppsed。

● 程式清單 6-15 Xposed.h 檔案修改

```
// 修改前
#define XPOSED_PROP_FILE "/system/xposed.prop"
#define XPOSED_LIB_ART XPOSED_LIB_DIR "libxposed_art.so"
#define XPOSED_JAR "/system/framework/XposedBridge.jar"
#define XPOSED_CLASS_DOTS_ZYGOTE "de.robv.android.xposed.XposedBridge"
#define XPOSED_CLASS_DOTS_TOOLS "de.robv.android.xposed.
XposedBridge$ToolEntryPoint"
// 修改後
#define XPOSED_PROP_FILE "/system/xppsed.prop"
#define XPOSED_LIB_ART XPOSED_LIB_DIR "libxppsed_art.so"
#define XPOSED_JAR "/system/framework/XppsedBridge.jar"
#define XPOSED_CLASS_DOTS_ZYGOTE "de.robv.android.xppsed.XposedBridge"
#define XPOSED_CLASS_DOTS_TOOLS "de.robv.android.xppsed.
XposedBridge$ToolEntryPoint"
```

（3）如程式清單 6-16 所示，修改 xposed_service.cpp。

⊘ 程式清單 6-16　xposed_service.cpp 檔案修改

```
// 修改前
IMPLEMENT_META_INTERFACE(XposedService, "de.robv.android.xposed.
IXposedService");
// 修改後
IMPLEMENT_META_INTERFACE(XposedService, "de.robv.android.xppsed.
IXposedService");
```

（4）如程式清單 6-17 所示，還需要修改 xposed_shared.h 檔案中包含 xposed 字串的地方。

⊘ 程式清單 6-17　xposed_shared.h 檔案修改

```
// 修改前
#define XPOSED_DIR "/data/user_de/0/de.robv.android.xposed.installer/"
#define XPOSED_DIR "/data/data/de.robv.android.xposed.installer/"
// 修改後
#define XPOSED_DIR "/data/user_de/0/de.robv.android.xppsed.installer/"
#define XPOSED_DIR "/data/data/de.robv.android.xppsed.installer/"
```

（5）修改編譯設定檔 ART.mk，修改處如程式清單 6-18 所示。

⊘ 程式清單 6-18　ART.mk 檔案修改

```
// 修改之前
libxposed_art.cpp
LOCAL_MODULE := libxposed_art
// 修改之後
libxppsed_art.cpp
LOCAL_MODULE := libxppsed_art
```

（6）將 xposed 資料夾下的 libxposed_art.cpp 檔案重新命名為 libxppsed_art.cpp 即可完成 Xposed 框架原始程式的修改工作。

最後，修改 XposedTools 編譯工具的原始程式。

事實上，這部分程式與最終生成的 ZIP 套件並無直接聯繫，只是在編譯過程中需要透過字串的方式連接各個模組並編譯而已，因此對這部分的修改原則在於：只要編譯過程不顯示出錯就行。

由於在之前的修改中存在 3 處修改生成檔案的地方：xppsed.prop、XppsedBridge.jar 和 libxppsed_art。因此，在修改過程中只要將 build.pl 和 zipstatic/_all/META-INF/com/google/android/flash-script.sh 這兩個檔案中的上述字串修改為魔改後的字串即可。

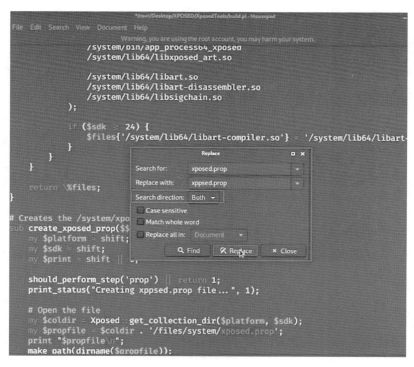

▲ 圖 6-34　查詢、替換功能

需要注意的是，在這部分修改的過程中，可充分利用編輯器的查詢、替換功能，保證不要有遺漏，如圖 6-34 所示。在修改完之後，可以到

XposedTools 根目錄下執行 grep 命令，保證找不到對應的字串，即全部替換完成。

```
# grep -ril "xposed.prop" *
```

到這一步，檔案修改的部分就完成了。接下來是透過 XposedTools 對各個模組編譯的步驟。注意，在最終編譯之前，需要把自訂修改後編譯生成的 XppsedBridge.jar 檔案放置到 $AOSP/out/java/ 目錄中，並刪除原有的 XposedBridge.jar 檔案。

與 6.2 節中編譯的方式一致，按照如下命令進行編譯即可：

```
# ./build.pl -t arm64:25
```

在成功編譯並刷入手機後，再次使用 XposedChecker 工具進行檢測，最終檢測效果如圖 6-35 所示，可以發現所有檢測 Xposed 的部分已經故障，魔改成功。

▲ 圖 6-35 XposedChecker 檢測結果

在使用魔改完畢的 Xposed 框架時要注意，市面上所有基於原版 Xposed 框架的模組都不會再生效，比如 GravityBox 在安裝到安裝了自訂修改的 Xposed 框架的手機上後，甚至都無法開啟（見圖 6-36），究其原因，正是因為 GravityBox 會在後台尋找 de.robv.android.xposed. IXposedHookZygoteInit 這個類別，但可惜的是這個類別被修改了。

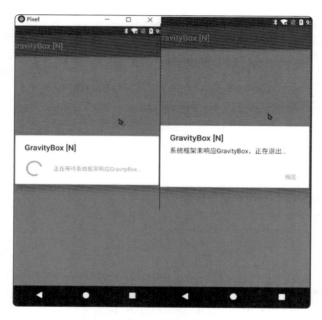

▲ 圖 6-36 GravityBox 無法開啟

最後介紹一下基於自訂修改的 Xposed 框架撰寫 Xposed 模組的方式，其對應的撰寫步驟如下：

首先，將在編譯 XposedBridge.jar 專案時編譯的兩個附屬品 api.jar 和 api-source.jar 檔案放置到 Android 專案的 libs/ 目錄中，並在 App 專案的 app/build.gradle 檔案的 dependencies 節點中，使用 compileOnly files('libs/api.jar') 指定使用剛剛編譯出來的函數庫，最終 dependencies

節點的內容如程式清單 6-19 所示。

⊘ 程式清單 6-19　build.gradle 檔案

```
dependencies {
    implementation 'androidx.appcompat:appcompat:1.1.0'
    implementation 'androidx.constraintlayout:constraintlayout:1.1.3'
    testImplementation 'junit:junit:4.12'
    androidTestImplementation 'androidx.test.ext:junit:1.1.1'
    androidTestImplementation 'androidx.test.espresso:espresso-core:3.2.0'
    compileOnly files('libs/api.jar')
    compileOnly files('libs/api-sources.jar')
}
```

此時再次撰寫 Hook 程式，即可正常使用快速鍵將需要的套件匯入，最終匯入的模組名如圖 6-37 所示。

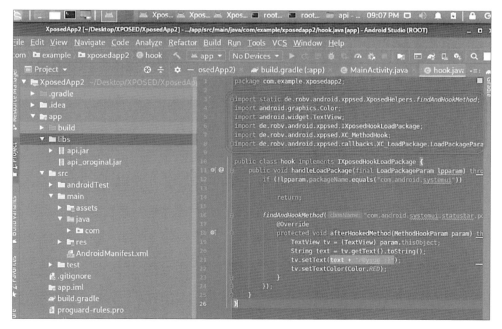

▲ 圖 6-37　自訂模組撰寫匯入

這裡撰寫了在系統時間欄後面加字串和顏色更換的 Hook 程式，具體內容如程式清單 6-20 所示。

◯ 程式清單 6-20 Hook 程式

```
public class hook implements IXposedHookLoadPackage {
    public void handleLoadPackage(final LoadPackageParam lpparam) throws
Throwable {
        if (!lpparam.packageName.equals("com.android.systemui"))
            return;

        findAndHookMethod("com.android.systemui.statusbar.policy.Clock",
lpparam.classLoader, "updateClock", new XC_MethodHook() {
            @Override
            protected void afterHookedMethod(MethodHookParam param) throws
Throwable {
                TextView tv = (TextView) param.thisObject;
                String text = tv.getText().toString();
                tv.setText(text + "r0ysue :)");
                tv.setTextColor(Color.RED);
            }
        });
    }
}
```

最終編譯執行後，Hook 效果以及 XposedChecker 檢測結果如圖 6-38 所示。

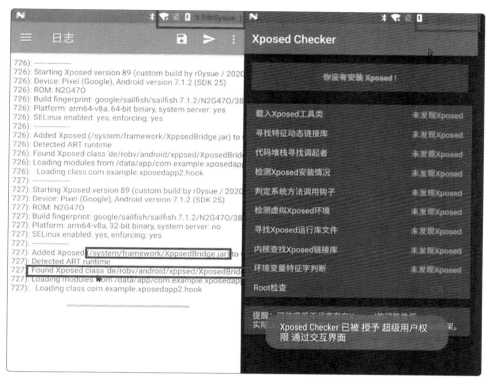

▲ 圖 6-38 自訂模組執行結果

6.3 本章小結

本章帶領讀者從簡單的 Android 原始程式的編譯工作開始一步一步介紹 Xposed 這一框架的編譯與自訂修改，並在 XposedChecker 這一開放原始碼 Xposed 檢測工具的檢測下成功隱藏了 Xposed 的特徵。但事實上，正 如筆者在介紹 XposedChecker 原始程式時所講，該工具檢測 Xposed 特徵的方式大部分只是透過簡單的字串進行檢測，而 Xposed 的特徵肯定不止有字串這種方式，比如被 Xposed Hook 的函數其 access_flags 屬性變成了 native 這一特徵。因此，要做到一個真正可用的、無法被檢測出來的自訂 Xposed，在本章中最終得到的「魔改」版 Xppsed 還任重道遠。當然，這裡僅僅是 磚引玉，如果讀者對這部分內容有興趣，可以自行深入研究。

07

Android 沙盒之加解密庫「自吐」

相信如果讀者對 Android 系統足夠了解，那麼一定對「沙盒」這個詞不陌生：每個 Android 應用都執行在獨立的沙盒中。但是從這一章開始要介紹的「沙盒」並不是指應用執行時期的沙盒，而是指系統級的沙盒。這裡所謂的「系統級的沙盒」是指：透過自訂系統原始程式編譯特定系統，使得執行在自訂系統上的 App 行為都曝露在系統的監控下，進而輔助完成後續逆向分析的任務。從本章開始，將從多個角度簡單介紹系統級沙盒的開發方式，帶領讀者領略沙盒的強大之處。

7.1 沙盒介紹

對於系統來說，由於 App 的全部程式都是依賴系統執行的，因此無論是保護 App 在執行時期的脫殼，還是 App 發送和接收資料封包，對於系統本身來說，App 的行為都是沒有隱私的。換句話說，如果在系統層或者更底層對 App 進行行為監控，App 的很多關鍵資訊就會曝露在「陽光」之下，一覽無餘。

基於這種從高維度對抗進而降維打擊應用的思路，DexHunter、Fupk、FART 等脫殼機從 ART 虛擬機器層面對執行在應用層的 App 進行記憶體資料的 dump，進而提出一、二代殼（整體保護和函數取出殼）的解決方案，甚至 FART 10 基於這種思路解決了部分三代殼的脫殼問題，更有 TinyTool 從核心中呼叫 JProbe 來監控 syscall 系統呼叫，這樣即使 App 應用使用靜態編譯的二進位檔案或透過 svc 組合語言指令在使用者態直接進行系統呼叫，最終導致使用者態 Hook 故障且沒有意義，還可以從核心中列印出一份日誌為分析 App 的行為提供依據。程式清單 7-1 是筆者在測試 TinyTool 時列印的部分日誌內容。

⬇ **程式清單 7-1 TinyTool 部分日誌**

```
[34728.283575] REHelper device open success!
[34728.285504] Set monitor pid: 3851
[34728.287851] [openat] dirfd: -100, pathname /dev/__properties__, flags:
a8000, mode: 0
[34728.289348] [openat] dirfd: -100, pathname /proc/stat, flags:20000,mode: 0
[34728.291325] [openat] dirfd: -100, pathname /proc/self/status, flags:
20000, mode: 0
[34728.292016] [inotify_add_watch]: fd: 4, pathname: /proc/self/mem, mask: 23
[34729.296569] PTRACE_PEEKDATA: [src]pid = 3851 --> [dst]pid = 3852, addr:
40000000, data: be919e38
```

除了以上這些開放原始碼的沙盒外，事實上各大安全或防毒公司都有自己的沙盒：將 App 安裝在沙盒上執行一次，從而快速得到 App 的執行流，最終得到一個成熟的安全分析報告。目前，筆者已知的公開提供這項服務的包括微步線上雲沙盒（官網位址：https://s.threatbook.cn/，透過模擬檔案執行環境來分析和收集檔案的靜態和動態行為資料，結合微步威脅情報雲，分鐘級發現未知威脅）、Cuckoo Sandbox（官網位址：https://cuckoosandbox.org/）等。

除了以上基於系統原始程式或者核心原始程式的沙盒外，還有一些其他類型的沙盒。比如基於 Hook 類型的沙盒：r0capture 雖然沒有直接修改系統原始程式，但是其基於 Hook 的思路對系統收發送封包函數進行二進位插樁，從而提出了應用層抓取封包的通殺方案；appmon（專案位址：https://github.com/dpnishant/appmon）更是基於 Frida Hook 對系統標準加密函數庫、檔案系統函數進行 Hook，從而追蹤 App 透過系統提供的API 執行的痕跡，最終輔助逆向工作人員對 App 進行分析。

但正如上面所説的，App 是因為依賴系統提供 API，進而導致本身行為曝露在系統監控中，那麼如果 App 想要對抗沙盒的分析，該怎麼做呢？筆者個人認為，首先 App 應當盡可能減少系統 API 的呼叫，其次關鍵函數的演算法儘量不直接使用系統提供的加密函數庫，而是盡可能自己實現一定量的演算法。除此之外，為了保護自實現演算法不被破解，可以採取對自實現演算法進行強混淆或者增加 VMP 保護等。當然，這裡提出的對抗思路僅能對抗部分沙盒，由於 App 開發的目的就是在系統上執行，不可避免地會有執行痕跡的存在，要完全對抗沙盒還需讀者們見仁見智。

基於此，本章將首先基於 appmon 這一 Hook 類型的沙盒提出針對加密函數庫進行分析的指令稿，從 Frida Hook 入手開發屬於我們自己的加密庫「自吐」沙盒。

7.2 雜湊演算法「自吐」

7.2.1 密碼學與雜湊演算法介紹

在資訊安全等級保護工作中,通常根據資訊系統的機密性(Confidentiality)、完整性(Integrity)、可用性(Availability)來劃分資訊系統的安全等級,簡稱為 CIA。其中機密性這一特性一般來說就是指一般人不可知曉。它的另一層意思是,只有被授權的主體才知道資訊的內容,要做到這一點就要依靠密碼學完成。以資料傳輸過程為例,在資料從發送方被發送出去之前需要對資料進行加密,以保證在傳輸過程中不被除接收方之外的協力廠商獲取。當然,僅僅是單一的加解密演算法並不能夠保證資料在傳輸過程中的機密性,因此在加密過程中還需要進一步保證只有接收方和發送方能夠得到真實資料的金鑰的參與,金鑰只掌握在發送方和接收方兩者手中,從而保障資料的機密性。

從上述介紹中,讀者會發現密碼學中最重要的概念其實是加解密演算法和金鑰,加解密演算法保證資料在傳輸過程中是加密狀態,僅掌握在通訊雙方手中的金鑰保障了即使協力廠商得知加解密的演算法和加密,也無法得到真實的原始資料。基於此,現代密碼學中出現了兩種流派,即對稱密碼和非對稱密碼。這裡的對稱和非對稱區別在於通訊雙方的金鑰上,對稱密碼通訊雙方持有的金鑰是一樣的,換句話說,同一金鑰不僅可以用於加密,還可以用於解密,其中經典的演算法有 DES、AES 等;相反,非對稱密碼通訊雙方持有的金鑰並不相同,分別稱為公開金鑰和私密金鑰,其中公開金鑰用於加密,私密金鑰用於解密,主要演算法有 RSA、ECC 等。

除了上述介紹的對稱密碼和非對稱密碼外，密碼學中還會有一類演算法一雜湊（Hash）演算法。與前兩者不同的是，雜湊演算法是一種能夠將任意長度的輸入轉化為固定長度輸出，且加密過程不可逆的演算法，訊息最終得到的輸出稱為雜湊值，因此雜湊演算法通常用於保障資料資訊不被篡改，又稱為訊息摘要演算法。另外，雜湊演算法還有另一個重要特徵，即訊息任意兩個不同的訊息，其 Hash 值一定不同，稱為雜湊演算法的抗衝突性。常用的雜湊演算法有 MD4/MD5 等 MD 系列演算法、SHA-1/SHA-256 等 SHA 系列演算法等，其中 MD5 演算法被廣泛使用，可以產生一個 128 位元（16 位元組，一位元組 8 位元）的雜湊值（常見的雜湊值是用 32 位的 16 進制字串表示的，比如 da00c473044a131e4c58e53b81187e9c）。

儘管密碼學本身的訊息演算法可能十分複雜，但如果想要使用目前已經公開的密碼學演算法對資料進行加解密或者獲取訊息摘要，還是存在很多的函數庫函數可以使用的。比如在 Android 領域，如果想要使用密碼學演算法，通常只需要呼叫系統 API，比如 Cipher 類別中的函數用於對稱 / 非對稱演算法，MessageDigest 類別中的函數用於雜湊函數的計算，Android 中如此便利的加解密函數庫封裝給逆向人員帶來了一定的幫助：可以直接透過 Hook 關鍵加解密函數庫函數對明文、加密甚至金鑰進行「自吐」操作。在接下來的章節中，將以 MD5 演算法為入口從 Hook 到沙盒分別介紹對加解密函數庫函數的「自吐」方法，帶領讀者切身體會沙盒的威力。

7.2.2 MD5 演算法 Hook「自吐」

如圖 7-1 所示，以 xianjianbang.apk 這一集合了許多標準加解密函數庫呼叫的 Demo 為例，針對其 Hash 演算法之一 MD5 進行測試研究。

▲ 圖 7-1　Demo 首頁面

在透過使用 Frida 將 hookEvent.js 指令稿注入應用並快速定位 JAVAMD5
按鈕回應函數的關鍵類別為 com.xiaojianbang.app.MainActivity 後，使
用 Jadx 開啟 App 並追蹤到該類別的 onClick 函數，會發現該函數中存在
著如圖 7-1 所示的許多控制項的回應入口，最終定位到實現 JAVAMD5 按
鈕的關鍵函數內容如程式清單 7-2 所示。

● 程式清單 7-2　JAVAMD5 按鈕回應函數

```
package com.xiaojianbang.app;
import java.security.MessageDigest;
public class MD5 {
    public static String md5_1(String arg2) throws Exception {
        MessageDigest v0 = MessageDigest.getInstance("MD5", "BC");
        v0.update(arg2.getBytes());
```

```
        return Utils.byteToHexString(v0.digest());
    }
}
```

需要注意的是，這個樣本中的 JAVAMD5 按鈕等利用 Java 標準加密函數庫的功能在新版本中是無法成功執行的，這是因為 MessageDigest 類別在加密時，如果指定 BC 模式，在 Android P 以上版本是會拋出 java.security.NoSuchAlgorithmException 異常，導致最終無法執行的，因此建議測試機版本最高選到 Android 9。

對 Android 開發相對了解的讀者，可能會發現程式清單 7-2 中 md5_1() 函數實現的 MD5 函數就是封裝的 Android 標準函數庫中的 Hash 加密函數，對應的系統關鍵類別為 java.security.MessageDigest 類別，在該類別中主要有 3 個函數：MessageDigest.getInstance() 函數用於初始化和設定 Hash 演算法類型，MessageDigest.update() 函數用於傳入待加密的明文，MessageDigest.digest() 函數用於計算輸入明文的 Hash 值。

基於上述分析並考慮到每個函數可能存在多個多載，這裡首先撰寫一個通用的用於 Hook 任意指定函數所有多載的指令稿，並針對 MessageDigest.getInstance 函數進行 Hook，具體程式如程式清單 7-3 所示。

⊙ 程式清單 7-3 hook.js

```
function hookMethod() {
    Java.perform(function () {
        // 指定要Hook的函數（套件名+類別名+函數名）
        var targetClassMethod = "java.security.MessageDigest.getInstance"
        // 獲取函數所在類別
        var delim = targetClassMethod.lastIndexOf(".");
        if (delim === -1) return;
        var targetClass = targetClassMethod.slice(0, delim)
        // 獲取函數名稱
```

```
        var targetMethod = targetClassMethod.slice(delim + 1,
targetClassMethod.length)
        var hook = Java.use(targetClass);
        // 獲取函數多載的數量
        var overloadCount = hook[targetMethod].overloads.length;
        for (var i = 0; i < overloadCount; i++) {
            // 對函數的每一個多載進行Hook
            hook[targetMethod].overloads[i].implementation = function () {
                console.warn("\n*** entered " + targetClassMethod);
                // 列印參數列表
                if (arguments.length >= 0) {
                    // 利用JS的特性：隱式參數arguments用於儲存參數列表
                    for (var j = 0; j < arguments.length; j++) {
                        console.log("arg[" + j + "]: " + arguments[j]);
                    }
                }
                // 主動呼叫該函數
                var retval = this[targetMethod].apply(this, arguments);
                // 列印呼叫堆疊
                console.log(Java.use("android.util.Log")
                            .getStackTraceString(Java.use("java.lang.
Throwable").$new()));
                // 列印返回值
                console.log("\nretval: " + retval);
                console.warn("\n*** leave " + targetClassMethod);
                return retval;
            }
        }
    })
}
setImmediate(hookMethod)
```

如圖 7-2 所示是最終 Hook 的結果，可以發現就如同普通函數的 Hook 一樣，Hash 演算法的具體類型及對應的 provider 都順利列印出來了。當然，加解密演算法在已知演算法的前提下，真實對我們有用的其實只有明文和對應的加密，因此還可以依樣畫葫蘆，將目標函數改成 digest 函數或者 update 函數，即可完成對其加密前明文和加密後內容的自吐，這裡不再展示。

```
[LGE Nexus 5X::HookTestDemo]->
[LGE Nexus 5X::HookTestDemo]->
*** entered java.security.MessageDigest.getInstance
arg[0]: MD5 => "MD5"
arg[1]: BC => "BC"

retval: MD5 Message Digest from BC, <initialized> => {"$handle":"0x244a","$weakRef":16}
java.lang.Throwable
        at java.security.MessageDigest.getInstance(Native Method)
        at com.xiaojianbang.app.MD5.md5_1(MD5.java:8)
        at com.xiaojianbang.app.MainActivity.onClick(MainActivity.java:81)
        at android.view.View.performClick(View.java:6294)
        at android.view.View$PerformClick.run(View.java:24770)
        at android.os.Handler.handleCallback(Handler.java:790)
        at android.os.Handler.dispatchMessage(Handler.java:99)
        at android.os.Looper.loop(Looper.java:164)
        at android.app.ActivityThread.main(ActivityThread.java:6494)
        at java.lang.reflect.Method.invoke(Native Method)
        at com.android.internal.os.RuntimeInit$MethodAndArgsCaller.run(RuntimeInit.java:438)
        at com.android.internal.os.ZygoteInit.main(ZygoteInit.java:807)

*** leave java.security.MessageDigest.getInstance
```

▲ 圖 7-2 Hook MessageDigest.getInstance 函數的結果

那麼當前已公開的成熟的沙盒是如何做到 Hash 演算法的「自吐」的？這裡選擇 appmon 進行測試。

由於 appmon 沙盒設計複雜，其整合了多個模組，甚至涉及一些 Python 與前端的知識，為了排除無關因素，只專注於功能本身，這裡僅僅使用單一的針對 Hash 函數進行 Trace 的指令稿（對應檔案路徑為 scripts/Android/Crypto/Hash.js）進行測試。

在將指令稿注入應用後，最終的 Hook 結果如圖 7-3 所示。

```
[Nexus 5X::com.xiaojianbang.app]->
[Nexus 5X::com.xiaojianbang.app]-> message: {'type': 'send', 'payload': '{"time":"2021-07-31T14:01:16.657Z","txnType":"Crypto","
thod":"update","artifact":[{"name":"Raw Data","value":"[object Object]","argSeq":0}]}'} data: None
message: {'type': 'send', 'payload': '{"time":"2021-07-31T14:01:16.660Z","txnType":"Crypto","lib":"java.security.MessageDigest",
"Algorithm","value":"MD5","argSeq":0},{"name":"Digest","value":"[object Object]","argSeq":0}]}'} data: None
message: {'type': 'send', 'payload': '{"time":"2021-07-31T14:01:17.213Z","txnType":"Crypto","lib":"java.security.MessageDigest",
"Raw Data","value":"[object Object]","argSeq":0}]}'} data: None
message: {'type': 'send', 'payload': '{"time":"2021-07-31T14:01:17.216Z","txnType":"Crypto","lib":"java.security.MessageDigest",
"Algorithm","value":"SHA-1","argSeq":0},{"name":"Digest","value":"[object Object]","argSeq":0}]}'} data: None
```

▲ 圖 7-3 appmon Hook 函數結果

在圖 7-3 中，根據日誌資訊發現，其內容都是關於函數 update 和 digest() 的 Hook 結果。但是進一步觀察日誌會驚奇地發現，日誌資訊中依舊存在關於演算法種類的資訊：MD5 以及 SHA-1 都被完美地辨識出來了，但是實際上如程式清單 7-4 中觀察 Hash.js 中的程式，會發現並未對上面介紹的 getInstance() 函數進行 Hook 以獲取演算法資訊，而是在 Hook digest 或者 update 函數時透過 getAlgorithm() 函數獲取對應演算法的種類。

◯ 程式清單 7-4　Hash.js

```
Java.perform(function() {
  var MessageDigest = Java.use("java.security.MessageDigest");

  if (MessageDigest.digest) {
    MessageDigest.digest.overloads[0].implementation = function() {
      var digest = this.digest.overloads[0].apply(this, arguments);
      // 獲取演算法資訊
      var algorithm = this.getAlgorithm().toString();

      /*   --- Payload Header --- */
      var send_data = {};
      send_data.time = new Date();
      send_data.txnType = 'Crypto';
      send_data.lib = 'java.security.MessageDigest';
      send_data.method = 'digest';
      send_data.artifact = [];

      /*   --- Payload Body --- */
      var data = {};
      data.name = "Algorithm";
      data.value = algorithm;
      data.argSeq = 0;
      send_data.artifact.push(data);

      /*   --- Payload Body --- */
```

```
      var data = {};
      data.name = "Digest";
      data.value = byteArraytoHexString(digest);
      data.argSeq = 0;
      send_data.artifact.push(data);

      send(JSON.stringify(send_data));
      return digest;
    }


  MessageDigest.digest.overloads[1].implementation = function(input) {
      ...
    }
}


if (MessageDigest.update) {
  MessageDigest.update.overloads[0].implementation = function(input) {
    //console.log("MessageDigest.update input: " + updateInput(input));
    /*   --- Payload Header --- */
    var send_data = {};
    send_data.time = new Date();
    send_data.txnType = 'Crypto';
    send_data.lib = 'java.security.MessageDigest';
    send_data.method = 'update';
    send_data.artifact = [];

    /*   --- Payload Body --- */
    var data = {};
    data.name = "Raw Data";
    data.value = updateInput(input);
    data.argSeq = 0;
    send_data.artifact.push(data);

    send(JSON.stringify(send_data));
```

```
    return this.update.overloads[0].apply(this, arguments);
  }

  MessageDigest.update.overloads[1].implementation = function(input,
offset, len) {
    ...
  }
  MessageDigest.update.overloads[2].implementation = function(input) {
    ...}
  MessageDigest.update.overloads[3].implementation = function(input) {
    //console.log("MessageDigest.update input: " + updateInput(input));
    /*   --- Payload Header --- */
    var send_data = {};
    send_data.time = new Date();
    send_data.txnType = 'Crypto';
    send_data.lib = 'java.security.MessageDigest';
    send_data.method = 'update';
    send_data.artifact = [];

    /*   --- Payload Body --- */
    var data = {};
    data.name = "Raw Data";
    data.value = updateInput(input);
    data.argSeq = 0;
    send_data.artifact.push(data);

    send(JSON.stringify(send_data));
    return this.update.overloads[3].apply(this, arguments);
  }
}

});
```

雖然 appmon 中的指令稿透過主動呼叫 getAlgorithm() 函數得到具體演算法種類的方式十分精妙，但是在測試過程中如圖 7-3 中著重標注的內容所示,其列印日誌資訊中,表示 data 資料的內容實際上只是 [object Object] 這種沒有真實內容的表示,這裡透過重寫 byteArraytoHexString() 這一方法修復這一問題後,測試結果如圖 7-4 所示,修改後的函數內容如程式清單 7-5 所示。

● 程式清單 7-5 byteArraytoHexString 函數

```
var byteArraytoHexString = function(byteArray) {
  if (!byteArray) { return 'null'; }
  if (byteArray.map) {
    return byteArray.map(function(byte) {
      return ('0' + (byte & 0xFF).toString(16)).slice(-2);
    }).join('');
  } else {
    return byteArray + "";
  }
}
```

▲ 圖 7-4 修改後的測試結果

在修復完成 appmon 的小 Bug 後，一個真正可用的 Hook「自吐」Hash 演算法的沙盒就暫且完成了。但考慮到 Hook 主要依賴於 Frida、Xposed 等 Hook 工具，要做到任意 Hash 演算法「自吐」也許會面臨需要繞過 Hook 工具在 App 中可能存在的對抗的情況，這種狀況的發生反而脫離了「自吐」的重點，因此接下來將化繁為簡，從系統原始程式層面直接修改原始程式，以期達到 Hook「自吐」的效果。

7.2.3 Hash 演算法原始程式沙盒「自吐」

雖然 Hook 沙盒和原始程式沙盒實現方式不同，但從本質上來講二者實際上都採取的是對原始程式插樁的方式實現的，只是 Hook 是針對二進位的動態程式插樁，而原始程式沙盒則是基於原始程式的插樁方式。另一方面，由於最終目的相同，都是針對特定類別和函數進行插樁，因此上述實現 Hook 沙盒的分析過程實際上已經為實現原始程式沙盒做了一些工作，包括插樁的目的類別和函數等，我們要實現原始程式沙盒，只需要針對目標函數內容進行修改即可。

但是在正式開始對原始程式進行修改之前，為了方便快捷地對系統原始程式進行修改，我們還需要做一些準備工作，即將原始程式匯入 Android Studio 等 IDE 工具。之所以要將原始程式匯入 IDE 而非單純地找到目的檔案直接進行修改，是因為 Android Studio 這些智慧編輯器帶給我們的良好體驗—能夠幫助我們在原始程式修改中避免一些拼字、語法上的錯誤。

當然，要實現原始程式匯入並不是簡單地直接使用 Android Studio 開啟原始程式資料夾即可，但 Android 原始程式也為原始程式匯入提供了支援，要實現原始程式匯入，依次執行如下步驟即可。

首先，在下載對應原始程式後（這裡筆者使用的 Android 版本為 8.1.0_
r1），切換到原始程式根目錄下並依次執行如下命令：

```
# source build/envsetup.sh
# mmm development/tools/idegen/
```

在 成 功 執 行 上 述 命 令 後，如 圖 7-5 所 示，會 在 out/host/linux-x86/
framework 目錄下生成一個 idegen.jar 檔案。

```
OUT_DIR=out
==========================================
ninja: no work to do.
[1/1] out/soong/.bootstrap/bin/soong_build out/soong/build.ninja
out/build-aosp_sailfish_development_tools_idegen_Android.mk.ninja is missing, regenerating...
PRODUCT_COPY_FILES device/generic/goldfish/data/etc/apns-conf.xml:system/etc/apns-conf.xml ignored.
No private recovery resources for TARGET_DEVICE sailfish
build/core/Makefile:34: warning: overriding commands for target `out/target/product/sailfish/system/lib64/libbcc.so'
build/core/base_rules.mk:390: warning: ignoring old commands for target `out/target/product/sailfish/system/lib64/libbcc.so'
build/core/Makefile:34: warning: overriding commands for target `out/target/product/sailfish/system/lib/libion.so'
build/core/base_rules.mk:390: warning: ignoring old commands for target `out/target/product/sailfish/system/lib/libion.so'
build/core/Makefile:34: warning: overriding commands for target `out/target/product/sailfish/system/lib64/libLLVM.so'
build/core/base_rules.mk:390: warning: ignoring old commands for target `out/target/product/sailfish/system/lib64/libLLVM.so'
[ 99% 144/145] glob test/vts/drivers/hal/common/include
[ 33% 1/3] host Java: idegen (out/host/common/obj/JAVA_LIBRARIES/idegen_intermediates/classes)
Picked up _JAVA_OPTIONS: -Dawt.useSystemAAFontSettings=on -Dswing.aatext=true
Picked up _JAVA_OPTIONS: -Dawt.useSystemAAFontSettings=on -Dswing.aatext=true
[100% 3/3] Install: out/host/linux-x86/framework/idegen.jar

#### build completed successfully (02:59 (mm:ss)) ####

[    ] -[~/aosp810_r1]
```

▲ 圖 7-5 生成 idegen.jar 檔案

在 生 成 idegen.jar 檔 案 後，只 需 繼 續 在 原 始 程 式 根 目 錄 下 執 行
development/tools/idegen/idegen.sh 命 令，即 可 在 根 目 錄 下 生 成
android.iml 和 android.ipr 這兩個檔案，用於匯入 Android Studio 的設
定檔。其中 android.iml 檔案包含原始程式匯入 Android Studio 時會被
匯入和排除的子目錄資料夾，android.ipr 則包含原始程式專案的具體設
定、程式以及依賴的 lib 等資訊。在生成上述檔案後，直接使用 Android
Studio 開啟 IPR 檔案，即可在等待一段時間後順利看到匯入成功的
Android 原始程式。

在完成上述準備工作後，我們正式開始開發 MessageDigest 相關演算法
的沙盒。

有了上述針對 appmon 相關程式的分析，要實現 MessageDigest 相關演算法「自吐」已經非常簡單，只需在 update 和 digest 的函數中插入「自吐」程式即可。

這裡需要注意的是，第一，用什麼方式實現「自吐」；第二，以上目的函數在 appmon 中雖然對多載函數進行了處理，但具體在原始程式碼中實現時還需要確定是哪一個多載函數以及是否存在互相呼叫的情況。

接下來依次解決以上問題。

針對第一個問題，事實上「自吐」方式很多，通常使用日誌列印或者檔案讀寫的方式實現。這裡使用的是日誌列印的方式，即呼叫 android.util.Log 類別中的日誌列印函數。

在具體實現時，筆者發現無法在開發 App 時直接使用 import 的方式匯入使用的 android.util.Log 類別，導致最終在編譯時出現如圖 7-6 所示的 cannot find symbol 錯誤，造成最終編譯失敗。

```
/bin/bash out/target/common/obj/JAVA_LIBRARIES/core-all_intermediates/classes-full-debug.jar.r
Picked up _JAVA_OPTIONS: -Dawt.useSystemAAFontSettings=on -Dswing.aatext=true
libcore/ojluni/src/main/java/java/security/MessageDigest.java:39: error: cannot find symbol
import android.util.Log;
                  ^
  symbol:   class Log
  location: package android.util
1 error
```

▲ 圖 7-6 cannot find symbol 錯誤

為了解決這個錯誤，這裡轉而使用反射方式實現呼叫 Log 類別中的函數，且由於反射呼叫可能會發生異常，因此需要在程式中對可能出現的異常進行處理，而不能簡單將異常拋出，最終日誌列印函式呼叫的程式如程式清單 7-6 所示。

⬤ **程式清單 7-6 反射呼叫日誌列印函數**

```
// update函數
Class logClass = null;
try {
    // 載入類別
    logClass = this.getClass().getClassLoader().loadClass("android.util.
Log");
} catch (ClassNotFoundException e) {
    e.printStackTrace();
}
Method loge = null;
try {
    // 獲取對應函數
    loge = logClass.getMethod("e",String.class,String.class);
} catch (NoSuchMethodException e) {
    e.printStackTrace();
}
try {
    // 呼叫函數
    loge.invoke(null,"r0ysue","input is => "+inputString);
} catch (IllegalAccessException e) {
    e.printStackTrace();
} catch (InvocationTargetException e) {
    e.printStackTrace();
}
```

針對第二個問題,這裡使用 Objection 命令首先確定 MessageDigest 類別中存在的目標函數,對應多載如圖 7-7 所示。

在找到所有多載後,直接使用原始程式分析每一個多載函數,最終發現只有 digest(byte[] input) 多載中又再次呼叫了 digest() 函數,其函數內容如程式清單 7-7 所示。

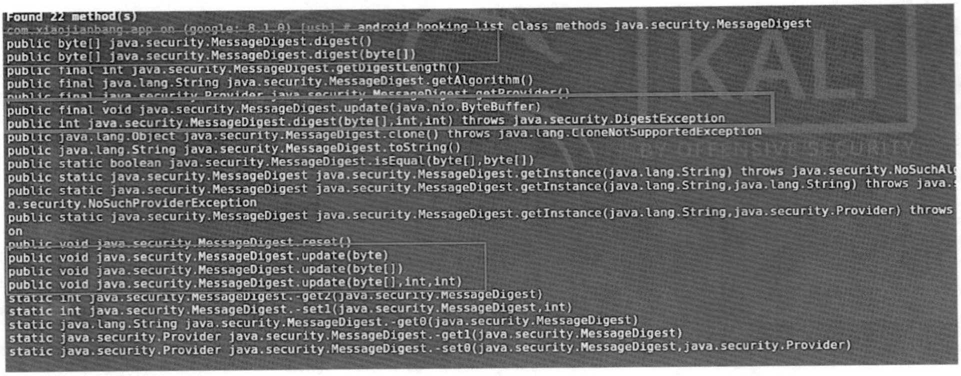

▲ 圖 7-7 MessageDigest 類別中的目標函數

⬤ 程式清單 7-7 digest(byte[] input) 多載

```
public byte[] digest(byte[] input) {
    update(input);
    return digest();
}
```

最終確認需要插樁的函數清單如下：

```
public byte[] java.security.MessageDigest.digest()
public int java.security.MessageDigest.digest(byte[],int,int) throws java.
security.DigestException
public void java.security.MessageDigest.update(byte)
public void java.security.MessageDigest.update(byte[])
public void java.security.MessageDigest.update(byte[],int,int)
public final void java.security.MessageDigest.update(java.nio.ByteBuffer)
```

因此，要實現 Hash 演算法「自吐」，只需依次在這些函數中插入「自吐」程式即可。比如最終 digest() 函數在增加「自吐」程式和呼叫堆疊列印後的程式如程式清單 7-8 所示，其他部分程式也類似，這裡為了節省篇幅，就不再列出了。

● 程式清單 7-8 digest() 函數自吐

```java
public byte[] digest(){
    /* Resetting is the responsibility of implementors. */
    byte[] result = engineDigest();
    state = INITIAL;
    // bytes 轉hex陣列
    String resultString = byteToHex(result);
    // Log.e("r0ysueDigest","result is => "+ resultString);
    Class logClass = null;
    try {
        logClass = this.getClass().getClassLoader().loadClass("android.
util.Log");
    } catch (ClassNotFoundException e) {
        e.printStackTrace();
    }
    Method loge = null;
    try {
        loge = logClass.getMethod("e",String.class,String.class);
    } catch (NoSuchMethodException e) {
        e.printStackTrace();
    }
    try {
        loge.invoke(null,"r0ysue","result is => "+resultString);
        // 列印呼叫堆疊
        Exception e = new Exception("r0ysueRESULT");
        e.printStackTrace();
    } catch (IllegalAccessException e) {
        e.printStackTrace();
    } catch (InvocationTargetException e) {
        e.printStackTrace();
    }

    return result;
}
```

細心的讀者可能會發現，在程式清單 7-8 中，其實還有一個並不存在的 byteToHex() 函數，事實上這個函數是為了處理 byte 陣列列印問題而手動增加的一個新函數，其具體內容如程式清單 7-9 所示。

⬤ 程式清單 7-9　bytes 陣列轉 hex

```
// 增加自用的API
/**
 * byte陣列轉hex
 * @param bytes
 * @return
 */
private static String byteToHex(byte[] bytes){
    String strHex = "";
    StringBuilder sb = new StringBuilder("");
    for (int n = 0; n < bytes.length; n++) {
        strHex = Integer.toHexString(bytes[n] & 0xFF);
        sb.append((strHex.length() == 1) ? "0" + strHex : strHex); // 每個
位元組由兩個字元表示，位元數不夠，高位元補0
    }
    return sb.toString().trim();
}
```

在原始程式修改完畢後，就可以正式開始編譯工作了。當然，讀者看到這裡可能有疑惑，為什麼第 6 章已經講解過原始程式編譯，這一章還要介紹。事實上，雖然第 6 章已經介紹過原始程式編譯，但是由於在進行原始程式編譯時增加了一個新函數 byteToHex()，會造成在編譯時出現如圖 7-8 所示的錯誤，導致無法編譯成功。

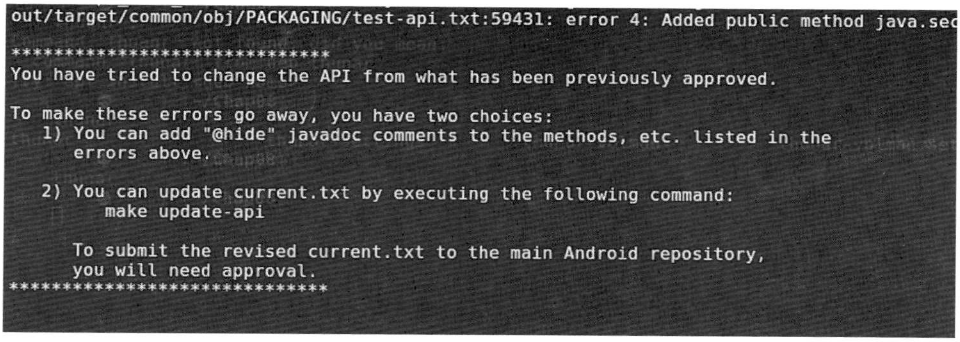

```
out/target/common/obj/PACKAGING/test-api.txt:59431: error 4: Added public method java.sec

******************************
You have tried to change the API from what has been previously approved.

To make these errors go away, you have two choices:
   1) You can add "@hide" javadoc comments to the methods, etc. listed in the
      errors above.

   2) You can update current.txt by executing the following command:
         make update-api

      To submit the revised current.txt to the main Android repository,
      you will need approval.
******************************
```

▲ 圖 7-8 增加新函數導致顯示出錯

這裡採取提示中的第二個解決方案，即執行 make update-api 解決這個增添新 API 導致的問題。

另一方面，在第 6 章編譯的過程中，細心的讀者會發現自編譯出來的鏡像是附帶 su 程式的，也就是擁有 Root 許可權，而這對於存在 Root 檢測的 App 是非常不友善的，因此這裡選擇編譯出一個不帶 Root 許可權的鏡像。要實現這一點，只需在執行 lunch 命令選擇編譯目標時進行修改即可。

根據官網（https://source.android.com/setup/build/building#choose-a-target）內容，圖 7-9 展示了在編譯時可以選擇的類型分為 user、userdebug 以及 eng，其中 user 類型是一個真正的 production，對應編譯出來的鏡像是沒有 Root 許可權的。

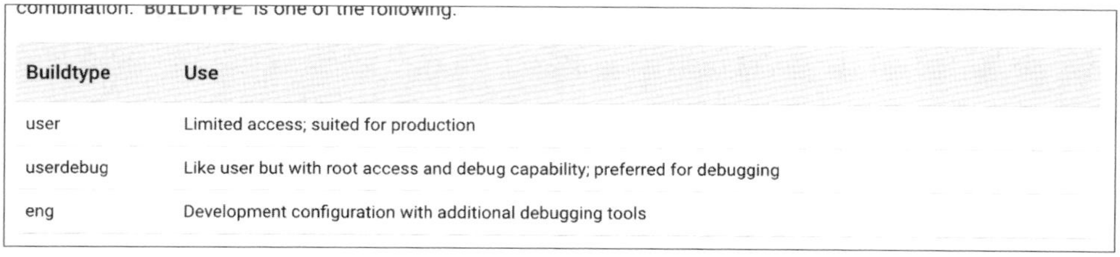

combination. BUILDTYPE is one of the following.

Buildtype	Use
user	Limited access; suited for production
userdebug	Like user but with root access and debug capability; preferred for debugging
eng	Development configuration with additional debugging tools

▲ 圖 7-9 編譯目標

但是，在執行 lunch 命令的時候，我們發現並沒有 user 結尾的編譯目標
存在（見圖 7-10），這是否意味著無法編譯出一個 user 類型的鏡像呢？

```
┌─(root💀kali)-[~/aosp810_r1]
└─# lunch

You're building on Linux

Lunch menu... pick a combo:
     1. aosp_arm-eng
     2. aosp_arm64-eng
     3. aosp_mips-eng
     4. aosp_mips64-eng
     5. aosp_x86-eng
     6. aosp_x86_64-eng
     7. full_fugu-userdebug
     8. aosp_fugu-userdebug
     9. aosp_car_emu_arm-userdebug
    10. aosp_car_emu_arm64-userdebug
    11. aosp_car_emu_x86-userdebug
    12. aosp_car_emu_x86_64-userdebug
    13. mini_emulator_arm64-userdebug
    14. m_e_arm-userdebug
    15. m_e_mips64-eng
    16. m_e_mips-userdebug
    17. mini_emulator_x86_64-userdebug
    18. mini_emulator_x86-userdebug
    19. uml-userdebug
    20. aosp_dragon-userdebug
    21. aosp_dragon-eng
    22. aosp_marlin-userdebug
    23. aosp_marlin_svelte-userdebug
    24. aosp_sailfish-userdebug
    25. aosp_walleye-userdebug
    26. aosp_walleye_test-userdebug
    27. aosp_taimen-userdebug
    28. aosp_angler-userdebug
    29. aosp_bullhead-userdebug
    30. aosp_bullhead_svelte-userdebug
    31. hikey-userdebug
    32. hikey960-userdebug

Which would you like? [aosp_arm-eng] 
```

▲ 圖 7-10 lunch 命令執行結果

當然不是，事實上筆者在研究後發現，如果在選擇編譯目標時將帶有
userdebug 字眼的對應目標中的 debug 字串移除，最終編譯出的鏡像就
是不帶 Root 許可權的鏡像，比如想要編譯適用於 sailfish 裝置的無 Root
鏡像，只需在執行 lunch 命令時輸入字串 aosp_sailfish-user 即可，具體
執行命令如圖 7-11 所示。

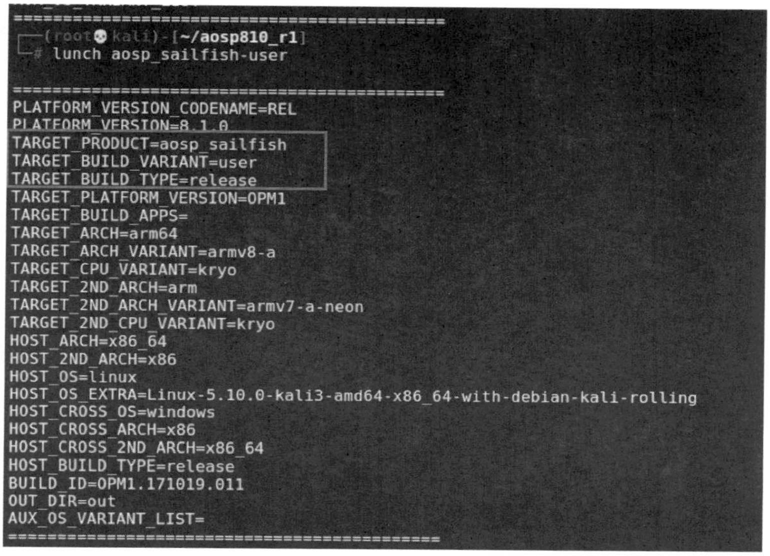

```
========================================
┌──(root💀kali)-[~/aosp810_r1]
└─# lunch aosp_sailfish-user

========================================
PLATFORM_VERSION_CODENAME=REL
PLATFORM_VERSION=8.1.0
TARGET_PRODUCT=aosp_sailfish
TARGET_BUILD_VARIANT=user
TARGET_BUILD_TYPE=release
TARGET_PLATFORM_VERSION=OPM1
TARGET_BUILD_APPS=
TARGET_ARCH=arm64
TARGET_ARCH_VARIANT=armv8-a
TARGET_CPU_VARIANT=kryo
TARGET_2ND_ARCH=arm
TARGET_2ND_ARCH_VARIANT=armv7-a-neon
TARGET_2ND_CPU_VARIANT=kryo
HOST_ARCH=x86_64
HOST_2ND_ARCH=x86
HOST_OS=linux
HOST_OS_EXTRA=Linux-5.10.0-kali3-amd64-x86_64-with-debian-kali-rolling
HOST_CROSS_OS=windows
HOST_CROSS_ARCH=x86
HOST_CROSS_2ND_ARCH=x86_64
HOST_BUILD_TYPE=release
BUILD_ID=OPM1.171019.011
OUT_DIR=out
AUX_OS_VARIANT_LIST=
========================================
```

▲ 圖 7-11　選擇 user 模式目標

在解決完上述編譯問題後，重新編譯並將編譯完成的鏡像重新刷入對應
裝置，執行 su 命令，所得到的結果如圖 7-12 所示。

```
┌──(root💀kali)-[~/aosp810_r1]
└─# adb shell
sailfish:/ $ su
/system/bin/sh: su: not found
127|sailfish:/ $ su
/system/bin/sh: su: not found
127|sailfish:/ $
```

▲ 圖 7-12　無 Root 許可權裝置

最終樣本的「自吐」日誌列印效果如圖 7-13 所示，這也標誌著一個簡單
的 Hash「自吐」沙盒順利開發完成。

▲ 圖 7-13 Hash 日誌「自吐」列印

7.3 crypto_filter_aosp 專案移植

在完成 Hash 函數的「自吐」後，我們再來研究一下針對對稱和非對稱加密的「自吐」方案。

同樣以 xianjianbang.apk 為例，這裡略過分析過程，直接將 DES、AES 等對稱密碼和 RSA 等非對稱密碼的程式從 Jadx-gui 的靜態反編譯結果中提取出來，對應實現如程式清單 7-10 所示。

○ 程式清單 7-10 標準對稱 / 非對稱加密實現

```
import javax.crypto.Cipher;
// AES
public static String aes(String args) throws Exception {
    SecretKeySpec key = new SecretKeySpec("1234567890abcdef1234567890abcd
ef".getBytes(), "AES");
    AlgorithmParameterSpec iv = new IvParameterSpec("1234567890abcdef".
getBytes());
```

```java
    Cipher aes = Cipher.getInstance("AES/CBC/PKCS5Padding");
    aes.init(1, key, iv);
    return Base64.encodeToString(aes.doFinal(args.getBytes("UTF-8")), 0);
}
// DES
public static String des_1(String args) throws Exception {
    SecretKey secretKey = SecretKeyFactory.getInstance("DES").
generateSecret(new DESKeySpec("12345678".getBytes()));
    AlgorithmParameterSpec iv = new IvParameterSpec("87654321".getBytes());
    Cipher cipher = Cipher.getInstance("DES/CBC/PKCS5Padding");
    cipher.init(1, secretKey, iv);
    cipher.update(args.getBytes());
    return Base64.encodeToString(cipher.doFinal(), 0);
}
// RSA加密
public static byte[] encrypt(byte[] plaintext) throws Exception {
    PublicKey publicKey = getPublicKey(pubKey);
    Cipher cipher = Cipher.getInstance("RSA/None/NoPadding", "BC");
    cipher.init(1, publicKey);
    return cipher.doFinal(plaintext);
}
// // RSA解密
public static byte[] decrypt(byte[] encrypted) throws Exception {
    PrivateKey privateKey = getPrivateKey(priKey);
    Cipher cipher = Cipher.getInstance("RSA/None/PKCS1Padding", "BC");
    cipher.init(2, privateKey);
    return cipher.doFinal(encrypted);
}
```

對比這 3 個密碼加解密的實現會發現，關鍵用於控制密碼加解密的類別
是 javax.crypto.Cipher，對應函數分別為：init 函數用於初始化加解密模
式並傳入加解密金鑰和向量，update 函數用於更新加解密輸入，doFinal
函數用於進行真正的加解密過程。

據此，讀者可以根據上一節中針對 Hash 函數的沙盒開發過程進行類似的分析，最終得到對應的「自吐」沙盒，但是在這一小節中不會講解這個過程，而是透過移植一個已有的項目（crypto_filter_aosp，專案位址為 https://github.com/icew4y/crypto_filter_aosp）來進行（非）對稱密碼的開發，並透過這個專案介紹一些關於沙盒開發的注意事項。

crypto_filter_aosp 專案是一個監控 Java 層的加密演算法的 ROM，可用於「自吐」標準函數庫的對稱或者非對稱密碼加解密、Hash 演算法加密以及 Mac 演算法加密。與前述介紹的 Hash 演算法「自吐」沙盒採用日誌列印的方式不同，這個專案採取的「自吐」方式是向應用私有目錄中寫檔案，相比之下，寫檔案的方式其實更加方便，不僅能夠對抗一些 Hook log 類別相關函數導致無日誌列印的情況，而且寫檔案的方式更加持久化。但是相對而言，寫檔案的方式對手機的儲存容量也做出了挑戰，因此採取讀取設定檔（設定檔為 /data/local/tmp/monitor_package）的方式實現只對目標應用進行監控的效果。但可惜的是，基於 Nexus 6p android 6.0.1 進行 ROM 的編譯和使用的，因此如果要使用專案提供的 backup 鏡像，還需要手中有一個 Nexus 6p 裝置，如果想要直接重複使用其原始程式碼，則需要重新下載 Android 6.0.1 的原始程式碼。因此，為了使得該專案能夠在 Android 8.1.0_r1 原始程式中重複使用，我們還需要做一些移植和調配的工作。

crypto_filter_aosp 專案的原始程式碼主要包括 6 個檔案：MessageDigest.java、Mac.java 和 Cipher.java 三個檔案是修改後的加解密相關檔案，MyUtil.java、ContextHolder.java 和 AndroidBase64.java 是三個新增加的檔案。

由於 crypto_filter_aosp 專案是基於 Android 6.0.1 程式的，因此在移植到 Android 8.1.0 上時，考慮到不同版本間程式的差異性，直接進行檔案的覆蓋編譯是非常危險且特別容易出錯的，因此還需要對比修改前後的加解密相關檔案到底修改了哪些函數，增加了哪些程式。這裡為了方便對比，從原始程式網站上下載了原版的 AOSP 6.0.1 的相關檔案，並透過檔案對比工具對對應檔案進行了對比，以 MessageDigest.java 檔案為例，這裡使用 VS Code 進行檔案的對比，效果如圖 7-14 所示。

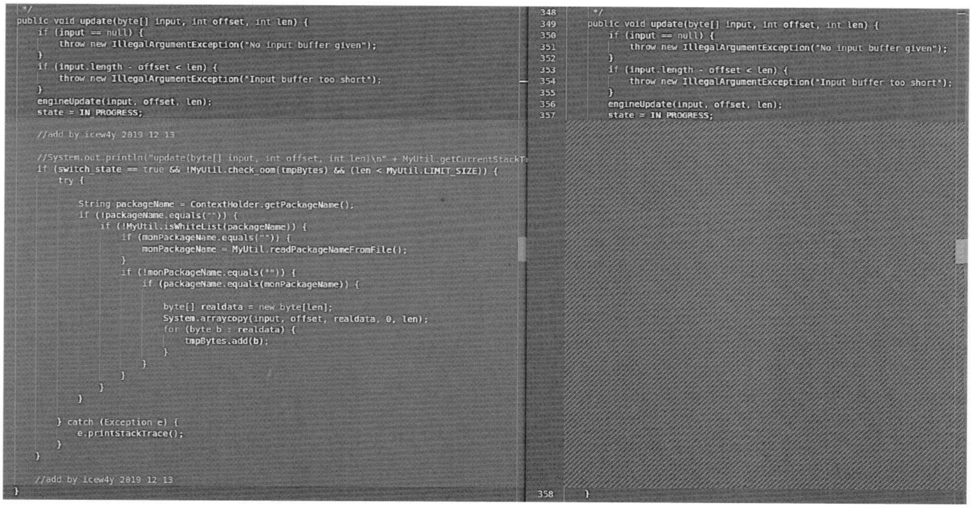

▲ 圖 7-14　MessageDigest.java 檔案對比

透過檔案對比可以清楚並快速地定位 crypto_filter_aosp 專案到底修改了哪些函數和程式，從而進行快速的移植工作。

具體的移植過程暫且不談，這裡還需要介紹一下 crypto_filter_aosp 專案的程式。透過專案中的介紹會發現最終寫入記錄檔中的資料其實都是 JSON 格式的，且與在上一節中開發的沙盒不同，該專案在輸出時是將一次加解密的過程只輸出一筆日誌，這是因為在進行沙盒開發時，為每一個目標類別增加了一個成員變數 jsoninfo，訊息用於存放一次加解密的訊息內容，且在 digest 這類最終進行加解密操作的函數中，呼叫寫檔案函數 priter() 將日誌輸出到檔案中，同時在輸出後就直接清空 jsoninfo 內容，以 digest 函數為例，其具體實現如程式清單 7-11 所示。

⬤ **程式清單 7-11 digest 函數**

```
public int digest(byte[] buf, int offset, int len) throws DigestException {
    ...
    //add by icew4y 2019 12 13
    //System.out.println("digest(byte[] buf, int offset, int len)");
    int result = engineDigest(buf, offset, len);
    if (switch_state == true && !MyUtil.check_oom(tmpBytes)) {
        try {
            // 獲取套件名
            String packageName = ContextHolder.getPackageName();
            if (!packageName.equals("")) {
                // 判斷是不是白名單應用，白名單應用不監控
                if (!MyUtil.isWhiteList(packageName)) {
                    if (monPackageName.equals("")) {
                        // 讀取/data/local/tmp/monitor_package檔案中的套件名
                        monPackageName = MyUtil.readPackageNameFromFile();
                    }
                    // 判斷是不是目標應用
                    if (!monPackageName.equals("")) {
                        if (packageName.equals(monPackageName)) {
                            // jsoninfo放置演算法類型
                            jsoninfo.put("Algorithm", getAlgorithm());
                            // provider
                            Provider provider_ = getProvider();
                            if (provider_ != null) {
                                jsoninfo.put("Provider", provider_.
getName());
                            }
                            StringBuffer tmpsb = new StringBuffer();
                            if (tmpBytes.size() > 0) {
                                int n = tmpBytes.size();
```

```
                                    byte[] resultBytes = new byte[n];
                                    for (int i = 0; i < n; i++) {
                                        resultBytes[i] = (byte) tmpBytes.get(i);
                                    }
                                    // hex格式原始資料
                                    jsoninfo.put("data", byteArrayToString(resu
ltBytes));

                                    // base64加密狀態資料
                                    jsoninfo.put("Base64Data", AndroidBase64.
encodeToString(resultBytes, AndroidBase64.NO_WRAP));
                                } else {
                                    jsoninfo.put("data", "");
                                }
                                //資料
                                byte[] readresult = new byte[len];
                                System.arraycopy(buf, offset, readresult, 0,
len);
                                // 加密結果資料
                                jsoninfo.put("digest", toHexString(readresult));
                                // base64格式的呼叫堆疊資料
                                jsoninfo.put("StackTrace", AndroidBase64.
encodeToString(MyUtil.getCurrentStackTrack(Thread.currentThread().
getStackTrace()).getBytes(), AndroidBase64.NO_WRAP));
                                // 寫檔案函數
                                priter("MessageDigestTag:" + jsoninfo.
toString(), packageName);

                                // 清空內容
                                jsoninfo = new JSONObject();
                                tmpBytes.clear();
                            }
                        }
                    }
                }

        } catch (Exception e) {
```

```
        e.printStackTrace();
    }
}

return result;
//add by icew4y 2019 12 13
}
```

另外，還需要介紹自己增加的 3 個檔案：ContextHolder.java 檔案用於獲取 Context 上下文，以獲取當前執行的應用套件名供外部呼叫；AndroidBase64.java 檔案用於計算輸入內容的 Base64 編碼結果，實際上該檔案中的內容與 Android 原始程式中附帶的 Base64 編碼的內容是一致的，但是為了避免在上一節中提過的無法使用 import 關鍵字匯入的情況，轉而直接複製一份程式並放置於與其他加解密相關檔案相同的目錄下，以實現直接使用 import 關鍵字匯入類別，進而呼叫其中函數的效果，這樣做就順利避免了反射呼叫的冗餘碼；而 MyUtil.java 檔案則包括幫助判斷是否記憶體溢位的函數 check_oom()、幫助判斷當前應用是不是白名單內的應用的函數 isWhiteList(String packageName) 以及其他用於讀寫檔案的相關函數 readPackageNameFromFile() 等。具體的程式在這裡不再贅述，讀者如果有興趣，可以閱讀原始程式進行研究。

在移植程式完成後，接下來進行最終的編譯過程。

由於這個專案中增加了 3 個原本 Android 原始程式中沒有的檔案，因此還需要在對應 libcore 子專案的編譯設定檔 libcore/obenjdk_java_files.mk 中增加對應檔案的全路徑，最終效果如圖 7-15 所示。

同時與上一小節相同的是，在設定檔中增加路徑成功後，正式編譯之前還需要執行 make update-api 命令以更新系統 API。

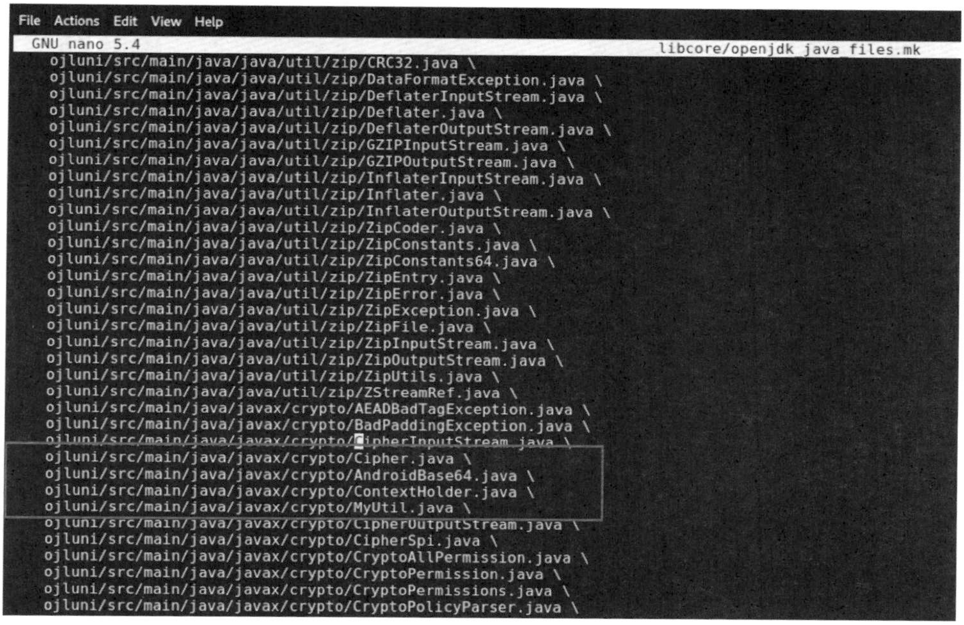

```
File Actions Edit View Help
GNU nano 5.4                                                    libcore/openjdk_java_files.mk
  ojluni/src/main/java/java/util/zip/CRC32.java \
  ojluni/src/main/java/java/util/zip/DataFormatException.java \
  ojluni/src/main/java/java/util/zip/DeflaterInputStream.java \
  ojluni/src/main/java/java/util/zip/DeflaterOutputStream.java \
  ojluni/src/main/java/java/util/zip/Deflater.java \
  ojluni/src/main/java/java/util/zip/GZIPInputStream.java \
  ojluni/src/main/java/java/util/zip/GZIPOutputStream.java \
  ojluni/src/main/java/java/util/zip/InflaterInputStream.java \
  ojluni/src/main/java/java/util/zip/Inflater.java \
  ojluni/src/main/java/java/util/zip/InflaterOutputStream.java \
  ojluni/src/main/java/java/util/zip/ZipCoder.java \
  ojluni/src/main/java/java/util/zip/ZipConstants.java \
  ojluni/src/main/java/java/util/zip/ZipConstants64.java \
  ojluni/src/main/java/java/util/zip/ZipEntry.java \
  ojluni/src/main/java/java/util/zip/ZipError.java \
  ojluni/src/main/java/java/util/zip/ZipException.java \
  ojluni/src/main/java/java/util/zip/ZipFile.java \
  ojluni/src/main/java/java/util/zip/ZipInputStream.java \
  ojluni/src/main/java/java/util/zip/ZipOutputStream.java \
  ojluni/src/main/java/java/util/zip/ZipUtils.java \
  ojluni/src/main/java/java/util/zip/ZStreamRef.java \
  ojluni/src/main/java/javax/crypto/AEADBadTagException.java \
  ojluni/src/main/java/javax/crypto/BadPaddingException.java \
  ojluni/src/main/java/javax/crypto/CipherInputStream.java \
  ojluni/src/main/java/javax/crypto/Cipher.java \
  ojluni/src/main/java/javax/crypto/AndroidBase64.java \
  ojluni/src/main/java/javax/crypto/ContextHolder.java \
  ojluni/src/main/java/javax/crypto/MyUtil.java \
  ojluni/src/main/java/javax/crypto/CipherOutputStream.java \
  ojluni/src/main/java/javax/crypto/CipherSpi.java \
  ojluni/src/main/java/javax/crypto/CryptoAllPermission.java \
  ojluni/src/main/java/javax/crypto/CryptoPermission.java \
  ojluni/src/main/java/javax/crypto/CryptoPermissions.java \
  ojluni/src/main/java/javax/crypto/CryptoPolicyParser.java \
```

▲ 圖 7-15　在 obenjdk_java_files.mk 檔案中增加新檔案的全路徑

完成以上設定後，就可以按照第 6 章中所介紹的原始程式編譯方式編譯出一個不帶 Root 的全新鏡像檔案，最終樣本測試的對稱 / 非對稱加解密以及 Mac 演算法的「自吐」效果分別如圖 7-16 和圖 7-17 所示。

```
oot@angler:/data/data/com.xiaojianbang.app # cat Cipher
ipherTag:{"opmode":"ENCRYPT_MODE","key":"1234567890abcdef1234567890abcdef","Key(Base64)":"MTIzNDU2Nzg5MGFiY2RlZj
yMzQ1Njc4OTBhYmNkZWY=","algorithm":"AES","SecureRandom":"SHA1PRNG","iv":"1234567890abcdef","Iv(Base64)":"MTIzNDU
Nzg5MGFiY2RlZg==","provider":"","transformation":"AES\/CBC\/PKCS5Padding","data":"xiaojianbang","Base64Data":"eG
hb2ppYW5iYW5n","doFinal":"GnT4C40I9oQb0BvimD9\/cA==","Base64Cipher":"GnT4C40I9oQb0BvimD9\/cA==","StackTrace":"ZG
sdmlrLnN5c3RlbS5SWTVN0YWNrLmdldFRocmVhZFN0YWNrVHJhY2UoKSAtMiA8LSAKamF2YS5sYW5nLlRocmVhZC5nZXRTdGGFja1RyYWNlKCkgNTg
IDWtIAPQYXZheC5jcnlwdG8uQ2lwaGVyLmRv_RmluYWwoKSAxLDc4NiA8LSAKY29tLnNpaW9qaWFuYmFuZy5hcHAuTWFpbkFjdGl2aXR5JGEub25DbGl
20ueGlhb2ppYW5iYW5nLmFwcC5NYWluQWN0aXZpdHkub25DcmVhdGUoKSAxNDQ8LSAKYW5kcm9pZC5hcHAuQWN0aXZpdHkucGVyZm9ybUNyZWF0ZSgp
IwNCA8LSAKYW5kcm9pZC5hcHAuQWN0aXZpdHlUaHJlYWQucGVyZm9ybUxhdW5jaEFjdGl2aXR5KCkgMzA8LSAKYW5kcm9pZC5hcHAuQWN0aXZpdHlUa
iYW5rCkgNzM5IDWtIAPhbmRyb2lkLmFwcC5BY3Rpdml0eVRocmVhZC5oYW5kbGVMYXVuY2hBY3Rpdml0eSgpIDE1NSA8LSAKYW5kcm9pZC5hcHAuQWN0a
NDggPC0gCmFuZHJvaWQuYXBwLkFjdGl2aXR5VGhyZWFkLm1haW4oKSA1LDDQyMiA8LSAKamF2YS5sYW5nLnJlZmxlY3QuTWV0aG9kLmludm9rZShT_gp
C0yIDwtIAPjb20uYW5kcm9pZC5pbnRlcm5hbC5vcy5aeWdvdGVJbml0JE1ldGhvZEFuZEFyZ3NDYWxsZXIucnVuKCkgNzI2IDwtIAPjb20uYW5kcm9pZ
9pZC5pbnRlcm5hbC5vcy5aeWdvdGVJbml0Lm1haW4oKSA2MTYgPC0gCmRhbHZpay5zeXN0ZW0uWHBvc2V2Vk0uPnJ2GdlLm1haW4oKS
xMDc="}
```

▲ 圖 7-16　對稱 / 非對稱加解密「自吐」效果

```
root@angler:/data/data/com.xiaojianbang.app # cat Mac
MacTag:{"key":"FridaHook","Key(Base64)":"RnJpZGFIb29r","Algorithm":"HmacSHA1","Provider":"AndroidOpenSSL"
,"data":"xiaojianbang","Base64Data":"eGlhb2pppYW5iYW5n","doFinal":"a878329db8a16027e1f6c9ebe99963f2974ed78
6","StackTrace":"ZGFsdmlrLnN5c3RlbS5WTVN0YWNrLmdldFRocmVhZFN0YWNrVHJhY2UoKSAtMiA8LSAKamF2YS5sYW5nLlRocmVVh
ZC5nZXRTdGFja1RyYWNlKCkgNTgwIDwtIApqYXZheC5jjcnlwdG8uTWFjLmRvRmluYWwoKSA4ODEgPC0gCmNvbS54aaWFvamlhbmJhbmcuY
XBwLk1BQy5tYWNfMSgpIDE0ODIgPC0gCmNvbS54aaWFvamlhbmJhbmcuYXBwLk1BQy5tTFwcC5NQW0aXZlX01uaXQoKSAtMSA8LSAKamF2Y
lkLnpZX0vbWlldy5ZXZXJmb3JtQ2xpY2soKSAtMSA8LSAKamF2YS5sYW5nLnJlZmxlY3QuTWV0aG9kLmludm9rZSgpIC0xIDwtIApjb20u
YW5kcm9pZC5pbnRlcm5hbC5vcy5aaWdvdGVJJbml0JE1ldGhvZEFuZFJyYgsjUN0YWNrW5kYW5kCkgNzI1IDwtIApjb20uYWaWZRyb2lkLkM
1NSA8LSAKYW5kcm9pZC5vcy5ZXW5kbGVyLmhhbmRsZU1lc3NhZ2UoKSAtMSA8LSAKYW5kcm9pZC5vcy5Iww5kbGVyLmRpc3BhdGNoTWVz
c2FnZSgpIC0xIDwtIApaW5kcm9pZC5vcy5Mb29wZXIubG9vcCgpIC0xIDwtIApjb20uYW5kcm9pZC5pbnRlcm5hbC5vcy5aaWdvdGVJbml0
JE1ldGhvZEFuZFJyYgsjUN0YWNrCkgNTU4IDwtIApjb20uYWaWZRyb2lkLmludGVybmFsLm9zLlpcZ290ZUlbml0Lm1haW4oKSAtMSA8LSA
KYW5kcm9pZC5pbnRlcm5hbC5vcy5ZXW5kbGVyLmhhbmRsZU1lc3NhZ2UoKSAtMSA8LSAKY29tLmFuZHJvaWQuaW50ZXJuYWwub3MuWmln
b3RlSW5pdCRNZXRob2RBbmRBcmdzQ2FsbGVyLnJ1bigpIC0xIDwtIApjb20uYW5kcm9pZC5pbnRlcm5hbC5vcy5aaWdvdGVJbml0Lm1haW
8Lm1haW4oKSA2MTYgPC0gCmRhLmVuJvYnYuYW5kcm9pZC54cC54cG9zZWQuWVBQanJpJpZGdlLm1haW4oKSAxMDc="}
```

▲ 圖 7-17 Mac 演算法「自吐」效果

7.4 本章小結

本章介紹了關於沙盒的概念並簡介了目前業記憶體在的一些沙盒情況，同時為了讓讀者體驗沙盒的效果，帶領讀者一起開發了一個簡單的 Hash 函數「自吐」沙盒，並透過移植 crypto_filter_aosp 專案的方式介紹了在進行沙盒開發時需要了解的基礎開發知識以及在增加新函數和新檔案後最終編譯需要注意的問題，希望從本章開始，讀者能夠真正領略到從高維降維完成低維應用的魅力。

08

Android 沙盒開發之網路函數庫與系統庫「自吐」

眾所皆知,抓取封包問題一直是困擾著許多逆向工程師的難題,無論是各種反 WiFi 代理、反 VPN 代理的對抗抓取封包方式,還是伺服器驗證用戶端、用戶端驗證伺服器這類 CA 證書層面的對抗,這些問題始終是阻礙逆向分析的攔路虎,讓人頭疼不已。雖然 r0capture 選擇從程式層面抓取資料封包資訊,徹底繞過了以上這些應對中間人抓取封包的對抗方式,但由於 r0capture 依賴於 Frida,導致一旦 Frida 被對抗,那麼接下來的抓取封包工作就無法進行,最終大大拖慢了逆向分析的工作處理程序。因此,本章將 r0capture 的功能完全移植到沙盒中,以避免因為環境問題最終導致抓取封包失敗的情況。另外,本章還會介紹一些黑色產業鏈與風控的基礎知識,並簡單地製作一個能夠隱藏部分裝置指紋資訊的「玩具級」沙盒,希望能夠幫助初學者了解在平靜的 App 執行過程中,黑色產業鏈和安全風控在洋流下的暗流湧動。

8.1 從 **r0capture** 到原始程式沙盒網路庫「自吐」

8.1.1 **App** 抓取封包分析

為了製作網路庫「自吐」沙盒，首先我們要明白對網路函數庫的「自吐」工作，從沙盒角度可以做到哪一步呢？

經過第 7 章密碼相關「自吐」沙盒的開發，相信讀者會發現：對於 Android 系統來說，如果要適用於沙盒，那麼首先其實現需要能夠在 Android 原始程式中找到相對應的位置，這樣才能進一步徹底修改其實現，使得修改後的原始程式能夠按照預期實現目標。簡而言之，脫離程式的本質談沙盒的建構是不切實際的。

既然沙盒充分依賴於程式本身，那麼在 OSI 七層模型和 TCP/IP 四層模型中，系統原始程式能夠實現的邊界在哪裡呢？

為了方便後續針對抓取封包的分析，這裡還要介紹一下 TCP/IP 四層模型。如圖 8-1 所示，當我們在討論 MAC 位址時，其實指的就是鏈路層的相關內容；而如果討論的是 IP 位址，則是指網路層的相關內容。通訊埠相關話題與傳輸層協定息息相關。一般來說，當我們討論傳輸內容時，往往是指在應用層中透過 HTTP/XMPP 等應用層協定傳輸資料內容本身。

在 App 的開發過程中，以應用許可權來講，筆者認為其只能針對傳輸層及以上層面進行控制修改。換言之，App 只能修改所使用的應用層協定類型，包括其中的資料格式等內容，或者修改傳輸所使用的通訊埠，甚至直接使用 TCP/UDP 進行通訊，而鮮少存在 App 可以修改 IP 等網路層相關內容，更不用提修改網路層以下網路層面的資料（VPN 應用只是利用下層提供介面 API，建立出一個新的網路介面，對應 IP 還是 VPN 伺服器分配的）。

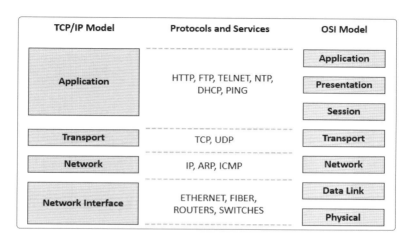

▲ 圖 8-1 OSI 七層模型和 TCP/IP 四層模型對比圖

（來源：https://www.electronicedu.com/2021/03/tcpip-model.html）

具體來說，與網頁應用不同，在應用層這個維度上，App 整體的邏輯和程式都是交由廠商全面控制的，通常來說，App 使用封裝完整的系統 API 或者呼叫更加好用的網路框架實現 HTTP(S) 等通用協定的互動與開發，App 開發者不需要考慮具體的協定方法與詳細內容，只需要關注 App 想要發送和接收的功能，這類成熟的框架包括存取網站的 OkHttp、播放視訊的 Exoplayer、非同步平滑圖片捲動載入框架 Glide 等。這類上層的成熟協定框架通常其底層協定封裝還是交由系統 API 處理，為 App 協定的安全性造成了很大困擾。基於此，大部分應用往往採取多種手段防止應用層面的抓取封包手段。比如，App 採用特定 API（Proxy.NO_PROXY、System.getProperty("http.proxyHost") 等）檢測，甚至繞過 WiFi 代理方式抓取封包，即使逆向人員使用 VPN 應用從網路層將資料流程程轉發到抓取封包軟體，繞過 WiFi 代理檢測方式，還是會存在如圖 8-2 所示的

getNetWorkCapabilities() 等 API 檢測當前網路介面，從而避免軟體本身被抓取封包。

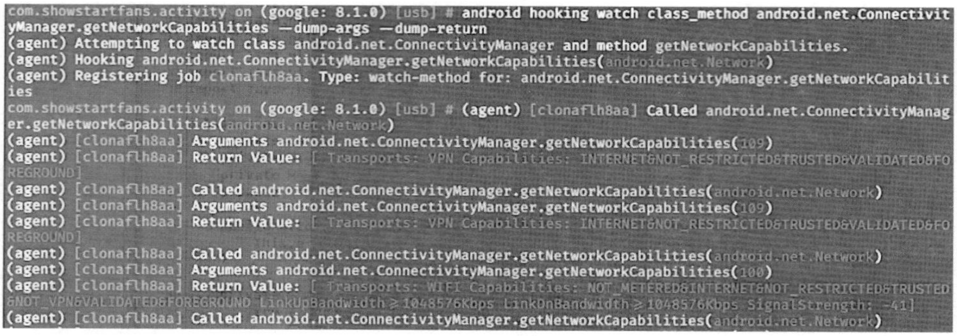

▲ 圖 8-2 VPN 檢測

除了以上透過檢測當前手機聯網方式來達到對抗抓取封包的手段，App還可以透過證書（CA）層面的對抗來達到對抗抓取封包的方式，比如用戶端驗證伺服器的方式：基於加密通訊過程中只需要證書合法即可完成通訊的已知事實，在用戶端和伺服器之間完成「握手」環節時驗證 CA 的 Hash 值，來達到只與持有相同 CA 的伺服器進行通訊的方式，從而進一步加強通訊的安全性。還會有利用服務端不在使用者手中的優勢，在服務端只與使用特定 CA 進行通訊的用戶端完成資料互動。

由於協定通用性問題，即使在應用層做到如此多的防護手段，只要攻擊者能夠成功繞過上述所有防護機制，其傳輸的具體資料最終還是能夠被攻擊者成功獲取和解析。因此，在用戶端開發過程中，開發者可能會採用一些小眾協定甚至自研應用層協定（比如騰訊的 JceStruct 協定等）來達到即使資料流程量被攻擊者成功獲取，依舊沒辦法從中得到有效資訊的效果。這種自研協定在傳輸層被利用得淋漓盡致，比如某廠商開創性地提出了自建代理長連通道的網路加速方案，將 App 中絕大部分的請求透過 CIP 通道中的 TCP 子通道與長連伺服器通訊，長連伺服器在收到請求後，將收到的請求代理再轉發到業務伺服器，從而大大提高了業務效率；更有某些公司在通訊標準演進的道路上大步快跑，在目前 HTTP/2 都沒有普及的情況下，甚至提前邁入 HTTP/3 的時代，在性能最佳化的 KPI 上一騎絕塵，從核心、演算法、傳輸層網路函數庫和服務端全部自研。在這樣的情況下，App 已經脫離了系統框架的限制，達到了真正的沙盒無法「自吐」的效果。

值得逆向工作者慶幸的是，限於開發能力，絕大多數的 App 實現 HTTP/SSL 的方案都非常直白，就是呼叫系統的 API 或者呼叫更加好用的網路框架，這些才是 Android 應用程式開發者的日常。作為逆向工作人員，基於從高維降維操作的思路，只需在應用層的下層：Socket 介面選擇必然經過的函數進行 Hook 之後，列印呼叫堆疊即可清晰地得出從肉眼可見的應用資料到被封裝成 HTTP 封包，進一步進入 SSL 進行加解密，再透過 Socket 與伺服器進行通訊的完整過程。如圖 8-3 所示是某非法應用視訊解析的示意圖。

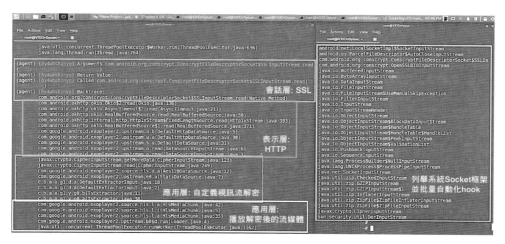

▲ 圖 8-3 Trace 系統 API 呼叫堆疊

基於上述關於 App 抓取封包問題的討論，接下來的兩小節將利用修改系統原始程式的優勢依次介紹如何從沙盒層面實現 App 無感知抓取封包以及沙盒能夠在對抗應用層防抓取封包方面所做的努力。

8.1.2 從 r0capture 到沙盒無感知抓取封包

如果讀者對 r0capture（https://github.com/r0ysue/r0capture）的開發過程十分了解就會知道，基於上述從底層監聽上層資料內容的方式，將應用層資料統一分為加密和非加密兩種類型，並分別針對多個應用層協定框架進行驗證和測試，最終得到明文資料協定所發送資料的過程必然會經過 java.net.SocketOutputStream 類別的 socketWrite0() 函數，接收到的資料則必然會經過 java.net.SocketInputStream 類別的 socketRead0() 函數，其對應的 r0capture 程式如程式清單 8-1 所示。

◉ 程式清單 8-1 r0capture 中關於明文協定的 Hook 指令稿

```
Java.use("java.net.SocketOutputStream").socketWrite0.overload('java.
io.FileDescriptor', '[B', 'int', 'int').implementation = function (fd,
bytearry, offset, byteCount) {
    var result = this.socketWrite0(fd, bytearry, offset, byteCount);
    var message = {};
    message["function"] = "HTTP_send";
    message["ssl_session_id"] = "";
    // 原位址和目的位址以及對應通訊埠資訊
    message["src_addr"] = ntohl(ipToNumber((this.socket.value.
getLocalAddress().toString().split(":")[0]).split("/").pop()));
    message["src_port"] = parseInt(this.socket.value.getLocalPort().
toString());
    message["dst_addr"] = ntohl(ipToNumber((this.socket.value.
getRemoteSocketAddress().toString().split(":")[0]).split("/").pop()));
```

```
    message["dst_port"] = parseInt(this.socket.value.
getRemoteSocketAddress().toString().split(":").pop());

    // 呼叫堆疊
    message["stack"] = Java.use("android.util.Log").
getStackTraceString(Java.use("java.lang.Throwable").$new()).toString();
    var ptr = Memory.alloc(byteCount);
    for (var i = 0; i < byteCount; ++i)
    Memory.writeS8(ptr.add(i), bytearry[offset + i]);
    send(message, Memory.readByteArray(ptr, byteCount))
    return result;
}
Java.use("java.net.SocketInputStream").socketRead0.overload('java.
io.FileDescriptor', '[B', 'int', 'int', 'int').implementation = function
(fd, bytearry, offset, byteCount, timeout) {
    var result = this.socketRead0(fd, bytearry, offset, byteCount,
timeout);
    var message = {};
    message["function"] = "HTTP_recv";
    message["ssl_session_id"] = "";
    // 原位址和目的位址以及對應通訊埠資訊
    message["src_addr"] = ntohl(ipToNumber((this.socket.value.
getRemoteSocketAddress().toString().split(":")[0]).split("/").pop()));
    message["src_port"] = parseInt(this.socket.value.
getRemoteSocketAddress().toString().split(":").pop());
    message["dst_addr"] = ntohl(ipToNumber((this.socket.value.
getLocalAddress().toString().split(":")[0]).split("/").pop()));
    message["dst_port"] = parseInt(this.socket.value.getLocalPort());

    // 呼叫堆疊
    message["stack"] = Java.use("android.util.Log").
getStackTraceString(Java.use("java.lang.Throwable").$new()).toString();
```

```
    if (result > 0) {
    var ptr = Memory.alloc(result);
    for (var i = 0; i < result; ++i)
        Memory.writeS8(ptr.add(i), bytearry[offset + i]);
    send(message, Memory.readByteArray(ptr, result))
    }
    return result;
}
```

基於上述 r0capture 指令稿內容會發現，dump 的關鍵內容包括位址資訊及傳輸的資料內容本身。當然，由於傳輸的資料內容本身可能處於加密狀態，因此還可以將呼叫堆疊資訊列印出來，以輔助後續定位關鍵加解密資訊。

讓我們來一一解決相關問題。

（1）對於位址相關資訊，在 r0capture 中其實是透過 this.socket.value 的方式得到實例中成員 Socket 的物件，進而呼叫相關函數以獲取資料來源位址和目的位址相關資訊。在移植沙盒時，如圖 8-4 所示，筆者透過 WallBreaker 查看 SocketOutputStream 和 SocketInputStream 物件中的成員結構，發現其中的成員 Socket 對應的內容就是目的位址資訊，因此在移植沙盒時要獲取目的位址資訊，對比 r0capture 中的實現，僅僅需要一句 this.socket.toString() 即可獲取，至於原位址資訊，也可以與 r0capture 類似，透過呼叫 getLocalAddress() 函數實現，這裡為了簡便起見，暫時不加。

▲ 圖 8-4　使用 WallBreaker 查看 SocketOutputStream 物件的成員結構

（2）對於傳輸資料的「自吐」，事實上對比 r0capture 和 AOSP 原始程式碼會發現其實資料內容就儲存在函數的第二個參數中，對於 socketWrite0 這一負責發送資料的函數來說，其有效資料的起始位置和長度分別由參數 off 和 len 決定，因此要「自吐」發送資料內容，只需將這部分資料列印出來即可；而對於 socketRead0 這一負責接收資料的函數來講，雖然其有效資料的起始位置由參數 off 決定，但是其真實傳輸的資料長度卻保留在返回值中，導致無法直接在 socketRead0 函數中獲取，而只能在上層函式呼叫這個函數結束後，再對資料內容進行「自吐」。幸運的是，經過 Android Studio 對 socketRead0 函數進行交叉引用，查詢發現只有處於同一個類別中的 socketRead 函式呼叫了該函數，如圖 8-5 所示。當然，由於 socketWrite0 和 socketRead0 都是 native 函數，其具體實現都在 Native 層，為了避免對 Native 層這個更複雜的部分進行修改，這裡採用在對應函數的上層函數進行「自吐」工作，而 socketWrite0 函數只對處於 SocketOutputStream 類別中的 socketWrite 函數進行呼叫，如圖 8-6 所示。

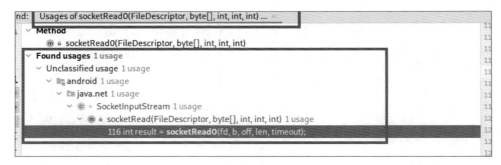

▲ 圖 8-5 查詢交叉引用

```
/*
private native void socketWrite0(FileDescriptor fd, byte[] b, int off,
                                 int len) throws IOException;

/**
 */
private native int socketRead0(FileDescriptor fd,
                               byte b[], int off, int len,
                               int timeout)
        throws IOException;
```

▲ 圖 8-6 native 函式宣告

（3）為了幫助定位資料封包的加密，我們還需要讓沙盒在資料「自吐」時列印函式呼叫堆疊。相比於 r0capture 中透過翻譯 Log.getStackTraceString(Throwable) 函數列印呼叫堆疊的方式，這裡採用更加簡單的方案完成呼叫堆疊的列印工作，其相關程式如程式清單 8-2 所示。

◯ 程式清單 8-2 列印呼叫堆疊

```
Exception e = new Exception("r0ysueSOCKETresponse");
e.printStackTrace();
```

在解決上述問題後，以 socketRead0 函數的「自吐」實現為例，其最終修改後的沙盒程式如程式清單 8-3 所示。

由於網路資料在 App 中是頻繁發生且資料量龐大的，為了避免手機儲存不足，「自吐」的方案仍舊是透過反射方式呼叫 Log.e() 函數列印日誌實現。

◯ 程式清單 8-3 接收資料自吐

```java
// socketInputStream.java
private int socketRead(FileDescriptor fd,
                       byte b[], int off, int len,
                       int timeout)
    throws IOException {
    int result = socketRead0(fd, b, off, len, timeout);

    if(result>0){
        byte[] input = new byte[result];
        // 將有效資料複製到新的 byte 陣列中
        System.arraycopy(b,off,input,0,result);

        String inputString = new String(input);
        Class logClass = null;
        try {
            logClass = this.getClass().getClassLoader().loadClass("android.
util.Log");
        } catch (ClassNotFoundException e) {
            e.printStackTrace();
        }
        Method loge = null;
        try {
            loge = logClass.getMethod("e",String.class,String.class);
        } catch (NoSuchMethodException e) {
            e.printStackTrace();
        }
        try {
```

```
        loge.invoke(null,"r0ysueSOCKETresponse","Socket is => "+this.
socket.toString());
        loge.invoke(null,"r0ysueSOCKETresponse","buffer is =>
"+inputString);
        Exception e = new Exception("r0ysueSOCKETresponse");
        e.printStackTrace();
    } catch (IllegalAccessException e) {
        e.printStackTrace();
    } catch (InvocationTargetException e) {
        e.printStackTrace();
    }
}
return result;
}
```

在移植完明文資料的沙盒「自吐」問題後，讓我們聚焦 r0capture 中對加密應用層資料的「自吐」部分。

如程式清單 8-4 所示，事實上在 r0capture 中針對 SSL 加密資料的「自吐」工作是繼承 r0capture 的原專案 frida_ssl_logger 在 Native 層（或者稱為 SO 層）所做的 Hook，而這與我們只想在 Java 層中對 AOSP 原始程式進行修改的初衷相互違背，且由於 Native 層的呼叫堆疊資訊難以列印，因此對於這部分內容，在移植到沙盒之前還需要重新尋找 Java 層相關 API。

● 程式清單 8-4　r0capture SSL 資料自吐

```
Interceptor.attach(addresses["SSL_read"],
  {
    onEnter: function (args) {
      var message = getPortsAndAddresses(SSL_get_fd(args[0]), true);
      message["ssl_session_id"] = getSslSessionId(args[0]);
      message["function"] = "SSL_read";
      this.message = message;
      this.buf = args[1];
    },
```

```
  onLeave: function (retval) {
    retval |= 0; // Cast retval to 32-bit integer.
    if (retval <= 0) {
      return;
    }
    send(this.message, Memory.readByteArray(this.buf, retval));
  }
});

Interceptor.attach(addresses["SSL_write"],
  {
    onEnter: function (args) {
      var message = getPortsAndAddresses(SSL_get_fd(args[0]), false);
      message["ssl_session_id"] = getSslSessionId(args[0]);
      message["function"] = "SSL_write";
      send(message, Memory.readByteArray(args[1], parseInt(args[2])));
    },
    onLeave: function (retval) {
    }
});
```

事實上，在快速定位關於 SSL 相關函數時，筆者採取的是透過 Objection
搜索所有與 socket 字串相關的類別，並利用 Objection 在執行注入時支
持執行檔案中所有命令的 -c 參數，對以上所有 socket 類別進行 Trace 進
而快速定位，最終考慮到 Android 8 和 Android 10 兩個系統的相容性，
選擇了圖 8-7 中的 com.android.org.conscrypt.ConscryptFileDescriptorSo
cket$SSLInputStream.read() 作為資料接收和 com.android.org.conscrypt.
ConscryptFileDescriptorSocket$SSLOutputStream.write() 作為資料發送
「自吐」函數。

在定位到關鍵 Java 函數後，與實現明文資料在沙盒中「自吐」一樣，仍
舊要解決關於資料內容、位址資訊與呼叫堆疊的問題。

▲ 圖 8-7 Trace Socket 相關函數得到 Hook 結果

首先，為了解決資料內容的「自吐」問題，我們可以與上述「自吐」明文協定資料內容一樣直接在找到的兩個函數中實現「自吐」，但是在研究時發現，二者的函數內容都會呼叫 SSL 成員所在類別中的函數，且傳遞的參數中包含一切和資料有關內容，如圖 8-8 和圖 8-9 所示。

▲ 圖 8-8 ConscryptFileDescriptorSocket$SSLInputStream.read() 函數

```java
    @Override
    public void write(byte[] buf, int offset, int byteCount) throws IOException {
        Platform.blockGuardOnNetwork();
        checkOpen();
        ArrayUtils.checkOffsetAndCount(buf.length, offset, byteCount);
        if (byteCount == 0) {
            return;
        }

        synchronized (writeLock) {
            synchronized (stateLock) {
                if (state == STATE_CLOSED) {
                    throw new SocketException("socket is closed");
                }

                if (DBG_STATE) {
                    assertReadableOrWriteableState();
                }
            }

            ssl.write(Platform.getFileDescriptor(socket), buf, offset, byteCount,
                    writeTimeoutMilliseconds);

            synchronized (stateLock) {
                if (state == STATE_CLOSED) {
                    throw new SocketException("socket is closed");
                }
            }
        }
    }
```

▲ 圖 8-9 ConscryptFileDescriptorSocket$SSLOutputStream.write() 函數

在追蹤 SSL 成員定義後，會發現該成員實際上是 sslWrapper 類型的物件，為了後續方便沙盒的移植工作，最終轉而在 sslWrapper 類別中實現沙盒的「自吐」工作。

其次，既然選擇在 sslWrapper 類別的 read() 和 write() 函數中實現傳輸資料的收發「自吐」，那麼該類別中是否有類似于明文協定中 Socket 這樣的成員用於表示位址資訊呢？為了驗證這一點，這裡同樣使用 WallBreaker 對處理程序堆積中的 sslWrapper 實例進行記憶體物件資訊的列印，會發現其 handshakeCallbacks 等多個成員的值都剛好與明文協定中的 Socket

成員造成的作用一致,因此在沙盒「自吐」中,筆者可從這幾個成員中
任意選擇進行列印,從而「自吐」出位址資訊,如圖 8-10 所示。

▲ 圖 8-10　sslWrapper 物件資訊

最後,在呼叫堆疊問題上,其實沙盒列印呼叫堆疊的方式與明文協定中
列印呼叫堆疊的方式相同,但是細心的讀者可能會發現,在程式清單 8-4
中,r0capture 並沒有列印呼叫堆疊相關程式,而之所以這麼做,是因為
Frida 列印的 Native 層呼叫堆疊資訊可能不準確,且參考意義不大,如果
一定要加呼叫堆疊資訊,怎麼辦呢?

這裡考慮到 Java 層的函數最終總是透過 Native 層的函數完成底層程式
呼叫,換言之,Java 上層函數總是在 Native 層相關函數被呼叫前開始執
行,因此這裡透過增加針對 com.android.org.conscrypt.ConscryptFileDe
scriptorSocket$SSLInputStream.read() 和 com.android.org.conscrypt.Co
nscryptFileDescriptorSocket$SSLOutputStream.write() 兩個函數的 Hook
後,在其中僅增加列印呼叫堆疊相關資訊,並在獲取到呼叫堆疊資訊後
保存到外部變數中,以便最終在對底層函數進行 Hook 時與其他資訊一起
列印出來,最終相關程式如程式清單 8-5 所示。

● 程式清單 8-5 帶呼叫堆疊的 r0capture

```
var SSLstackwrite = null;
var SSLstackread = null;
Interceptor.attach(addresses["SSL_read"],
  {
    onEnter: function (args) {
      var message = getPortsAndAddresses(SSL_get_fd(args[0]), true);
      message["ssl_session_id"] = getSslSessionId(args[0]);
      message["function"] = "SSL_read";
      // 保存的呼叫堆疊資訊
      message["stack"] = SSLstackread;
      this.message = message;

      this.buf = args[1];
    },
    onLeave: function (retval) {
      retval |= 0; // Cast retval to 32-bit integer.
      if (retval <= 0) {
        return;
      }
      send(this.message, Memory.readByteArray(this.buf, retval));
    }
  });

Interceptor.attach(addresses["SSL_write"],
  {
    onEnter: function (args) {
      var message = getPortsAndAddresses(SSL_get_fd(args[0]), false);
      message["ssl_session_id"] = getSslSessionId(args[0]);
      message["function"] = "SSL_write";
      //  保存的呼叫堆疊資訊
      message["stack"] = SSLstackwrite;
      send(message, Memory.readByteArray(args[1], parseInt(args[2])));
    },
```

```
  onLeave: function (retval) {
  }
});
Java.use("com.android.org.conscrypt.ConscryptFileDescriptorSocket$SSLOut
putStream").write.overload('[B', 'int', 'int').implementation = function
(bytearry, int1, int2) {
    var result = this.write(bytearry, int1, int2);
    SSLstackwrite = Java.use("android.util.Log").getStackTraceString(Java.
use("java.lang.Throwable").$new()).toString();
    return result;
}
Java.use("com.android.org.conscrypt.ConscryptFileDescriptorSocket$SSLIn
putStream").read.overload('[B', 'int', 'int').implementation = function
(bytearry, int1, int2) {
    var result = this.read(bytearry, int1, int2);
    SSLstackread = Java.use("android.util.Log").getStackTraceString(Java.
use("java.lang.Throwable").$new()).toString();
    return result;
}
```

在解決上述問題後，最終修改後的 SslWrapper.java 相關內容（以 write
函數為例）如程式清單 8-6 所示，完整檔案見附件。

○

○ 程式清單 8-6 SslWrapper.java 訂製部分內容

```
// SslWrapper.java
// TODO(nathanmittler): Remove once after we switch to the engine socket.
void write(FileDescriptor fd, byte[] buf, int offset, int len, int
timeoutMillis)
        throws IOException {
    if(len>0){
        byte[] input = new byte[len];
        System.arraycopy(buf,offset,input,0,len);
```

```
        String inputString = new String(input);
        Class logClass = null;
        try {
            logClass = this.getClass().getClassLoader().loadClass("android.
util.Log");
        } catch (ClassNotFoundException e) {
            e.printStackTrace();
        }
        Method loge = null;
        try {
            loge = logClass.getMethod("e",String.class,String.class);
        } catch (NoSuchMethodException e) {
            e.printStackTrace();
        }
        try {
            loge.invoke(null,"r0ysueSSLrequest","SSL is => "+this.
handshakeCallbacks.toString());
            loge.invoke(null,"r0ysueSSLrequest","buffer is =>
"+inputString);
            Exception e = new Exception("r0ysueSSLrequest");
            e.printStackTrace();
        } catch (IllegalAccessException e) {
            e.printStackTrace();
        } catch (InvocationTargetException e) {
            e.printStackTrace();
        }
    }

    NativeCrypto.SSL_write(ssl, fd, handshakeCallbacks, buf, offset, len,
timeoutMillis);
}
```

在完成上述所有沙盒程式的訂製撰寫後，按照前面介紹的編譯方法，最終編譯出不帶 Root 許可權的系統。在更新軔體後，測試某應用抓取封包的效果，如圖 8-11 所示。

```
bullhead:/sdcard/Download $ cat ██████.txt |grep 186213
12-27 22:00:34.917 11519 12004 E r0ysueSSLrequest: app_version=11.6.5&device_type=AOSP%2Bon%
BBullHead&lon=&device_name=AOSP+on+BullHead&qyidv2=E8E1219732A33BD761F3DDBD2CAF27745&agenttyp
=21&sdk_version=11.6.5&is_reg_confirm=1&hui_version=&hfvc=95&lang=zh_CN&lat=&s2=wd&s3=&s4=WD
login&fromSDK=21&device_id=eb4774dbbbbb7e2ec15d58b469b6b0bfal108&area_code=86&qd_sg=eb4774dbb
b7e2ec15d58b469b6b0bfal108-11.6.5-1609077634430-null&ptid=02022001010000000000&dfp=14e1af8d3
f30843dcb3c1bc290cafaff3236d4b0e12d9b854b9b2f434d8b3f8&iqid=79b323be1353a0909d07ab89172f707
407100009&biqid=eb4774dbbbbb7e2ec15d58b469b6b0bfal108&QC005=E8E1219732A33BD761F3DDBD2CAF27451
09077605245&app_lm=cn&account=18621300999&qd_sc=318823aa326790f2f51a8198f5baeb69
bullhead:/sdcard/Download $
```

▲ 圖 8-11　無感知抓取封包效果

8.1.3　使用沙盒輔助中間人抓取封包

雖然前面說明的方式能夠在 App 無感知狀態下達到抓取封包的效果，但正如 r0capture 本身的侷限一樣，對於部分採用自訂 SSL 框架的方式（比如 WebView、小程式、Flutter 等）進行資料通信的 App 來說，由於其資料通信的方式已經不依賴於系統本身，而是收發送封包本身都交由 App 自身進行處理，因此即使是從目前開發的沙盒角度依舊無法得到這類 App 的傳輸資料。此時中間人抓取封包的優點就表現出來了：綁架所有從系統中發出去的資料封包，即使是自訂資料傳輸方式也不例外。

但是，正如在 8.1.1 節中所介紹的那樣，中間人抓取封包的方式存在很多對抗方式阻礙資料的抓取，比如利用 Proxy.NO_PROXY 等 API 對抗 WiFi 代理抓取封包，或者利用 getNetWorkCapabilities() 等 API 檢測 VPN 代理從而對抗抓取封包。當然，即使在繞過上述檢測系統代理的方法後，還是可能會遇到其他問題，比如伺服器驗證用戶端，或者用戶端驗證伺服器等。

面對如此多的對抗手段，是否能夠反制這些對抗手段呢？

答案是可以的，甚至在系統沙盒中也能夠做到這一點，這也正是本小節存在的意義—透過修改系統原始程式幫助排除阻礙中間人抓取封包的「元兇」。

在正式開始介紹沙盒在中間人抓取封包中的作用前,我們先了解一下中間人抓取封包的流程。

我們知道要實現中間人抓取封包,如果是抓取明文的 HTTP 協定,只需要在將手機和電腦放置於同一網路環境中並確定二者能夠相互 Ping 通後,再將系統 WiFi 代理或者 VPN 代理設定為電腦系統的 IP 以及對應代理軟體的通訊埠即可完成資料封包的抓取,如圖 8-12 所示。

▲ 圖 8-12 HTTP 代理抓取封包

但是如果使用 HTTPS 等需要 CA 認證成功才能正常通訊的協定,仍舊使用這樣的方法抓取封包就會發現存取網頁時出現如圖 8-13 所示的「您的連接不是私密連接」提示。

要解決這個問題,需要逆向人員將 Charles 等代理軟體的對應證書檔案放置到使用者信任的證書甚至系統信任的證書空間中,才能順利抓取 HTTPS 資料。要將 Charles 證書內建到系統信任證書的位置,需要透過 mount 命令臨時將系統磁碟分割設定為可寫後才能成功移動,而如果想要將系統磁碟分割設定為可寫，又需要 Root 許可權，這就幾乎斷絕了逆向人員在非 Root 環境中將證書移植到系統信任證書空間的可能性。那麼作為無 Root 沙盒，能否做到這一點呢？

▲ 圖 8-13 HTTPS 協定抓取封包錯誤訊息

答案是可以的，要做到這一點，我們只需將一個 Charles 證書檔案轉換為 Android 系統能夠辨識的形式放置到系統證書在原始程式環境的對應目錄下即可。

那麼如何將 Charles 匯出的證書檔案轉換成系統所辨識的 .0 這種格式呢？

如圖 8-14 所示，這裡將證書檔案安裝到手機的使用者信任證書中，並從使用者信任證書所在的系統位置 /data/misc/user/0/cacerts-added/ 複製出來供後續使用。

需要注意的是，由於每一個 Charles 軟體生成的證書都是不一致的，因此這裡移植的證書只能用於這個虛擬機器中的 Charles 抓取封包。

```
┌──(root💀kali)-[~/aosp810_r1]
└─# adb shell
bullhead:/ $ su
bullhead:/ # cd /data/misc/user/0/cacerts-added
bullhead:/data/misc/user/0/cacerts-added # ls
bf630c44.0
bullhead:/data/misc/user/0/cacerts-added # cp  bf630c44.0 /sdcard/
bullhead:/data/misc/user/0/cacerts-added # exit
bullhead:/ $ exit
┌──(root💀kali)-[~/aosp810_r1]
└─# adb pull /sdcard/bf630c44.0
/sdcard/bf630c44.0: 1 file pulled. 0.1 MB/s (1330 bytes in 0.025s)
┌──(root💀kali)-[~/aosp810_r1]
```

▲ 圖 8-14 將 Charles 證書匯出備用

在取出證書後，接下來的工作就只剩下定位系統證書在原始程式環境的對應目錄，並將證書放置到該位置即可。最終透過在原始程式目錄中搜索 certificates 關鍵字發現在原始程式的 /system/ca-certificates/google/files/ 目錄下存在著大量的證書檔案，在與真實執行的系統中的證書進行對比後，最終確認原始程式的 /system/ca-certificates/google/files/ 目錄即為系統執行時期系統信任證書所在的位置。在移動完畢後，還需要確認所移動的證書所屬的使用者、使用者群組以及對應的許可權都與其他已有證書一致，如圖 8-15 所示。

```
┌──(root💀kali)-[~/aosp810_r1/system/ca-certificates/files]
└─# ls -alit
total 1036
1999017 drwxr-xr-x 2 root root 4096 Aug  7 08:55 .
 649668 -rw-r--r-- 1 root root 1330 Aug  7 08:39 bf630c44.0
1992786 drwxr-xr-x 5 root root 4096 Jun 30 04:43 ..
1992874 -rw-r--r-- 1 root root 4819 Jun 30 04:43 91739615.0
1992875 -rw-r--r-- 1 root root 7245 Jun 30 04:43 9282e51c.0
1992876 -rw-r--r-- 1 root root 8002 Jun 30 04:43 9339512a.0
1992877 -rw-r--r-- 1 root root 2971 Jun 30 04:43 9479c8c3.0
1992878 -rw-r--r-- 1 root root 7064 Jun 30 04:43 9576d26b.0
1992879 -rw-r--r-- 1 root root 4511 Jun 30 04:43 95aff9e3.0
1992880 -rw-r--r-- 1 root root 4150 Jun 30 04:43 9685a493.0
1992881 -rw-r--r-- 1 root root 4405 Jun 30 04:43 9772ca32.0
1992882 -rw-r--r-- 1 root root 7204 Jun 30 04:43 9c3323d4.0
1992883 -rw-r--r-- 1 root root 7223 Jun 30 04:43 9d6523ce.0
1992884 -rw-r--r-- 1 root root 4369 Jun 30 04:43 9dbefe7b.0
1992885 -rw-r--r-- 1 root root 8390 Jun 30 04:43 9f533518.0
1992886 -rw-r--r-- 1 root root 5461 Jun 30 04:43 a0bc6fbb.0
1992887 -rw-r--r-- 1 root root 7188 Jun 30 04:43 a2c66da8.0
1992888 -rw-r--r-- 1 root root 4284 Jun 30 04:43 a3896b44.0
1992889 -rw-r--r-- 1 root root 4123 Jun 30 04:43 a7605362.0
1992890 -rw-r--r-- 1 root root 2878 Jun 30 04:43 a7d2cf64.0
1992891 -rw-r--r-- 1 root root 4395 Jun 30 04:43 a81e292b.0
1992892 -rw-r--r-- 1 root root 4333 Jun 30 04:43 ab5346f4.0
```

▲ 圖 8-15 移動後的證書

在完成上述步驟後,重新編譯原始程式並刷入手機會發現 Charles 的證書
已經內建在系統信任證書中,如圖 8-16 所示。此時再次測試抓取 HTTPS
資料,就會發現原本無法存取的 HTTPS 網站已經能夠正常存取且資料在
Charles 軟體中都能夠正常顯示。

▲ 圖 8-16 編譯後的 AOSP 沙盒系統信任證書

在解決基本的 HTTPS 協定抓取封包問題後,讓我們聚焦沙盒幫助對抗伺
服器驗證用戶端和 SSL Pinning 的問題。

所謂伺服器驗證用戶端,是指伺服器在與用戶端進行通訊時,會在握手
階段驗證用戶端使用證書的公開金鑰,但是當使用中間人進行抓取封包
時,實際上與伺服器進行通訊的對方是 Charles 等抓取封包軟體,其使用
的證書就不再是服務端認可的證書檔案,因此在解決這種問題時,通常
是在 App 中找到對應證書檔案和對應密碼,並將之轉換為 P12 格式的證
書,最終匯入 Charles 以達到欺騙伺服器進行通訊的效果。

通常來說，用戶端要實現使用特定證書與伺服器進行通訊，就不可避免地會使用證書密碼開啟證書。對應地，以筆者的經驗，開發者通常會使用 KeyStore.load(InputStream, char[]) 函數開啟證書。筆者曾經寫過一個指令稿用於 Hook 該函數，從而 dump 證書檔案和對應密碼，具體 Hook 指令稿如程式清單 8-7 所示。

⊘ 程式清單 8-7 saveClientCer.js

```
Java.perform(function () {
  var StringClass = Java.use("java.lang.String");
  var KeyStore = Java.use("java.security.KeyStore");
  KeyStore.load.overload('java.io.InputStream', '[C').implementation =
function (arg0, arg1) {
    // 列印堆疊資訊
    console.log(Java.use("android.util.Log").getStackTraceString(Java.
use("java.lang.Throwable").$new()));
    // arg1即為證書金鑰
    console.log("KeyStore.load2:", arg0, arg1 ? StringClass.$new(arg1) :
null);

    if (arg0){
      // 將證書保存到/sdcard/Download/目錄下
        var file =  Java.use("java.io.File").$new("/sdcard/Download/"+
String(arg0)+".p12");
        var out = Java.use("java.io.FileOutputStream").$new(file);
        var r;
        while( (r = arg0.read(buffer)) > 0){
            out.write(buffer,0,r)
        }
        console.log("save success!")
        out.close()
    }
    this.load(arg0, arg1);
  };
}
```

在測試某伺服器驗證用戶端證書的樣本中，測試結果如圖 8-17 所示。

```
java.lang.Throwable
        at java.security.KeyStore.load(Native Method)
        at cn.██████.android.net.l.<init>(TLSSocketFactory.java:10)
        at cn.██████.android.net.k.a(SoulNetStorages.java:1)
        at cn.██████.android.net.g.a(OkHttpClientHelper.java:15)
        at cn.██████.android.net.SoulNetworkSDK.a(SoulNetworkSDK.java:93)
        at cn.██████.android.net.winter.b.b.c.a(NetProxy.java:1)
        at cn.██████.android.net.winter.b.b.a.a(ChinaProxy.java:8)
        at cn.██████.android.net.SoulNetworkSDK.a(SoulNetworkSDK.java:27)
        at cn.██████.android.net.i.<init>(SoulApiManager.java:20)
        at cn.██████.android.net.i.<init>(SoulApiManager.java:1)
        at cn.██████.android.net.i$b.<clinit>(SoulApiManager.java:1)
        at cn.██████.android.net.i.b(SoulApiManager.java:1)
        at cn.██████.android.net.ab.b.a(ABTestSDK.java:2)
        at cn.██████.android.net.sip.wraper.a.a(SIPRequestWraper.java:3)
        at cn.██████.android.net.sip.wraper.a.b(SIPRequestWraper.java:10)
        at cn.██████.android.net.SoulNetworkSDK.a(SoulNetworkSDK.java:14)
        at g.a██████e.c.b(SNetWorkKit.kt:3)
        at g.a██████b.a(SACommonKit.kt:21)
        at cn.██████.android.n.d(InitializeUtils.java:13)
        at cn.██████.android.SoulApp.c(SoulApp.java:37)
        at cn.██████.lib.basic.app.MartianApp.onCreate(MartianApp.java:6)
        at android.app.Instrumentation.callApplicationOnCreate(Instrumentation.java:1119)
        at android.app.ActivityThread.handleBindApplication(ActivityThread.java:5740)
        at android.app.ActivityThread.handleBindApplication(Native Method)
        at android.app.ActivityThread.-wrap1(Unknown Source:0)
        at android.app.ActivityThread$H.handleMessage(ActivityThread.java:1656)
        at android.os.Handler.dispatchMessage(Handler.java:106)
        at android.os.Looper.loop(Looper.java:164)
        at android.app.ActivityThread.main(ActivityThread.java:6494)
        at java.lang.reflect.Method.invoke(Native Method)
        at com.android.internal.os.RuntimeInit$MethodAndArgsCaller.run(RuntimeInit.java:438)
        at com.android.internal.os.ZygoteInit.main(ZygoteInit.java:807)
KeyStore.load2: android.content.res.AssetManager$AssetInputStream@c0d05fe }%2██ ██████ pP!w%X  密碼
save success!
```

▲ 圖 8-17 keyStore.load 函數 Hook 結果

最終透過 KeyStore Explorer 等證書查看軟體使用 Hook 得到的密碼成功開啟 dump 下來的證書檔案後，即可完整地看到該證書的私密金鑰等資訊，如圖 8-18 所示。將證書檔案匯入 Charles 作為用戶端證書，樣本即可在中間人環境下順利上網，且代理軟體能夠順利攔截到資料詳情。

既然是透過系統 API 使用密碼開啟證書檔案，那麼非常明顯，我們可以將其移植到系統原始程式中，最終在找到 java/security/KeyStore.java 檔案（系統中還會有一個 android/security/KeyStore.java 檔案，注意別弄錯）後，其最終修改的程式如程式清單 8-8 所示。

▲ 圖 8-18 Charles 匯入證書

◯ 程式清單 8-8 KeyStore.load 函數

```
// libcore/ojluni/src/main/java/java/security/KeyStore.java
public final void load(InputStream stream, char[] password)
        throws IOException, NoSuchAlgorithmException,
CertificateException {
    if (password != null) {
        String inputPASSWORD = new String(password);
        Class logClass = null;
        try {
            logClass = this.getClass().getClassLoader().loadClass("android.
util.Log");
        } catch (ClassNotFoundException e) {
            e.printStackTrace();
        }
        Method loge = null;
        try {
            loge = logClass.getMethod("e", String.class, String.class);
        } catch (NoSuchMethodException e) {
            e.printStackTrace();
```

```
        }
        try {
            loge.invoke(null, "r0ysueKeyStoreLoad", "KeyStore load PASSWORD
is => " + inputPASSWORD);
            Exception e = new Exception("r0ysueKeyStoreLoad");
            e.printStackTrace();
        } catch (IllegalAccessException e) {
            e.printStackTrace();
        } catch (InvocationTargetException e) {
            e.printStackTrace();
        }

        Date now = new Date();
        String currentTime = String.valueOf(now.getTime());
        // 寫檔案
        FileOutputStream fos = new FileOutputStream("/sdcard/Download/" +
inputPASSWORD + currentTime);
        byte[] b = new byte[1024];
        int length;
        while ((length = stream.read(b)) > 0) {
            fos.write(b, 0, length);
        }
        fos.flush();
        fos.close();

    }

    keyStoreSpi.engineLoad(stream, password);
    initialized = true;
}
```

在修改完畢後,重新編譯原始程式並更新軔體進行測試後,查看日誌會
發現最終證書密碼「自吐」如圖 8-19 所示。此時將沙盒「自吐」的證書
開啟,發現其內容與 Hook 得到的結果一致。

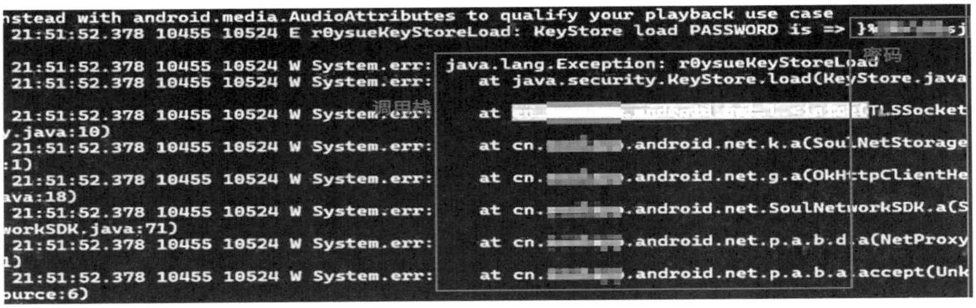

▲ 圖 8-19 沙盒「自吐」證書密碼

但是，在測試樣本時發現，只要在這個函數中對證書進行 dump 操作，就會導致樣本崩潰退出，因此在這裡選擇使用 Objection Trace 與字串 keystore 相關的類別，從而幫助找到其他更加通用的函數用於 dump 證書和密碼，最終 Trace 後發現，圖 8-20 中的函數在樣本開啟時會被呼叫。

▲ 圖 8-20 Trace keystore 相關類別結果

在使用 WallBreaker 搜索並查看對應的物件結構時，發現 java.security. KeyStore$PrivateKeyEntry 物件中存在很多證書相關資訊，如圖 8-21 所示。

那麼是否可以透過這些內容獲取到一個真正的證書檔案呢？

研究後發現，在獲取到證書的 chain 和 privateKey 後，確實可以得到最終的證書檔案，具體保存證書的程式如程式清單 8-9 所示，其中函數的第一個參數是私密金鑰；第二個參數是憑證連結的字串形式；第三個參數是保存的憑證路徑，可指定；第四個參數是保存的證書密碼，這個參數也是由使用者自訂的。

▲ 圖 8-21 PrivateKeyEntry 物件結構

◯ 程式清單 8-9 保存證書相關程式

```
public static void storeP12(PrivateKey pri, String p7, String p12Path,
String p12Password) throws Exception {
    CertificateFactory factory = CertificateFactory.getInstance("X509");
    //初始化憑證連結
    X509Certificate p7X509 = (X509Certificate) factory.
generateCertificate(new ByteArrayInputStream(p7.getBytes()));
    Certificate[] chain = new Certificate[]{p7X509};
    // 生成一個空的p12證書
    KeyStore ks = KeyStore.getInstance("PKCS12", "BC");
    ks.load(null, null);
```

```
    // 將伺服器返回的證書匯入到p12中去
    ks.setKeyEntry("client", pri, p12Password.toCharArray(), chain);
    // 加密保存p12證書
    FileOutputStream fOut = new FileOutputStream(p12Path);
    ks.store(fOut, p12Password.toCharArray());
}
```

為了幫助確定 KeyStore$PrivateKeyEntry 類別中哪些函數在服務端驗證用
戶端機制中會被呼叫，這裡使用 Objection 對該類別進行 Trace 後發現，
圖 8-22 中的 getPrivateKey() 和 getCeritficateChain() 函數都會被呼叫，
因此最終具體的 Hook 得到證書的程式如程式清單 8-10 所示。

▲ 圖 8-22 trace KeyStore$PrivateKeyEntry 類別

◯ 程式清單 8-10 saveClientCer2.js

```javascript
function storeP12(pri, p7, p12Path, p12Password) {
  var X509Certificate = Java.use("java.security.cert.X509Certificate")
  var p7X509 = Java.cast(p7, X509Certificate);
  var chain = Java.array("java.security.cert.X509Certificate", [p7X509])
  var ks = Java.use("java.security.KeyStore").getInstance("PKCS12", "BC");
  ks.load(null, null);
  ks.setKeyEntry("client", pri, Java.use('java.lang.
String').$new(p12Password).toCharArray(), chain);
  try {
    var out = Java.use("java.io.FileOutputStream").$new(p12Path);
    ks.store(out, Java.use('java.lang.String').$new(p12Password).
toCharArray())
  } catch (exp) {
    console.log(exp)
```

```
  }
}
Java.use("java.security.KeyStore$PrivateKeyEntry").getPrivateKey.
implementation = function () {
  var result = this.getPrivateKey()
  storeP12(this.getPrivateKey(), this.getCertificate(), '/sdcard/
Download/'  + uuid(10, 16) + '.p12', 'r0ysue');
  return result;
}
Java.use("java.security.KeyStore$PrivateKeyEntry").getCertificateChain.
implementation = function () {
  var result = this.getCertificateChain()
  storeP12(this.getPrivateKey(), this.getCertificate(), '/sdcard/
Download/'  + uuid(10, 16) + '.p12', 'r0ysue');
  var message = {};
  return result;
}
```

最終在 Hook 測試後會發現，以這樣的方式保存的證書同樣能夠開啟且
其中的資訊與在 KeyStore.load() 函數中 dump 得到的證書檔案內容是
一致的，只是可以明顯地發現，在 KeyStore$PrivateKeyEntry 這個類別
中得到的證書檔案無須得知原本證書的密碼，相比於上一種方式更加優
雅，最終在測試成功後，同樣將相關程式移植到沙盒中，最終修改後的
getPrivateKey() 函數部分內容如程式清單 8-11 所示。

○ 程式清單 8-11 getPrivateKey() 函數內容

```
// libcore/ojluni/src/main/java/java/security/KeyStore.java
public PrivateKey getPrivateKey() {

    String p12Password = "r0ysue";
    Date now = new Date();
    String currentTime = String.valueOf(now.getTime());
    String p12Path = "/sdcard/Download/"  + currentTime + ".p12";
```

```
    X509Certificate p7X509 = (X509Certificate) chain[0];
    Certificate[] mychain = new Certificate[]{p7X509};
    // 生成一個空的p12證書
    KeyStore myks = null;
    try {
        myks = KeyStore.getInstance("PKCS12", "BC");
        myks.load(null, null);
        myks.setKeyEntry("client", privKey, p12Password.toCharArray(),
mychain);
    } catch (KeyStoreException e) {
        // ...
    } catch (CertificateException e) {
        e.printStackTrace();
    }
    // 加密保存p12證書
    FileOutputStream fOut = null;
    try {
        fOut = new FileOutputStream(p12Path);
    } catch (FileNotFoundException e) {
        e.printStackTrace();
    }
    try {
        myks.store(fOut, p12Password.toCharArray());
    } catch (KeyStoreException e) {
        // ...
    }

    return privKey;
}
```

在透過沙盒解決完服務端驗證用戶端的問題後，讓我們聚焦於用戶端驗證伺服器之 SSL Pinning 的部分。事實上，由於 SSL Pinning 是一種用戶端驗證當前使用的證書是不是特定證書的方式，其全部邏輯都是在 App 層面對證書進行驗證，因此無論是 Objection 中的 SSL Pinning Bypass 功能還是 FridaContainer（https://github.com/deathmemory/FridaContainer）、

DroidSSLUnpinning（https://github.com/WooyunDota/DroidSSLUnpinning）中針對這類對抗手段的 bypass 方法，都是透過對每一種不同的框架證書驗證的程式分別進行處理從而達到繞過的效果，因此就系統本身來説，其實無法直接干預這個過程，但是由於 SSL Pinning 驗證的方法總是以開啟證書加上 Hash 驗證兩個部分組成，因此沙盒所能做的就是在開啟檔案這一操作上進行監控，透過呼叫堆疊的方式幫助定位可能進行證書驗證的函數。

為了驗證上述想法，這裡選擇了一個存在 SSL Pinning 驗證手段的樣本（實際上是一個混淆的 OkHttp3 樣本）進行測試，透過 Objection hook Java.io.File 類別的建構函數 $init 並列印其參數和呼叫堆疊，最終發現函式呼叫堆疊中出現已知證書驗證函數的 File 類別建構函數總是 java.io.File.$init(File,String) 這個多載建構函數，且第二個參數值永遠是 .0 格式的證書名，如圖 8-23 所示。

▲ 圖 8-23 java.io.File.$init(File,String) 函數 Hook 結果

根據上述 Hook 結果，這裡重新寫了針對 File 類別這一特定建構函數進行 Hook 的指令稿，其程式如程式清單 8-12 所示。

🔻 **程式清單 8-12 sslHelper.js**

```
setImmediate(function(){
    Java.perform(function(){
        Java.use("java.io.File").$init.overload('java.io.File', 'java.lang.
String').implementation = function(file,cert){
            var result = this.$init(file,cert)
            console.log("path,cart",file.getPath(), cert)
            return result;
        }
    })
})
```

在測試後發現，當第二個參數是證書格式檔案名稱時，對應路徑總是包含 cacert 字串，在進一步根據路徑進行過濾後，透過列印呼叫堆疊會發現並不是所有開啟證書的呼叫堆疊中都存在已知的證書驗證程式。

為了進一步增加程式清單 8-12 中程式在 SSL Pinning Bypass 中的有效性，這裡透過將多個已知是證書驗證呼叫鏈的呼叫堆疊保存為檔案並進行檔案對比，發現 X509TrustManagerExtensions.checkServerTrusted 函數總是存在於這些呼叫堆疊中，從而獲得了最終幫助定位 SSL Pinning 的指令稿內容如程式清單 8-13 所示。

🔻 **程式清單 8-13 sslHelper.js**

```
// (agent) [90m[5208352434982] [39mArguments [32mjava.io.File[39m.
[92mFile[39m([31m"<instance: java.io.File>", "ab1f3027.0"[39m
// (agent) [90m[5208352434982] [39mArguments [32mjava.io.File[39m.
[92mFile[39m([31m"<instance: java.io.File>", "ac1595c4.0"[39m)
// (agent) [90m[5208352434982] [39mArguments [32mjava.io.File[39m.
[92mFile[39m([31m"<instance: java.io.File>", "ab1f3027.0"[39m)
```

```
// (agent) [90m[5208352434982] [39mArguments [32mjava.io.File[39m.
[92mFile[39m([31m"<instance: java.io.File>", "35105088.1"[39m)
setImmediate(function(){
    Java.perform(function(){
        Java.use("java.io.File").$init.overload('java.io.File', 'java.lang.
String').implementation = function(file,cert){
            var result = this.$init(file,cert)
            var stack = Java.use("android.util.Log").
getStackTraceString(Java.use("java.lang.Throwable").$new());
            // 根據多個呼叫堆疊分析發現呼叫堆疊中總是出現這些情況
            if(
            // 開啟的檔案一定是證書
            file.getPath().indexOf("cacert")>0
            //
            && stack.indexOf("X509TrustManagerExtensions.
checkServerTrusted")> 0){
                console.log("path,cart",file.getPath(), cert)
                console.log(stack);

            }
            return result;
        }
    })
})
```

與之前的流程一致，在測試完畢後，將這部分程式翻譯為 Java 程式並增加到原始程式中，最終修改後的 File 類別對應建構函數程式如程式清單 8-14 所示。

◎ 程式清單 8-14 File 類別

```
public File(File parent, String child) {
    if (child == null) {
        throw new NullPointerException();
    }
```

```
    if (parent != null) {
        if (parent.path.equals("")) {
            this.path = fs.resolve(fs.getDefaultParent(),
                    fs.normalize(child));
        } else {
            this.path = fs.resolve(parent.path,
                    fs.normalize(child));
        }
    } else {
        this.path = fs.normalize(child);
    }


    Class logClass = null;
    Method getStackTraceString = null;
    try {
        logClass = this.getClass().getClassLoader().loadClass("android.
util.Log");
        getStackTraceString = logClass.getMethod("getStackTraceString",Thro
wable.class);
    } catch (ClassNotFoundException e) {
        e.printStackTrace();
    } catch (NoSuchMethodException e) {
        e.printStackTrace();
    }
    try {
      // 反射呼叫Log.getStackTraceString(Thrwoable)函數
        String stack = (String)getStackTraceString.invoke(null,new
Throwable());
        if (parent.getPath().indexOf("cacert") >= 0 &&
            // 獲取呼叫堆疊字串
                stack.indexOf("X509TrustManagerExtensions.
checkServerTrusted") >= 0) {
                // 列印呼叫堆疊幫助定位SSL Pinning具體函數
            Exception e = new Exception("r0ysueFileSSLpinning");
            e.printStackTrace();
```

```
        }
    } catch (IllegalAccessException e) {
        e.printStackTrace();
    } catch (InvocationTargetException e) {
        e.printStackTrace();
    }

    this.prefixLength = fs.prefixLength(this.path);
}
```

在修改完程式並重新編譯系統刷入手機後，一個新的針對網路庫「自吐」和輔助中間人抓取封包的沙盒環境就新鮮出爐了。

雖然在這一小節中並沒有介紹如何透過修改原始程式完成對 WiFi/VPN 代理的檢測，但相信對完成上述修改原始程式過程的讀者來説，要完成檢測實際上已經非常簡單了，這裡限於篇幅，不再繼續介紹。

另外，讀者要清楚的是在輔助中間人抓取封包這一塊，沙盒能夠做到的工作是有限的，比如在輔助繞過 SSL Pinning 的指令稿其基點是基於 App 會透過開啟證書檔案及其呼叫堆疊中包含的 X509TrustManagerExtensions. checkServerTrusted 函數等特徵來幫助定位的，有特徵就一定存在繞過手段；再比如，在輔助繞過服務端驗證用戶端的措施中，如果證書內容被強制寫入在程式中，就有可能繞過 KeyStore.load 函數。筆者對這部分所做的講解僅是 磚引玉，希望讀者能夠了解其中的原理，並且在遇到問題時懂得如何去思考。

8.2 風控對抗之簡單實現裝置資訊的篡改

8.2.1 風控對抗基礎介紹

在金融安全領域，「風控」一詞頻繁地被提及，這是用於金融領域中風險控制的專業術語。在當今的網際網路時代，由於金融和網際網路深度結合，「風控」一詞又被廣泛應用于網際網路安全中，與一般風控用途類似，電腦領域的風控同樣用於控制在電子支付或者其他相關場景保護甲方產品免於利益損失。

那麼既然存在風險控制，說明風險本身是存在的，那麼是什麼導致風險的存在呢？

這個答案並不確定。以金融投資來說，如果花幾百萬買了一家公司的股票，未來的盈利是不能預測的，其結果可能是虧損、回本、盈利，其中虧損的機率就是風險存在的可能性，之所以存在風險控制，就是為了儘量減少虧損的機率，保證投資人和公司的利益在盈利和回本之間。

與風控相對，在網際網路中存在著一類團體專門針對網際網路產品業務的漏洞或者福利鑽漏洞，從而發現低成本獲得利益的機會，甚至採取一定的手段從中獲取更大的經濟利益，這類團體通常被稱為黑色產業鏈或灰產。

與電影中駭客輕鬆駭進別人電腦獲取資訊類似，真實的黑色產業鏈團體通常也有一群掌握電腦關鍵技術的人，利用自身的技術優勢繞過或破解已知業務產品的機制從中獲利。以過去非常頻繁的 QQ 盜號事件為例，黑色產業鏈團體透過病毒或其他非法手段盜取 QQ 使用者和密碼，並從獲取的 QQ 資訊中選擇等級較高的號碼高價轉賣以獲取利益，再比如存在黑色產業鏈團體透過某些 App 應用「手機號 + 驗證碼」弱驗證的方式獲取使用者身份證字號等個人重要資訊，利用使用者個人資訊更改手機服務密碼等，同時還會利用話術欺騙誘導電信企業客服人員將已掛失的電話卡進行解掛，利用部分網貸平台「找回使用者密碼」漏洞重置使用者支付密碼騙取網貸資金，最終造成使用者和企業信用與財產損失。事實上，不僅存在上述利用某些技術實現黑色/灰色產業鏈目的的黑色產業鏈團體，還有一些與普通線民息息相關的黑色/灰色產業鏈事件，這類事件的發生往往沒有任何專業技術的要求，只需要能夠操作手機即可，比如眾所皆知的兼職刷單等，這部分黑色/灰色產業鏈從事者往往並非特定黑色/灰色產業鏈團體，相反學生、無穩定工作的人都可能在不知情的情況下成為其中的一員。而追究這類群眾形成的原因，最終往往逃不過「利益」二字。

黑色產業鏈與風控就像是宿命敵人，黑色產業鏈作為攻擊方從各種角度繞過風控系統的防禦達到非法利益的獲取，作為防守方的風控團隊在防禦黑色產業鏈攻擊的過程中，會逐漸完善自身的防禦系統，從而保護產品的合法利益，甚至反客為主，追蹤並幫助打擊黑色產業鏈團體。當然，作為防守方，不僅需要防禦和阻止攻擊方的各種手段，而且需要在修復漏洞的同時做到業務系統本身的相容性與穩定性，同時還不能影響業務的正常進行。因此，相對于作為攻擊方的黑色產業鏈，一個完整的、可靠性高的風控防禦系統的建立其難度相對更高。

8.2.2 原始程式改機簡單實現

在以裝置為核心資源的網際網路時代，風控判斷使用者的真實性往往透過使用者是否使用真實的裝置來進行。接下來將透過沙盒的方式實現不需要切換真實裝置，但是 App 卻將修改前和修改後的系統認為是兩台裝置的效果，從而簡單介紹黑色產業鏈團體實現一鍵「新機」的原理。

那麼什麼是標識裝置身份的相關資訊呢？接下來將兩個獲取裝置資訊的應用安裝到手機上進行驗證。

如圖 8-24 和圖 8-25 所示，會發現關於裝置存在很多標識資訊，比如裝置指紋、裝置名稱、裝置型號、製造商和主機板、平台等，事實上如果對 Android 比較了解，就會發現以上這些資訊都是從 android.os.Build 這個類別的成員值獲取到的。這裡使用 Objection 注入「設定」應用並使用 WallBreaker 這一利器查看 android.os.Build 類別的類別結構，發現其類別成員都是靜態成員，且每一個成員的值都與圖 8-24 和圖 8-25 中的內容一一對應，比如圖 8-24 中的「裝置指紋」對應 Build 類別中的成員 FINGERPRINT，圖 8-24 和圖 8-25 中的「主機板」則對應 Build 類別中的 BOARD 成員的值等，具體 Build 類別中的成員值如程式清單 8-15 所示。

▲ 圖 8-24 裝置資訊 1

▲ 圖 8-25 裝置資訊 2

⬇ 程式清單 8-15 Build 類別結構

```
package android.os;

class Build {

    /* static fields */
    static String TAG; => Build
    static String TAGS; => release-keys
    static boolean IS_ENG; => false
    static String MODEL; => Nexus 5X
    static String CPU_ABI; => arm64-v8a
    static String CPU_ABI2; =>
    static String SERIAL; => 00abebcbe2a8ca31
    static String FINGERPRINT; => google/bullhead/bullhead:8.1.0/
OPM1.171019.011/4448085:user/release-keys
    static String USER; => android-build
    static String DEVICE; => bullhead
    static String BOARD; => bullhead
    static String BOOTLOADER; => BHZ31a
    static boolean IS_EMULATOR; => false
    static String RADIO; => M8994F-2.6.40.4.04
    static String[] SUPPORTED_ABIS; => arm64-v8a,armeabi-v7a,armeabi
    static long TIME; => 1510540425000
    static String[] SUPPORTED_32_BIT_ABIS; => armeabi-v7a,armeabi
    static String TYPE; => user
    static String[] SUPPORTED_64_BIT_ABIS; => arm64-v8a
    static boolean IS_USERDEBUG; => false
    static String ID; => OPM1.171019.011
    static boolean IS_DEBUGGABLE; => false
    static String HARDWARE; => bullhead
    static String HOST; => wphp10.hot.corp.google.com
    static boolean IS_CONTAINER; => false
    static boolean IS_USER; => true
    static String UNKNOWN; => unknown
    static boolean IS_TREBLE_ENABLED; => false
```

```
static String BRAND; => google
static String MANUFACTURER; => LGE
static boolean PERMISSIONS_REVIEW_REQUIRED; => false
static String PRODUCT; => bullhead
static String DISPLAY; => OPM1.171019.011

/* instance fields */

/* constructor methods */
Build();

/* static methods */
static boolean isBuildConsistent();
static String getString(String);
static String getRadioVersion();
static void ensureFingerprintProperty();
static String[] -wrap0(String, String);
static String getSerial();
static String deriveFingerprint();
static String -wrap1(String);
static String[] getStringList(String, String);
static long getLong(String);

/* instance methods */

}
```

那麼如何在 Android 開發過程中獲取這部分內容呢？透過反編譯以上兩個樣本發現，要獲取該類別中大部分成員的值，只需要直接透過類別存取即可，如程式清單 8-16 所示。

◯ 程式清單 8-16 樣本反編譯結果

```
package com.zhanhong.deviceinfo;
import android.os.Build;
```

```
...
public String getBOARD() {
    return Build.BOARD;
}
public String getBOOTLOADER() {
    return Build.BOOTLOADER;
}
public String getBRAND() {
    return Build.BRAND;
}
public String getBuilder() {
    return Build.USER + "@" + Build.HOST;
}
public String getDevice() {
    String v1 = Build.MANUFACTURER;
    String v0 = Build.MODEL;
    if(!v0.startsWith(v1)) {
        v0 = v1 + " " + v0;
    }
    return v0;
}
```

因此，如果要實現這部分資訊的改變，只需要修改對應的成員值即可，
讓我們直接來看對應 android.os.Build 類別的原始程式。

如圖 8-26 所示，在原始程式中，該類別的一些比較重要的成員實際
上都是在該類別被初始化時呼叫了 getString() 函數並傳遞對應的屬性
名稱得到的。進一步，如程式清單 8-16 所示，利用線上原始程式網
站追蹤 Android 8.1.0_r1 版本中 getString() 函數的實現會發現，該函
數具體是呼叫 SystemProperties 類別的 get() 函數獲取對應實現，而
SystemProperties.get() 函數具體又是如何進行函式呼叫的呢？這裡將其呼
叫鏈直接展示在程式清單 8-17 中，具體追蹤過程就不展開敘述了。

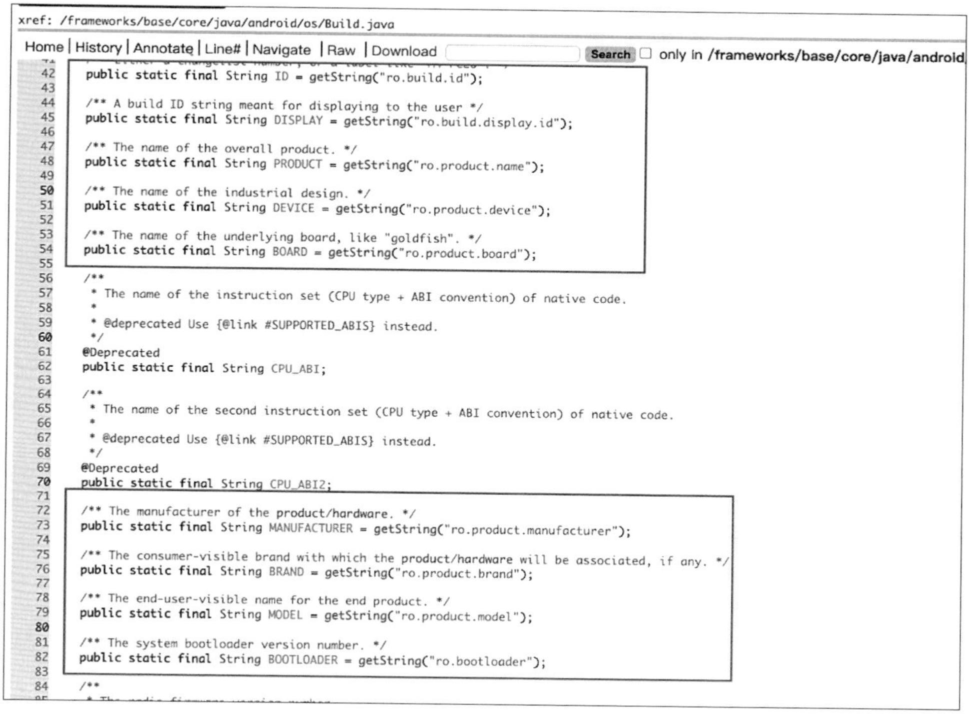

```
xref: /frameworks/base/core/java/android/os/Build.java

Home | History | Annotate | Line# | Navigate | Raw | Download [          ]  Search ☐ only in /frameworks/base/core/java/android
42    public static final String ID = getString("ro.build.id");
43
44    /** A build ID string meant for displaying to the user */
45    public static final String DISPLAY = getString("ro.build.display.id");
46
47    /** The name of the overall product. */
48    public static final String PRODUCT = getString("ro.product.name");
49
50    /** The name of the industrial design. */
51    public static final String DEVICE = getString("ro.product.device");
52
53    /** The name of the underlying board, like "goldfish". */
54    public static final String BOARD = getString("ro.product.board");
55
56    /**
57     * The name of the instruction set (CPU type + ABI convention) of native code.
58     *
59     * @deprecated Use {@link #SUPPORTED_ABIS} instead.
60     */
61    @Deprecated
62    public static final String CPU_ABI;
63
64    /**
65     * The name of the second instruction set (CPU type + ABI convention) of native code.
66     *
67     * @deprecated Use {@link #SUPPORTED_ABIS} instead.
68     */
69    @Deprecated
70    public static final String CPU_ABI2;
71
72    /** The manufacturer of the product/hardware. */
73    public static final String MANUFACTURER = getString("ro.product.manufacturer");
74
75    /** The consumer-visible brand with which the product/hardware will be associated, if any. */
76    public static final String BRAND = getString("ro.product.brand");
77
78    /** The end-user-visible name for the end product. */
79    public static final String MODEL = getString("ro.product.model");
80
81    /** The system bootloader version number. */
82    public static final String BOOTLOADER = getString("ro.bootloader");
83
84    /**
```

▲ 圖 8-26　Build 類別原始程式

�‣ 程式清單 8-17　getString() 函數實現關鍵函數追蹤

```
-> /frameworks/base/core/java/android/os/Build.java
private static String getString(String property) {
        return SystemProperties.get(property, UNKNOWN);
}


-> /frameworks/base/core/java/android/os/SystemProperties.java
public static String get(String key, String def) {
    if (TRACK_KEY_ACCESS) onKeyAccess(key);
    return native_get(key, def);
}
```

```
-> /frameworks/base/core/java/android/os/SystemProperties.java
private static native String native_get(String key, String def);

// 接下來就是Native層部分內容
-> /frameworks/base/core/jni/android_os_SystemProperties.cpp
static jstring SystemProperties_getS(JNIEnv *env, jobject clazz,
                                     jstring keyJ)
{
    return SystemProperties_getSS(env, clazz, keyJ, NULL);
}

-> /frameworks/base/core/jni/android_os_SystemProperties.cpp
static jstring SystemProperties_getSS(JNIEnv *env, jobject clazz,
                                      jstring keyJ, jstring defJ)
{
    int len;
    const char* key;
    char buf[PROPERTY_VALUE_MAX];
    jstring rvJ = NULL;
    // 這裡略去無關部分
    ...
    key = env->GetStringUTFChars(keyJ, NULL);
    // 關鍵獲取屬性函數
    len = property_get(key, buf, "");
    // ...
        rvJ = env->NewStringUTF(buf);
    } else {
        rvJ = env->NewStringUTF("");
    }
    env->ReleaseStringUTFChars(keyJ, key);

error:
    return rvJ;
}
```

```
-> /frameworks/rs/cpp/rsCppUtils.cpp
int property_get(const char *key, char *value, const char *default_value) {
    int len;
    // 呼叫__system_property_get函數
    len = __system_property_get(key, value);
    // ...
    if (default_value) {
        len = strlen(default_value);
        memcpy(value, default_value, len + 1);
    }
    return len;
}
```

從程式清單 8-17 中發現到最後實際上追蹤到了 __system_property_
get() 函數，這個函數的具體實現實際上是在 bionic/libc/bionic/system_
properties.cpp 檔案中，該檔案屬於 libc 基礎函數庫內容，而在追蹤該函
數實現時，發現該函數已經是關於裝置屬性資訊獲取的最底層函數—該
函數透過與 property_service_socket 裝置進行 Socket 通訊獲取具體屬性
值，這也是某些黑色產業鏈團夥在修改 ROM 時選擇該函數進行修改的原
因。如圖 8-27 所示是「永安線上」安全團隊在 2020 年黑色 / 灰色產業
鏈研究報告中提及的黑色產業鏈 ROM 改機的部分程式。

當然，這裡並不會直接對該函數進行修改，相反選擇最上層的 Build 類別
中的 getString() 函數進行修改。一方面，因為在這裡只是為了簡單地修
改 Build 類別中的成員值，而在對 Build 成員給予值的函數中，大都是透
過 getString() 函數獲取到的；另一方面，這部分改機部分實在過於敏感，
因此這裡僅介紹 Demo 等級的改機實現。這裡以部分成員值為例，最終
Build 類別中的 getString() 函數內容如程式清單 8-18 所示。

```
1  int __fastcall system_property_get(int a1, _BYTE *a2)
2  {
3    _BYTE *buffer1; // r4
4    const char *key; // r5
5    int result; // r0
6    _DWORD *v5; // r0
7
8    buffer1 = a2;
9    key = (const char *)a1;
10   if ( !shouldGjInLibc() )
11     goto LABEL_15;
12   if ( !j_strcmp((int)"init.svc.mtpd", (int)key) )
13   {
14 LABEL_11:
15     result = 0;                    rom作者新增的程式碼，原本android官方沒有這些
16     *buffer1 = 0;
17     return result;
18   }
19   if ( !j_strcmp((int)"sys.usb.config", (int)key) || !j_strcmp((int)"sys.usb.state", (int)key) )
20   {
21     buffer1[2] = 112;
22     *(_WORD *)buffer1 = 29805;
23     buffer1[3] = 0;
24     return 3;
25   }
26   if ( j_strcmp((int)"persist.sys.usb.config", (int)key) )
27   {
28 LABEL_15:
29     if ( isWgProp((int)key) )               判斷是否是自定義的屬性，如果呼叫的是自定的屬性獲取函
30       return wgzs_property_get(key, buffer1);  數，則獲取屬性值。這段程式碼也是作者新增的
31     v5 = (_DWORD *)j___system_property_find(key);
32     if ( v5 )
33       return j_j___system_property_read(v5, 0, (int)buffer1);  非自定義屬性獲取
34     goto LABEL_11;
35   }
36   *(_DWORD *)buffer1 = 0x656E6F6E;
37   buffer1[4] = 0;
38   return 4;
39 }
```

▲ 圖 8-27 某黑色產業鏈團體改機 ROM __system_property_get
函數的虛擬程式碼

● 程式清單 8-18 Build 類別

/** The consumer-visible brand with which the product/hardware will be
associated, if any. */
// 裝置名
public static final String BRAND = getString("ro.product.brand");
// 裝置製造商
public static final String MANUFACTURER = getString("ro.product.
manufacturer");
// 裝置主機板
public static final String BOARD = getString("ro.product.board");

```
private static String getString(String property) {
    String result = SystemProperties.get(property, UNKNOWN) ;
    if(property.equals("ro.product.brand")){
        result = new String("r0ysueBRAND");
    }else if(property.equals(("ro.product.manufacturer"))){
        result = new String("r0ysueMANUFACTUERER");
    }else if(property.equals("ro.product.board")){
        result = new String("r0ysueBOARD");
    }
    // 列印呼叫堆疊幫助定位在何處呼叫獲取裝置資訊
    Exception e = new Exception("r0ysueFINGERPRINT");
    e.printStackTrace();
    return result;
}
```

對比過多個相同類型裝置 Build 類別相關參數的讀者會發現，以上 Build 類別中大部分成員表示的裝置資訊非常有可能相互衝突，因此依賴上述資訊來標記裝置的唯一性實際上是略失偏頗的。在 Android 中，什麼資訊可以用於標識裝置的唯一性呢？

熟悉 Android 開發的讀者一定會脫口而出：IMEI（International Mobile Equipment Identity，國際行動裝置辨識碼）、IMSI（International Mobile Subscriber Identity，國際行動使用者辨識碼）、Android_id、SN，讓我們一一來介紹這些唯一性標識符。

（1）IMEI 是由 **15 位數字 ** 組成的「電子序號」，它與每台行動裝置一一對應，且該碼是全世界唯一的。對比每個人出生就有一個唯一的身份證字號，IMEI 由製造生產裝置的廠商所記錄，在裝置出場後無法再進行修改。如圖 8-28 所示，裝置 ID 可以透過在撥號鍵盤上輸入 *#06# 數字串查看。

▲ 圖 8-28　撥號鍵盤獲取裝置唯一 ID

如果協力廠商開發者要獲取裝置的唯一 ID，則需要在 AndroidManifest.
xml 檔案中宣告 android.permission.READ_PHONE_STATE 許可權（從
Android 10 開始，更是要求應用必須擁有 android.permission.READ_
PRIVILEGED_PHONE_STATE 隱私許可權），才可以在向使用者動態申請許
可權後，透過 TelephonyManager.getDeviceID() 或者 TelephonyManager.
getImei() 函數（Android 8 以上使用此函數）獲取唯一裝置 ID，具體在
Android 開發中獲取裝置唯一 ID 的方式如程式清單 8-19 所示。

◯ 程式清單 8-19　獲取裝置唯一 ID

```
import android.telephony.TelephonyManager;

final TelephonyManager telephonyManager = (TelephonyManager) context.
getSystemService(Context.TELEPHONY_SERVICE);
String imei="";
```

```
// 如果系統高於Android 8.0
if (android.os.Build.VERSION.SDK_INT >= 26) {
    imei=telephonyManager.getImei();
}
else
{
    imei=telephonyManager.getDeviceId();
}
```

確定在開發中如何修改 IMEI 後，如果不分析其底層實現，要實現對應原始程式的修改十分簡單，所需要做的只是找到目標 API 函數，然後直接修改該函數實現即可，這裡同樣僅簡單修改對應的函數返回值，最終對應函數修改後的內容如程式清單 8-20 所示。

需要注意的是，這裡修改的 IMEI 實際上是不符合 IMEI 格式標準的，具體這裡不再展開。

❍ 程式清單 8-20 IMEI 修改實現

```
//對應原始程式路徑：frameworks/base/telephony/java/android/telephony/
TelephonyManager.java
@Deprecated
@RequiresPermission(android.Manifest.permission.READ_PHONE_STATE)
public String getDeviceId() {
    try {
        ITelephony telephony = getITelephony();
        String result = telephony.getDeviceId(mContext.getOpPackageName());
        if (telephony == null)
            return null;
        //列印呼叫堆疊幫助定位在何處呼叫獲取IMEI
        Exception e = new Exception("r0ysueDeviceID");
        e.printStackTrace();
        //直接修改返回值
        return "r0ysueIMEI";
```

```
    } catch (RemoteException ex) {
        return null;
    } catch (NullPointerException ex) {
        return null;
    }
}
@RequiresPermission(android.Manifest.permission.READ_PHONE_STATE)
public String getImei() {
    String result = getImei(getSlotIndex());
    //列印呼叫堆疊幫助定位在何處呼叫獲取IMEI
    Exception e = new Exception("r0ysueDeviceID");
    e.printStackTrace();
    //直接修改返回值
    return "r0ysueIMEI";
}
```

（2）IMSI 用於區分行動網路中不同使用者在所有行動網路中不重複的辨識碼。IMSI 儲存在 SIM 卡中，可用於區別行動使用者的有效資訊，其總長度不超過 15 位，且同樣使用 0 ～ 9 的數字。與 IMEI 可以作為裝置的唯一辨識號類似，IMSI 同樣可以作為 SIM 卡的唯一身份標識，因為 SIM 卡總是要插入裝置啟用，一般來説 IMSI 和 IMEI 號會被一同組合用於辨識唯一裝置和使用者。

IMSI 由 行 動 國 家 號 碼（Mobile Country Code，MCC）、行 動 網 路 號碼（Mobile Network Code，MNC）和 行 動 使 用 者 辨 識 號 碼（Mobile Subscription Identification Number，MSIN）依次連接而成。其中 MCC 的長度為 3 位，在中國 MCC 的值為 460；MNC 的長度由 MCC 的值決定，可以是 2 位（歐洲標準）或 3 位數字（北美標準），比如中國移動的 MNC 值採用歐洲標準，其值為 00；MSIN 的值則由營運商自行分配，用以辨識某一行動通訊網中的行動使用者。以 IMSI 為 460001357924680

的 SIM 卡為例,其 MCC 值為 460,MNC 值為 00,剩下的 1357924680 則為中國移動為該 SIM 卡分配的唯一 ID。

與協力廠商 App 中獲取 IMEI 類似,如果要在協力廠商 App 中獲取 IMSI 號,同樣需要在 AndroidManifest.xml 檔案中宣告 android.permission. READ_PHONE_STATE 許 可 權(Android 10 上 也 需 要 宣 告 READ_ PRIVILEGED_PHONE_STATE 許可權),在動態申請許可權後,經過使 用者同意即可透過 telephonyManager.getSubscriberId() 函數獲取對應的 IMSI,具體程式如程式清單 8-21 所示。

◯ 程式清單 8-21 獲取 IMSI

```
import android.telephony.TelephonyManager;

final TelephonyManager telephonyManager=(TelephonyManager)context.
getSystemService(Context.TELEPHONY_SERVICE);
 //獲取IMSI
String imsi=telephonyManager.getSubscriberId();
```

最終 getSubscriberId() 函數原始程式被修改成如程式清單 8-22 所示。

◯ 程式清單 8-22 IMSI 函數修改

```
// 對應原始程式路徑:frameworks/base/telephony/java/android/telephony/
TelephonyManager.java
@RequiresPermission(android.Manifest.permission.READ_PHONE_STATE)
public String getSubscriberId() {
    String imsi =  getSubscriberId(getSubId());
    // 列印呼叫堆疊幫助定位在何處呼叫獲取IMSI
    Exception e = new Exception("r0ysueIMSI");
    e.printStackTrace();
    return "AAAABBBBIMSI";
}
```

（3）與上述兩者不同，android_id 是裝置第一次啟動時產生和儲存的一個 64bit 的數，也叫 SSAID（Settings.Secure.ANDROID_ID 的縮寫），在裝置正常執行過程中，其值是固定的，因此透過它可以知道裝置的壽命資訊。但是，一旦當裝置被更新軔體或者恢復出廠設定後，這個數就會被重置。在 Android 8 及更高的版本中，每一個應用在第一次安裝時都會根據應用簽名，系統使用者 ID（一般 ID 為 0）以及裝置資訊共同組合，根據一定演算法生成一個新的 android_id，也正因此，每一個不同的應用獲取到的 android_id 都是不同的（具體解釋見 https://developer.android.com/about/versions/oreo/android-8.0-changes?hl = zh-cn#privacy-all）。

暫且不論其唯一性問題，相較於獲取 IMEI 和 IMSI 需要獲取一定許可權，android_id 的獲取完全不需要任何許可權，協力廠商應用獲取 android_id 的具體方式如程式清單 8-23 所示。

⬇ 程式清單 8-23 獲取 android_id

```
import android.provider.Settings;

String android_id = Settings.Secure.getString(getContentResolver(),
                                Settings.Secure.ANDROID_ID);
```

雖然在開發中要獲取 ANDROID_ID 只需程式清單 8-23 中的一行程式即可，但實際上在找對應原始程式實現時，讀者首先要明白實際上獲取 ANDROID_ID 是透過 Settings 類別中的子類別 Secure 類別的 getString() 函數實現的，因此要找到對應原始程式實現，首先需要找到 Settings 類別的路徑，在找到該類別實現後，才能直接在對應檔案中找到 Secure.getString() 函數的實現，最終對應修改的程式如程式清單 8-24 所示。

⬤ 程式清單 8-24 android_id 獲取實現

```
// 路徑：frameworks/base/core/java/android/provider/Settings.java
public static String getString(ContentResolver resolver, String name) {
    // 如果是獲取ANDROID_ID
    if(name.equals(ANDROID_ID)){
        // 列印呼叫堆疊幫助定位在何處呼叫獲取IMSI
        Exception e = new Exception("r0ysueAndroid_id");
        e.printStackTrace();
        // 返回虛假的android_id
        return "r0syueAndroid_id";
    }
    return getStringForUser(resolver, name, UserHandle.myUserId());
}
```

（4）最後介紹一個在前面曾經一筆帶過的裝置標識符：SN（Serial Number）。可能讀者會奇怪，明明前文並未出現過，為什麼這裡還説在前面曾經介紹過呢？事實上，這裡所要介紹的 SN 對應的就是前面介紹的 android.os.Build 類別中的 SERIAL 成員，通常是手機生產廠商提供的裝置序號，是為了驗證產品合法而存在的，用來保障使用者的正版權益和合法服務，其具體格式由生產廠商自訂，每一個生產廠商的 SN 的格式都可能不同。

要獲取 SN，不同系統版本的獲取方式不盡相同。Android 8 以下獲取裝置序號無須申請任何許可權，只需要透過 Build.SERIAL 的方式獲取即可；而在 Android 8～10 的裝置中，如果要獲取 SN，則需要事先在 AndroidManifest.xml 檔案中宣告 READ_PHONE_STATE 許可權，在使用者同意授權後，才能透過 Build.getSerial() 函數獲取到裝置序號；在 Android 10 以上的裝置中，透過 Build.getSerial() 函數獲取到的裝置序號甚至可能是 unknown 或者直接抛出安全異常崩潰。暫且不論無法獲取到

SN 的情況，程式清單 8-25 僅介紹在 Android 10 以下的裝置中獲取 SN 的方法。

● 程式清單 8-25　獲取 SN

```
import android.os.Build;

if (Build.VERSION.SDK_INT >= Build.VERSION_CODES.O) {
    return Build.getSerial();
} else {
    return Build.SERIAL;
}
```

當然，SN 除了可以透過程式方式獲取外，還可以利用 ADB Shell 使用如下兩種方式獲取，如圖 8-29 所示。

```
$ getprop ro.serialno
$ getprop ro.boot.serialno
```

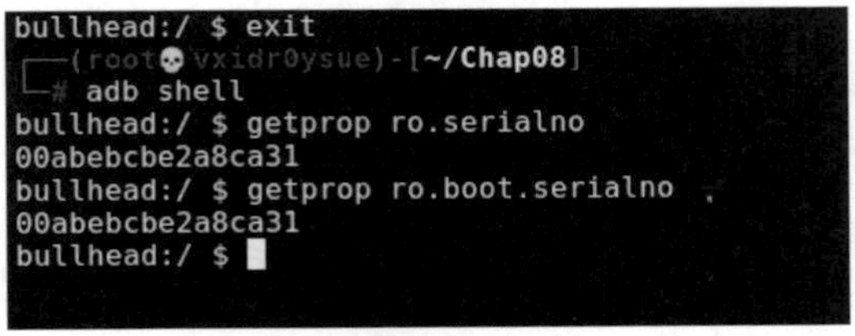

▲ 圖 8-29　ADB Shell 獲取 SN

對比透過 getprop 方式獲取到的 SN 結果和程式清單 8-15 中對應 SERIAL 變數的值，會發現其結果完全一致。

暫且不討論透過 ADB Shell 獲取 SN 的方式，這裡僅討論 Java 程式的實現，與前文介紹的裝置資訊的修改相同，SN 的篡改同樣需要修改 Build 類別實現，最終 SN 修改相關程式如程式清單 8-26 所示。

◆ 程式清單 8-26　SN 修改實現

```java
// 程式路徑：/frameworks/base/core/java/android/os/Build.java

@Deprecated
public static final String SERIAL = getString("no.such.thing");

@RequiresPermission(Manifest.permission.READ_PHONE_STATE)
public static String getSerial() {
    IDeviceIdentifiersPolicyService service =
IDeviceIdentifiersPolicyService.Stub
            .asInterface(ServiceManager.getService(Context.DEVICE_
IDENTIFIERS_SERVICE));
    try {
        String result =service.getSerial();
        // 列印呼叫堆疊幫助定位在何處呼叫獲取Serial
        Exception e = new Exception("r0ysueSerial");
        e.printStackTrace();
        // 直接返回虛假的Serial
        return "r0ysueserial1234";
    } catch (RemoteException e) {
        e.rethrowFromSystemServer();
    }
    return UNKNOWN;
}
private static String getString(String property) {
    String result = SystemProperties.get(property, UNKNOWN) ;
    // 事實上在Android 8.1系統版本中，這個if分支永遠不會進入
    if(property.equals("no.such.thing")){
        result = new String("r0ysueAAAABBBBCCCCDDDD");
    }
    Exception e = new Exception("r0ysueFINGERPRINT");
    e.printStackTrace();
    return result;
}
```

在修改完畢 SN 的原始程式後，改機的簡單實現暫且告一段落。接下來對原始程式進行編譯並更新韌體，這個過程在前面的章節中已經介紹過多次，此處略過，這裡直接舉出最終在改機後再次使用上述樣本測試的結果，如圖 8-30 所示。

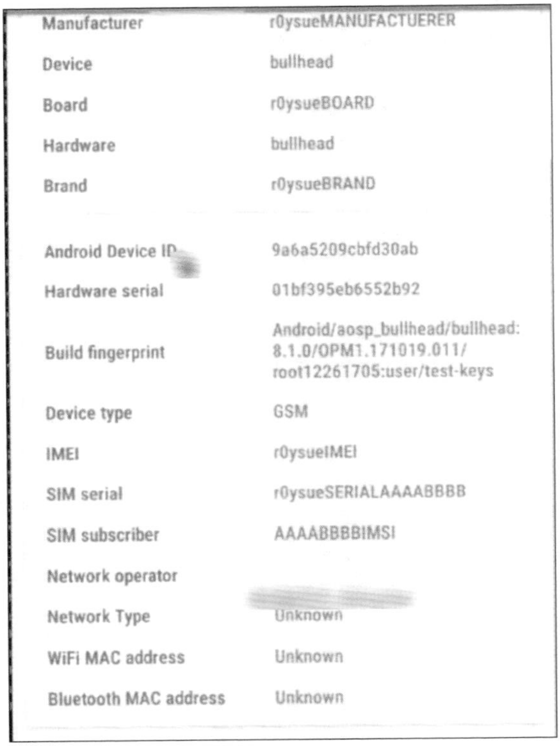

Manufacturer   r0ysueMANUFACTUERER

Device   bullhead

Board   r0ysueBOARD

Hardware   bullhead

Brand   r0ysueBRAND

Android Device ID   9a6a5209cbfd30ab

Hardware serial   01bf395eb6552b92

Build fingerprint   Android/aosp_bullhead/bullhead:
                      8.1.0/OPM1.171019.011/
                      root12261705:user/test-keys

Device type   GSM

IMEI   r0ysueIMEI

SIM serial   r0ysueSERIALAAAABBBB

SIM subscriber   AAAABBBBIMSI

Network operator

Network Type   Unknown

WiFi MAC address   Unknown

Bluetooth MAC address Unknown

▲ 圖 8-30「改機」效果

顯而易見，真實黑色產業鏈的改機工具不可能如此簡單，如果使用在這一小節中修改後編譯的系統，甚至只要 App 檢測獲取到的資訊是否合法即可辨識出 ROM 修改的痕跡，更不用説繞過風控檢測了。當然，真實改機實現肯定更加複雜、智慧，在這裡僅是做粗略的技術討論，更深入的學術研究留待讀者朋友自行完成，毋庸置疑，筆者不建議讀者利用以上相關技術做任何黑色 / 灰色產業鏈行為。

8.3 本章小結

經過本章的學習，相信讀者會發現事實上在系統函數庫中能夠實現的事情很多：幫助實現抓取封包、輔助中間人抓取封包、原始程式改機等，甚至如果在 Java.io.File 等檔案類別、Java.lang.String 關鍵字串類別的相關函數中列印日誌，對於系統來說上層 App 所有 Java 層的檔案和字串相關內容幾乎毫無隱私可言。當然，如果覺得這種直接在 File 檔案類別列印日誌的方式得到的結果太過容錯，還可以只專注 SharedPreferences 等特定檔案類別中的函數進行日誌列印，從而減少無關日誌資訊干擾。如果 App 想要實現不被系統底層窺探，就需要如本章介紹的協力廠商函數庫實現 SSL Pinning 證書綁定的方式一樣，儘量將所有關鍵功能完全交由應用自身實現，以減少對系統的依賴，而非簡單地呼叫系統提供的 API，這也是如今 Frida 應用逆向難（自實現了虛擬機器，不依賴 art 虛擬機器解析指令）、自訂 OpenSSL 庫無法使用 r0capture 抓取封包的原因。當然，這對應用程式開發者的水準提出了一定的挑戰。

同時，對比使用 Frida 對系統 API 進行 Hook 和直接在系統對應函數原始程式中增加日誌列印兩種方式，其實後者才是更加穩定且不易檢測的方式。比如使用 Frida Hook String 類別的 toString() 函數，大機率對如此基礎的函數進行 Hook 會引起 App 崩潰，但如果選擇在原始程式中列印日誌的方式，其帶來的副作用最多就是系統執行較慢而已。但相對的，由於原始程式修改本身針對的是整個系統，而非單一 App，導致如果修改的是頻繁被呼叫的函數，那麼對應產生的日誌資訊會十分容錯且不容易定位；同時，原始程式修改後要生效，每次都需要重新編譯和更新韌體，其過程相對複雜，而 Frida 這類只針對單一 App、靈活注入的方式更加亮眼，因此讀者在不同場景使用時建議因時而定、因地制宜。

Android 協定分析之收費
直播間逆向分析

前面的章節中我們已經學習了 Frida 和 Xposed 工具的使用方式，
並介紹了二者在逆向過程中的作用；同時，還介紹了逆向「大殺
器」—沙盒。但俗話説，「讀萬卷書，不如行萬里路」，本章我們透過一
些實戰案例來深入理解前面介紹的理論，做到真正的知行合一。

9.1 VIP 功能繞過

作為逆向工程師，通常在拿到一個大部分功能需要付費才能正常使用的
樣本時，第一步想要做的就是完成樣本限制功能的繞過，這裡使用的樣
本也不例外。

由於樣本違法成分過多，這裡僅做文字功能描述：A 品牌 App 是一個直
播類型的應用樣本，但每次開啟直播間都會發現存在兩個限制：第一，

彈出視窗提示 VIP 才能正常使用觀看收費視訊功能;第二,非 VIP 使用者觀看視訊存在 15 秒的時間限制。

由於樣本 App 是經過保護處理的,這裡為了後續的靜態分析過程順利,先使用 frida-dexdump 完成脫殼處理。使用時將電腦透過 ADB 連接上手機並將 frida-server 啟動,在 frida-server 執行成功後,執行目標 App 並使之處於前臺,然後透過 Python 執行對應脫殼指令稿,即可得到脫殼完畢的 DEX 檔案,最終脫殼效果如圖 9-1 所示。

▲ 圖 9-1 dexdump 脫殼

在脫殼完成後,即可使用 Jadx-gui 將脫殼得到的全部 DEX 檔案匯入備用。

回到正題,首先繞過第一個限制:彈出視窗提示。

相信看過第 4 章的讀者都知道,彈出視窗這種類型的限制是非常容易繞過的。由於彈出視窗本身是透過 Android 系統提供的 API 實現的,因此彈出視窗的程式中不可避免地會呼叫系統提供的 API。採用系統 API 實現的彈出視窗方式無非就幾種,其中通用的彈出視窗 API 為 android.app.

Dialog 類別，要透過這個類別實現視窗的彈出，有一定開發經驗的讀者一定知道是透過呼叫該類別的 show 函數實現的。這裡為了驗證是不是透過 Dialog 類別的 show 函數實現的，我們直接使用如下 Objection 命令進行測試（注意，如果是使用 Objection 1.8.4 版本 Hook 特定函數，一定要加上對返回值、參數或呼叫堆疊的列印，否則會顯示出錯。另外，這裡為了定位 App 業務層實現彈出視窗的部分，一定要加上對呼叫堆疊的列印，也就是 --dump-backtrace 參數），最終在開啟新的收費直播間時，Objection 的 Hook 結果如圖 9-2 所示。

```
android hooking watch class_method android.app.Dialog.show --dump-backtrace
--dump-return
```

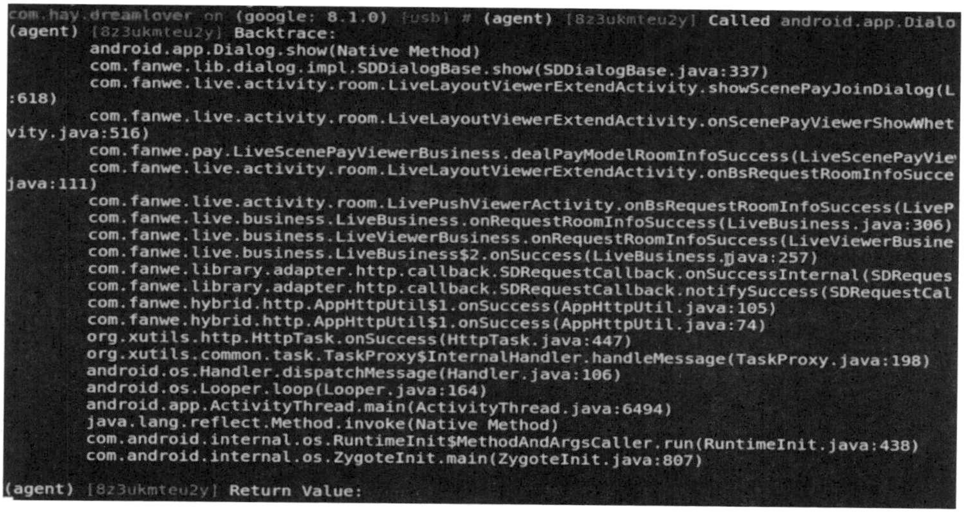

▲ 圖 9-2 Objection 的 Hook 結果

根據圖 9-2 的 Hook 結果，最終定位到業務層實現彈出視窗的函數為 SDDialogBase.show()，進而得到移除彈出視窗的 Frida 指令稿程式如程式清單 9-1 所示。

● 程式清單 9-1 hookVIP.js

```
setImmediate(function(){
    Java.perform(function(){
        console.log("Entering hook")
        // 移除彈框
        Java.use("com.fanwe.lib.dialog.impl.SDDialogBase").show.
implementation = function(){
            console.log("hook show ")
        }
    })
})
```

在繞過 VIP 彈出視窗限制後,讓我們繼續完成倒計時限制的繞過工作。

事實上,在繞過 VIP 彈出視窗限制的過程中,如果一步一步地使用 Jadx-gui
查看呼叫堆疊中 App 相關的程式,就會發現其呼叫堆疊由下至上依次用於
接收網路資料、處理資料以及彈出視窗提示相關函數,如圖 9-3 所示。

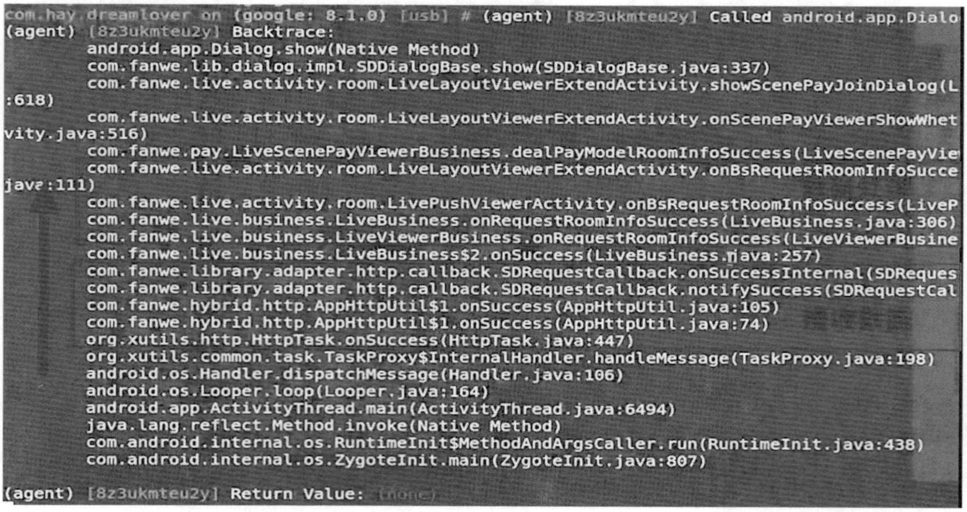

▲ 圖 9-3 Objection Hook 結果

在分析 onBsRequestRoomInfoSuccess() 函數時發現，該函數中呼叫的函數 dealPayModelRoomInfoSuccess(actModel) 又再次呼叫了 onTimePayViewerShowCoveringPlayeVideo() 函數用於設定 是否僅播放聲音（is_only_play_voice）、視訊播放倒計時（countdown）以及視訊網址（preview_play_url）等視訊相關參數，其具體程式如程式清單 9-2 所示。

⬡ 程式清單 9-2 設定時間

```
@Override
public void onBsRequestRoomInfoSuccess(App_get_videoActModel actModel) {
    super.onBsRequestRoomInfoSuccess(actModel);
    // 倒計時控制函數
    getTimePayViewerBusiness().dealPayModelRoomInfoSuccess(actModel);
    getScenePayViewerBusiness().dealPayModelRoomInfoSuccess(actModel);
}
public void dealPayModelRoomInfoSuccess(App_get_videoActModel actModel) {
    UserModel user = UserModelDao.query();
    if (user != null) {
        // 判斷是否付費
        if (!user.getUser_id().equals(actModel.getUser_id()) && actModel.
getIs_live_pay() == 1 && actModel.getLive_pay_type() == 0) {
            LogUtil.i("is_pay_over:" + actModel.getIs_pay_over());
            if (actModel.getIs_pay_over() == 1) {
                agreePay();
                return;
            }
            setCanJoinRoom(false);
            isPreViewPlaying(actModel);
            this.businessListener.onTimePayViewerShowCoveringPlayeVideo(act
Model.getPreview_play_url(), actModel.getCountdown(), actModel.getIs_only_
play_voice()); // 設定觀看時間函式呼叫
            startMonitor();
        }
    }
```

```
}
public void onTimePayViewerShowCoveringPlayeVideo(String preview_play_url,
int countdown, int is_only_play_voice) {
    showPayModelBg();
    this.payLiveBlackBgView.setIs_only_play_voice(is_only_play_voice);
    // 設定觀看時間
    this.payLiveBlackBgView.setProview_play_time(countdown * 1000);
    this.payLiveBlackBgView.startPlayer(preview_play_url);
}
```

進一步觀察 onTimePayViewerShowCoveringPlayeVideo() 函數會發現，關於視訊的相關參數實際上都是透過 payLiveBlackBgView 實例中的函數去控制的，該實例對應的完整類別名為 com.fanwe.pay.appview.PayLiveBlackBgView。

此時，使用 Objection 中關於 Hook class 的命令對該類別中的所有函數進行 Hook，以驗證在開啟直播間時其中的函數被呼叫的情況。最終可以發現其 Hook 結果其實和靜態分析結果一致，如圖 9-4 所示。

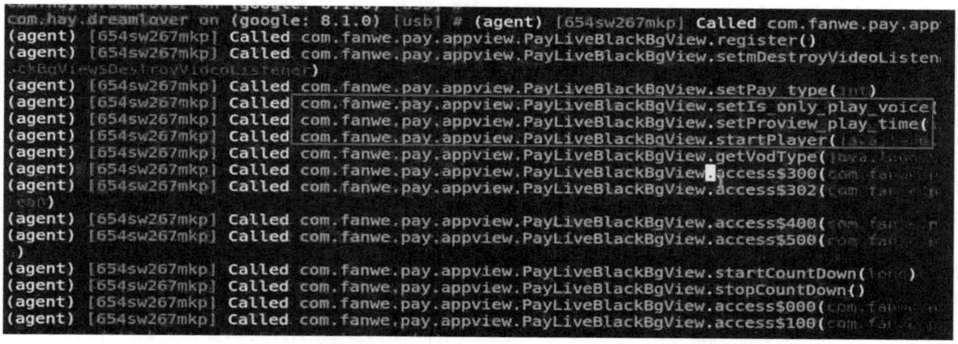

▲ 圖 9-4 PayLiveBlackBgView Hook 結果

繼續 Hook 這 3 個相關函數時，會發現 setProview_play_time 函數是最終用於設定剩餘觀看時長的關鍵函數，因此如果要繞過視訊觀看時長的

限制，只需 Hook 這個函數，並將其參數設定成相對大的數字即可，這裡
最終繞過時間限制的 Hook 指令稿如程式清單 9-3 所示。

◉ 程式清單 9-3 hookVIP.js

```
setImmediate(function(){
    Java.perform(function(){
        console.log("Entering hook")
        // 移除彈框
        ...
        // 移除倒計時限制，第一種方法
        Java.use("com.fanwe.pay.appview.PayLiveBlackBgView")
            .setProview_play_time.implementation = function(x){
                console.log("Calling setProview_play_time ")
                // 設定倒計時為1000* 3600秒
                return this.setProview_play_time(1000*3600)
            }
    })
})
```

至此，其實上述 VIP 相關限制都已經繞過了。但是實際上在繞過播放時
長的限制時，選擇的 Hook 函數—setProview_play_time() 只是用於設
定所在實例的一個成員變數而已，並不是真實用於計時的函數，在後續
的分析中發現，真實用於計時的函數實際上是圖 9-4 中倒數第 4 個函數
startCountDown()（這裡透過對比 startCountDown 函數中的相關中文字
串以及 App 中直播間顯示的文字可以確定）。startCountDown() 函數程式
具體內容如程式清單 9-4 所示。

◉ 程式清單 9-4 startCountDown 函數

```
public void startCountDown(long time) {
        stopCountDown();
        if (time > 0) {
            this.countDownTimer = new CountDownTimer(time, 1000) {
```

```
            public void onTick(long leftTime) {
                String time;
                if (PayLiveBlackBgView.this.pay_type == 1) {
                    time = "該直播按場收費，您還能預覽倒計時:" +
(leftTime / 1000) + "秒";
                } else {
                    time = "1分鐘內重複進入，不重複扣款，請能正常預覽視訊
後，點擊進入，以免扣款後不能正常進入，您還能預覽倒計時:"+(leftTime / 1000) +
"秒";
                }
                PayLiveBlackBgView.this.tv_time.setText(time);
            }

            public void onFinish() {
                String time;
                if (PayLiveBlackBgView.this.pay_type == 1) {
                    time = "該直播按場收費，您還能預覽倒計時:0秒";
                } else {
                    time = "1分鐘內重複進入，不重複扣款，請能正常預覽視訊
後，點擊進入，以免扣款後不能正常進入，您還能預覽倒計時:0秒";
                }
                PayLiveBlackBgView.this.tv_time.setText(time);
                if (PayLiveBlackBgView.this.mDestroyVideoListener !=
null) {
                    PayLiveBlackBgView.this.mDestroyVideoListener.
destroyVideo();
                }
                PayLiveBlackBgView.this.destroyVideo();
            }
        };
        this.countDownTimer.start();
    }
}
```

因此，秉持著 Hook 時離資料越近越好的原則，真正最後用於繞過觀看視

訊時間限制的 Hook 程式如程式清單 9-5 所示。

⬛ 程式清單 9-5 hookVIP.js

```
setImmediate(function(){
    Java.perform(function(){
        console.log("Entering hook")
        // 移除彈框
        ...
        // 移除倒計時，第一種方法
        ...
        // 移除倒計時，第二種方法
        Java.use("com.fanwe.pay.appview.PayLiveBlackBgView").
startCountDown.implementation = function(x){
            console.log("Calling countdown ")
            return this.startCountDown(1000*3600)
        }
    })
})
```

9.2 協定分析

在進行協定分析之前，首先要解決的是抓取封包問題。為此，這裡首先將虛擬機器設定為「橋接模式」並將手機和電腦連線同一網路環境下，然後設定手機端的 VPN 代理，使得執行在電腦中的 Charles 軟體能夠順利完成手機上網路流量的抓取。

但是在測試時發現，無論在 App 啟動前還是在 App 登入成功後的任意時刻，只要手機上開啟 VPN 代理，就會出現如圖 9-5 所示的無法正常使用 App 的狀態。當然，如果使用 WiFi 代理，經過筆者測試，也會出現一樣的結果。

▲ 圖 9-5 開啟 VPN 狀態下無法正常使用 App

遇到這種情況，大機率可以認為是因為 App 對抓取封包做了對抗，比如透過 java.net.NetworkInterface.getName() 函數判斷當前處於活躍狀態的網路卡是不是 tun0，從而檢測當前是不是透過 VPN 代理方式上網的。

但這裡並未仔細研究其中的對抗方式，而是轉而使用 r0capture 這種 Hook 抓取封包方式完成後續的抓取封包。

這裡順帶提一下，由於 r0capture 工具本身基於 Frida，如果使用 r0capture 對應用進行 spawn 模式抓取封包時，發現應用無法正常啟動或者異常崩潰，可以採取以下 3 種方式解決：

- 放棄使用 spawn 模式，轉而透過 attach 模式對應用進行注入抓取封包。
- 透過 -w 參數設定 spawn 模式的延遲時間。
- 放棄 r0capture 外層的 Python 包裝，直接透過命令列將 script.js 指令稿注入應用，實現 spawn 模式抓取封包。

最終使用 r0capture 成功完成資料封包的抓取，請求的 URL 為 hhy2. hhyssing.com:46288/mapi/index.php，使用的通訊協定為 HTTP 協定，存在 requestData、i_type、ctl 等參數資訊，其中 requestData 欄位資訊明顯處於加密狀態，抓取到的資料封包內容如圖 9-6 所示。

```
SSL Session:
[HTTP_send] 127.0.0.1:48858 ──→ 127.0.0.1:46288
00000000: 50 4F 53 54 20 2F 6D 61   70 69 2F 69 6E 64 65 78   POST /mapi/index
00000010: 2E 70 68 70 20 48 54 54   50 2F 31 2E 31 0D 0A 43   .php HTTP/1.1..C
00000020: 6F 6F 6B 69 65 3A 20 79   75 6E 73 75 6F 5F 73 65   ookie: yunsuo_se
00000030: 73 73 69 6F 6E 5F 76 65   72 69 66 79 3D 62 66 39   ssion_verify=bf9
00000040: 64 33 33 33 33 63 64 63   32 38 35 65 39 62 34 31   d3333cdc285e9b41
00000050: 35 37 61 35 30 31 66 38   35 64 35 66 66 3B 20 50   57a501f85d5ff; P
00000060: 48 50 53 45 53 53 49 44   3D 68 63 76 72 67 74 6C   HPSESSID=hcvrgtl
00000070: 35 73 39 69 69 74 34 6F   6A 68 71 37 75 65 74 65   5s9iit4ojhq7uete
00000080: 6B 66 30 0D 0A 58 2D 4A   53 4C 2D 41 50 49 2D 41   kf0..X-JSL-API-A
00000090: 55 54 48 3A 20 73 68 61   31 7C 31 36 31 31 32 38   UTH: sha1|161128
000000A0: 31 34 32 34 7C 35 35 4F   34 30 69 34 4B 76 58 30   1424|55040i4KvX0
000000B0: 30 6F 7C 35 30 63 38 33   66 34 35 32 39 64 36 61   0o|50c83f4529d6a
000000C0: 31 31 63 31 64 31 30 62   63 61 38 30 35 35 66 35   11c1d10bca8055f5
000000D0: 31 34 30 63 30 35 32 37   34 65 63 0D 0A 43 6F 6E   140c05274ec..Con
000000E0: 74 65 6E 74 2D 54 79 70   65 3A 20 61 70 70 6C 69   tent-Type: appli
000000F0: 63 61 74 69 6F 6E 2F 78   2D 77 77 77 2D 66 6F 72   cation/x-www-for
00000100: 6D 2D 75 72 6C 65 6E 63   6F 64 65 64 3B 63 68 61   m-urlencoded;cha
00000110: 72 73 65 74 3D 55 54 46   2D 38 0D 0A 43 6F 6E 74   rset=UTF-8..Cont
00000120: 65 6E 74 2D 4C 65 6E 67   74 68 3A 20 34 39 38 0D   ent-Length: 498.
00000130: 0A 55 73 65 72 2D 41 67   65 6E 74 3A 20 44 61 6C   .User-Agent: Dal
00000140: 76 69 6B 2F 32 2E 31 2E   30 20 28 4C 69 6E 75 78   vik/2.1.0 (Linux
00000150: 3B 20 55 3B 20 41 6E 64   72 6F 69 64 20 31 30 3B   ; U; Android 10;
00000160: 20 50 69 78 65 6C 20 32   20 58 4C 20 42 75 69 6C    Pixel 2 XL Buil
00000170: 64 2F 51 51 33 41 2E 32   30 30 38 30 35 2E 30 30   d/QQ3A.200805.00
00000180: 31 29 0D 0A 48 6F 73 74   3A 20 68 68 79 32 2E 68   1)..Host: hhy2.h
00000190: 68 79 73 73 69 6E 67 2E   63 6F 6D 3A 34 36 32 38   hyssing.com:4628
000001A0: 38 0D 0A 43 6F 6E 6E 65   63 74 69 6F 6E 3A 20 4B   8..Connection: K
000001B0: 65 65 70 2D 41 6C 69 76   65 0D 0A 41 63 63 65 70   eep-Alive..Accep
000001C0: 74 2D 45 6E 63 6F 64 69   6E 67 3A 20 67 7A 69 70   t-Encoding: gzip
000001D0: 0D 0A 0D 0A 72 65 71 75   65 73 74 44 61 74 61 3D   ....requestData=
000001E0: 25 32 46 54 44 55 25 32   46 48 6E 37 41 75 58 50   %2FTDU%2FHn7AuXP
000001F0: 70 66 56 51 79 43 49 30   71 49 72 68 73 51 25 32   pfVQyCI0qIrhsQ%2
00000200: 46 49 69 6B 79 25 32 42   35 39 63 58 50 62 30 4D   FIiky%2B59cXPb0M
00000210: 54 53 63 72 31 68 67 61   79 4A 38 49 37 33 4F 62   TScr1hgayJ8I73Ob
00000220: 36 55 6E 63 79 5A 69 58   63 33 47 32 51 73 36 63   6UncyZiXc3G2Qs6c
00000230: 78 63 79 37 25 30 41 25   32 42 62 74 50 6C 53 6C   xcy7%0A%2BbtPlSl
```
× Find: login   ↓ Next   ↑ Previous   ⠿ Highlight All   M

▲ 圖 9-6　抓取的 r0capture 資料封包內容

圖 9-6 中存在一個很奇怪的資訊：觀察資料封包 IP 資訊會發現，竟然是透過 127.0.0.1 本地回路位址的某個通訊埠發給另一個通訊埠的。

經過研究發現，之所以出現本地與本地的通訊，實際上是因為某大型公司的遊戲盾的作用—透過將資料連接當地語系化以減少被網路攻擊的風險並防止 DDOS/CC 攻擊。

此時，如果想知道真實的伺服器位址，可以透過將測試手機刷入 Kali NetHunter 並透過其帶來的完整 Linux 環境中的 jnettop 命令獲取對應資料伺服器位址。如圖 9-7 所示是透過 SSH 連接手機上的 Kali 環境，最終透過 jnettop 命令定位的真實播放視訊的 IP 位址。

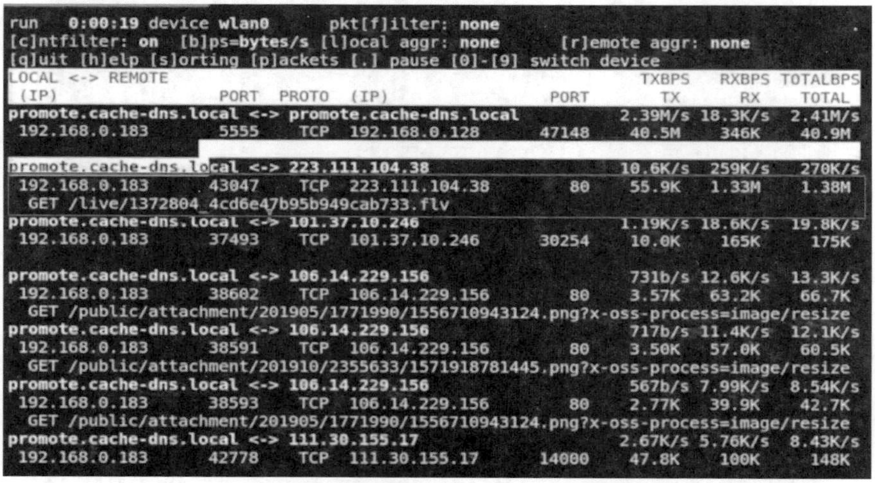

▲ 圖 9-7 jnettop 查看真實通訊伺服器位址

除此之外，還可以透過 VNC 連接刷入 Kali NetHunter 的手機系統，並透過手機上的 Wireshark 抓取本地回路位址的資料封包確定；或者使用將手機連接到透過無線網路卡自製的路由器，這樣就能夠掌握所有經由該無線網路卡的資料封包。

在解決了抓取封包問題後，讓我們正式開始針對該樣本的協定分析，這裡同樣以獲取直播間相關資訊的介面為例。

如圖 9-2 所示，在分析 9.1 節中彈出視窗限制的呼叫堆疊中的函數時，發現在接收網路通訊資料時有一個關鍵類別 AppHttpUtil。但奇怪的是，之前使用 dexdump 脫下的 DEX 檔案並不存在該類別的宣告，轉而使用 frida_fart 重新脫殼，最終得到 AppHttpUtil 類別的內容。

結合圖 9-2 中的 Hook 結果，靜態分析 AppHttpUtil 類別中的函數會發現真實用於發送資料封包的函數為 getImpl() 和 postImpl()，而以上兩個函數在執行過程中都會呼叫 parseRequestParams() 函數，parseRequestParams() 函數的內容如程式清單 9-6 所示。

🔻 **程式清單 9-6 處理請求資料函數 parseRequestParams()**

```
public RequestParams parseRequestParams(SDRequestParams params) {
    String ctl = String.valueOf(params.getCtl());
    String act = String.valueOf(params.getAct());
    StringBuilder url = new StringBuilder(params.getUrl());
    if (!"http://www.xxx.com/app.php?act=init".equals(url.toString())) {
        String otherUrl = AppRuntimeWorker.getApiUrl(ctl, act);
        if (!TextUtils.isEmpty(otherUrl)) {
            url = new StringBuilder(otherUrl);
        }
    }
    RequestParams request = new RequestParams(url.toString());
    if (SDResourcesUtil.getResources().getInteger(R.integer.is_open_auth_
token) == 1) {
        //  header關鍵參數
        request.addHeader("X-JSL-API-AUTH", LvBokeAuthUtils.getToken(url.
toString()));
    }
    printUrl(params);
    try {
        // signqt參數形成關鍵函數，分析後發現實際上是固定值
```

```
        params.put("signqt", MD5Util.MD5(this.settings.
getString("angeloip098", "") + "&*" + this.settings.getString("6969dolkoh",
"") + "()" + ctl + "+_" + act + "@!@###@"));
        // timeqt參數，獲取當前時間
        params.put("timeqt", System.currentTimeMillis() + "");
    } catch (Exception e) {
        Log.e("sodsb", "01");
    }
    Map<String, Object> data = params.getData();
    if (!data.isEmpty()) {
        String encryptData = null;
        int requestDataType = params.getRequestDataType();
        switch (requestDataType) {
            case 0:
                for (Map.Entry<String, Object> item : data.entrySet()) {
                    String key = item.getKey();
                    Object value = item.getValue();
                    if (value != null) {
                        request.addBodyParameter(key, String.valueOf(value));
                    }
                }
                break;
            case 1:
                String json = SDJsonUtil.object2Json(data);
                long lrtt = Long.parseLong(ApkConstant.getAeskeyHttp()) +
234512;
                // 呼叫AESUtils.encrypt()函數對上述資料進行加密，key為
8648754518945235
                encryptData = AESUtil.encrypt(json, "8648754518945235");
                break;
        }
        if (requestDataType != 0) {
            request.addBodyParameter("requestData", encryptData);
            request.addBodyParameter("i_type", String.
```

```
valueOf(requestDataType));
            request.addBodyParameter("ctl", ctl);
            request.addBodyParameter("act", act);
            if (data.containsKey("itype")) {
                request.addBodyParameter("itype", String.valueOf(data.
get("itype")));
            }
        }
    }
    Map<String, SDFileBody> dataFile = params.getDataFile();
    if (!dataFile.isEmpty()) {
        request.setMultipart(true);
        for (Map.Entry<String, SDFileBody> item2 : dataFile.entrySet()) {
            SDFileBody fileBody = item2.getValue();
            request.addBodyParameter(item2.getKey(), fileBody.getFile(),
fileBody.getContentType(), fileBody.getFileName());
        }
    }
    List<SDMultiFile> listFile = params.getDataMultiFile();
    if (!listFile.isEmpty()) {
        request.setMultipart(true);
        for (SDMultiFile item3 : listFile) {
            SDFileBody fileBody2 = item3.getFileBody();
            request.addBodyParameter(item3.getKey(), fileBody2.getFile(),
fileBody2.getContentType(), fileBody2.getFileName());
        }
    }
    return request;
}
```

為了確定該類別在發生網路請求的過程中是否被呼叫，可以使用 Objection Hook AppHttpUtil 類別，最終 Hook 結果如圖 9-8 所示。

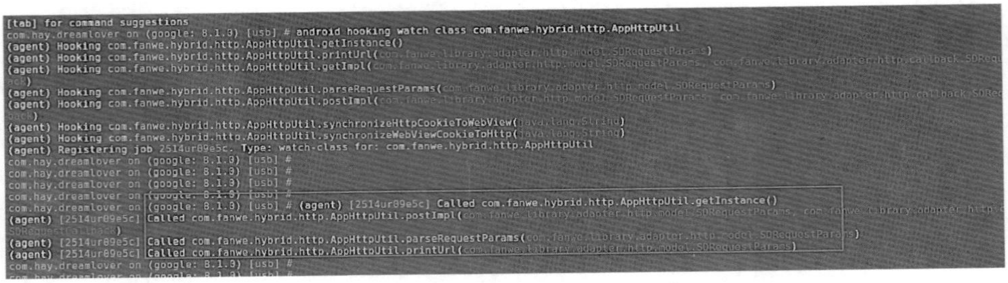

▲ 圖 9-8 Objection Hook AppHttpUtil 類別

在確定 parseRequestParams 函數在網路請求過程中被呼叫後，透過 Hook parseRequestParams 函數得到對應的請求呼叫鏈，如圖 9-9 所示。

▲ 圖 9-9 Objection Hook parseRequestParams 函數

對比圖 9-9 中 parseRequestParams 函數中的參數和返回值與 r0capture 抓取的資料封包，會發現其實該函數的返回值就是請求的資料封包內容，因此要實現最終的離線請求指令稿，就要從 parseRequestParams 函數入手。

仔細分析 parseRequestParams 函數的內容，會發現發送的資料封包中的 requestData 欄位是由多個欄位組合後，透過呼叫 AESUtil.encrypt 函數加密組成最後的 Base64 格式的發送封包資料的。

如程式清單 9-7 所示,追蹤 AESUtil.encrypt 函數的實現,會明顯發現這是一個標準的 AES ECB 模式加密。透過查看前幾章中的密碼沙盒的記錄檔也可以進一步確定,其內容如圖 9-10 所示。

◉ 程式清單 9-7 AESUtil.encrypt 函數內容

```java
public static String encrypt(String content, String key) {
    byte[] encryptResult = null;
    try {
        byte[] contentBytes = content.getBytes("UTF-8");
        SecretKeySpec skeySpec = new SecretKeySpec(key.getBytes("UTF-8"),
"AES");
        // 呼叫標準的AES加密,模式選用ECB模式,Padding選用PKCS5Padding方式
        Cipher cipher = Cipher.getInstance("AES/ECB/PKCS5Padding");
        cipher.init(1, skeySpec);
        encryptResult = cipher.doFinal(contentBytes);
    } catch (Exception ex) {
        ex.printStackTrace();
    }
    if (encryptResult != null) {
        // Base64編碼處理
        return Base64.encodeToString(encryptResult, 0);
    }
    return null;
}
```

▲ 圖 9-10 沙盒日誌分析

分析圖 9-10 中加密之前的資料，會發現除去 signqt 參數和 timeqt 參數對應的值未知，其餘參數的值實際上都是確定的。

分析程式清單 9-6 中的 parseRequestParams 函數，發現實際上 timeqt 參數是一個與當前時間有關的時間戳記資訊，而 signqt 參數則是透過 MD5Util.MD5 這個標準 MD5 加密的封裝函數加密後得到的，對應的函數內容如程式清單 9-8 所示。

◯ 程式清單 9-8　signqt 參數形成過程

```
params.put("signqt", MD5Util.MD5(this.settings.getString("angeloip098", "")
+ "&*" + this.settings.getString("6969dolkoh", "") + "()" + ctl + "+_" +
act + "@!@###@"));
public static String MD5(String value) {
    try {
        MessageDigest digest = MessageDigest.getInstance(ISecurity.SIGN_
ALGORITHM_MD5);
        digest.update(value.getBytes());
        byte[] bytes = digest.digest();
        StringBuilder sb = new StringBuilder();
        for (byte b : bytes) {
            String hex = Integer.toHexString(b & 255);
            if (hex.length() == 1) {
                sb.append('0');
            }
            sb.append(hex);
        }
        return sb.toString();
    } catch (NoSuchAlgorithmException e) {
        return null;
    }
}
```

此時，透過 Hook MD5Util.MD5 函數最終確定傳入該函數的參數內容如圖 9-11 所示。

▲ 圖 9-11 Hook MD5Util.MD5 函數

分析完對應參數的實現後,便能夠順利得到寫出離線請求資料的指令稿,其內容如程式清單 9-9 所示。

◎ 程式清單 9-9 離線指令稿

```python
import time
from hashlib import md5
import requests
import base64
import binascii
import re
from Crypto.Cipher import AES
import json


# aes 加密/解密
class AESECB:
    def __init__(self, key):
        self.key = key
        self.mode = AES.MODE_ECB
        self.bs = 16  # block size
```

```python
        self.PADDING = lambda s: s + \
            (self.bs - len(s) % self.bs) * chr(self.bs - len(s) % self.bs)


    def encrypt(self, text):
        generator = AES.new(self.key, self.mode)   # ECB模式無須向量
        crypt = generator.encrypt(self.PADDING(text))
        crypted_str = base64.b64encode(crypt)
        result = crypted_str.decode()
        return result


    def decrypt(self, text):
        generator = AES.new(self.key, self.mode)   # ECB模式無須向量
        text += (len(text) % 4) * '='
        decrpyt_bytes = base64.b64decode(text)
        meg = generator.decrypt(decrpyt_bytes)
        # 去除解碼後的非法字元
        try:
            result = re.compile(
                '[\\x00-\\x08\\x0b-\\x0c\\x0e-\\x1f\n\r\t]').sub('', meg.
decode())
        except Exception:
            result = '解碼失敗，請重試!'
        return result


# com.fanwe.hybrid.http.AppHttpUtil.parseRequestParams(SDRequestParams
params)
if __name__ == '__main__':
    ctl = "app"
    act = "init"
    signqt = md5(("&*()" + ctl + "+_" + act +
                "@!@###@").encode('utf8')).hexdigest() #
    timeqt = str(round(time.time() * 1000))
    # 請求標頭
    headers = {
```

```
    "X-JSL-API-AUTH": "sha1|1611928510|693SMeR0H|8fe0b019e47e9d09be043c
e85f0e7cf0582b50f2"}
   # 請求資料內容
   body = {
       "screen_width": "1440",
       "screen_height": "2392",
       "sdk_type": "android",
       "sdk_version_name": "1.3.0",
       "sdk_version": "2020031801",
       "ctl": ctl,
       "act": act,
       "signqt": signqt,
       "timeqt": timeqt
   }
   a = AESECB("8648754518945235")
   requestDATA = a.encrypt(str(body))
   url = "http://hhy2.hhyssing.com:46288/mapi/index.php?requestData=" + \
       requestDATA+"i_type=1&ctl="+ctl+"&act"+act
   rsp = requests.post(url, headers=headers)
   result = json.loads(rsp.text).get("output")
   d = AESECB("7489148794156147")
   print(d.decrypt(result))
```

但是當真實進行測試時，發現會始終出現如圖 9-12 所示的顯示出錯：
Connection refused，但是明明主機網路狀態良好，這是什麼原因導致的
呢？

```
    raise ConnectionError(e, request=request)
requests.exceptions.ConnectionError: HTTPConnectionPool(host='hhy2.hhyssing.com', port=47232): Max retries exceeded wit
/mapi/index.php?requestData=V+5FrkMLpGbskBLAbCmmn1wix7fDIqY3ZlPW9r8C+s7maNBQE7XdhCcnGuS5cyT40PDUnwyd+ztIBPLRtTEi7N9n5yz
5Kf6HoyOtlOs6DyTbFNhNENRCSsgw1LSoyTm6aYHnHgajoyA+QOoZrfQuyJuBffynMHp0QFny4P8HaRH+LGi9Vg18SM09IvpCjiJIbkBlzNEA4Wrh1VRwOt
qcQM6RlNOTlGjIuYLIv92xqjNAAFCjVx/3OQA5vNCfxD+5dKF5rRkOMGKftnH7h/RwnxZLbzjmXRKCty5v2JYuWs0efqfIDFOqVzkYiYgEd1U7dUNYwybpS
ELnX17IrKXIOKjIRVMxBWFfTLk6DIPLwcGi type=1&ctl=index&act=index (Caused by NewConnectionError('<urllib3.connection.HTTPC
on object at 0x7fe954372610>: Failed to establish a new connection: [Errno 111] Connection refused'))
```

▲ 圖 9-12　發送資料封包錯

如圖 9-13 所示，這裡最終透過 nslookup 命令查看 hhy2.hhyssing.com 域
名對應的 IP，發現該域名竟然指向本地回路位址，甚是神奇。在查閱資
料後找到了原因，之所以出現這樣的情況，是因為 App 整合遊戲盾的作
用。

```
Non-authoritative answer:
Name:    hhy2.hhyssing.com
Address: 127.0.0.1
```

▲ 圖 9-13 nslookup 命令

那麼如何解決這個問題呢？

這裡採用的是 ADB 通訊埠轉發的方法。使用資料線連接上手機並開啟
App，使用如下命令使得主機上發送到本地 46288 通訊埠的資料被 ADB
轉發到手機上的 46288 通訊埠，最終透過該 SDK 轉發到伺服器，從而獲
得最終返回資料。

```
adb forward tcp:46288 tcp:46288
```

這裡還需要注意的是，每個 App 開啟後，對應的通訊埠實際上都會變
化，因此還需要每次開啟 App 後重新獲取對應通訊埠，這個過程可以透
過 r0capture 確定。最終執行指令稿就可以實現「離線」呼叫，進而獲取
我們想要的資料資訊。

當然，讀者可能會發現，在這裡並未介紹對應的解密接收資料封包的函
數，實際上這部分解密過程筆者也並未進行分析，而是查看加解密沙盒
的記錄檔與對應的抓取封包資料時，發現資料封包的解密方式也是 AES/
ECB/PKCS5Padding 的標準模式，只是使用了不同的金鑰而已。當然，如
此草率地判定解密方式是不大可取的，筆者在獲取到這些資訊後也進行
了驗證，限於篇幅，這裡就不再介紹具體的驗證過程了。

9.3 主動呼叫分析

在完成了上述協定分析後，還要透過獲取該樣本中的首頁面獲取批次房
間資訊和具體直播間的房間資訊，以介紹一些關於主動呼叫的內容。

如圖 9-14 所示，在對 postImpl 函數進行 Hook 分析時，會發現始終存在
一個屬於 CommonInterface 類別中的函數的呼叫，後續經過靜態分析會
發現，該類別中存在很多用於獲取直播間內容的函數。

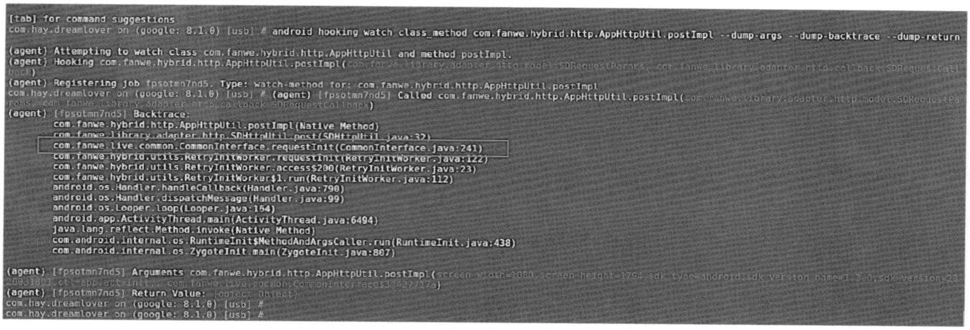

▲ 圖 9-14 Objection Hook postImpl 呼叫堆疊

為了確認在使用 App 開啟直播間時 CommonInterface 類別中有函數會被
呼叫，這裡直接透過 Objection 對該類別進行 Trace 分析，其結果如圖
9-15 所示。仔細研究後發現，其中的函數依次用於頁面中獲取簡略的房
間資訊（requestIndex 函數）、請求具體房間視訊（requestNewVideo 函
數）、請求具體房間資訊（requestRoomInfo 函數）。

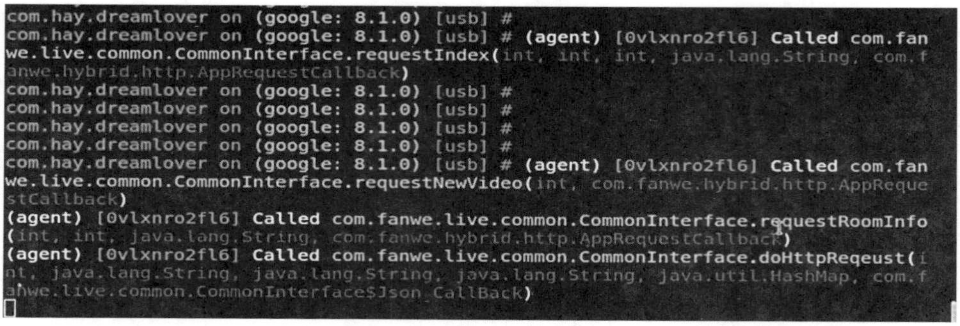

▲ 圖 9-15 Objection Hook CommonInterface 類別

9.3.1 簡單函數的主動呼叫

溫故知新，這裡首先介紹在首頁面滑動視窗獲取房間簡略資訊的主動呼叫方式，即主動呼叫 requestIndex 函數的方式。

正如在介紹 Frida 的三劍客時所說的，要實現主動呼叫，首先要進行 Hook。為此，這裡先透過 Objection Hook 該函數，進一步確認該函數被呼叫，其結果如圖 9-16 所示。

▲ 圖 9-16 Hook requestIndex() 函數

在確認該函數被呼叫後，還需要透過 Frida 指令稿進行 Hook，得到對應的 roomId，但是在實現 Hook 的過程中發現，如果僅僅是 Hook requestIndex() 函數，無法直接得到真實的 roomId，經過一番 Hook 與靜態函數的分析，最終發現真實的房間資訊隱藏在 requestIndex() 的第 4 個參數中，該參數實際上是一個範型類別，用於處理網路請求的結果，其中 onSuccess() 函數的參數 resp 包含著網路返回的加密資料封包。因此，要獲得最終的 roomId，還需要分析 SDResponse 類型的參數 resp 的類別結構，這裡透過 WallBreaker 查看該類別中的成員及函數，結果如圖 9-17 所示。

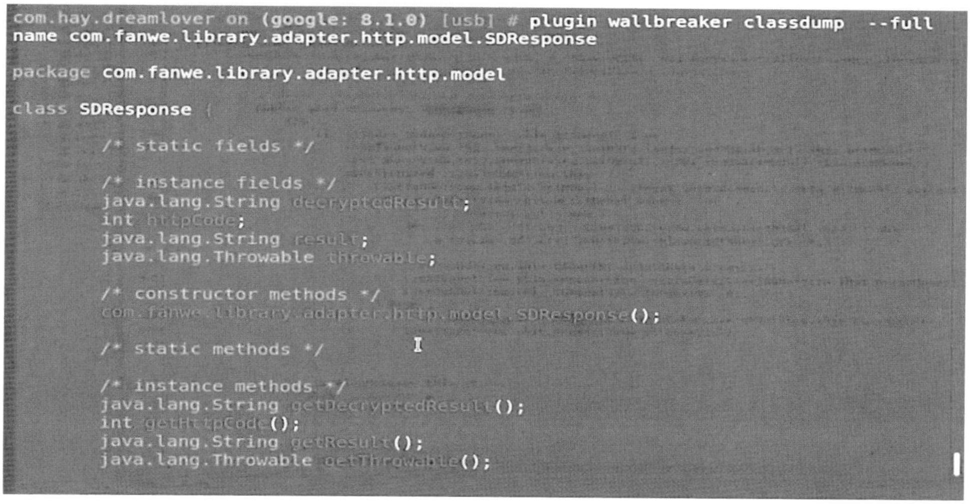

▲ 圖 9-17　SDResponse 類別

進一步分析 onSuccess 函數的內容會發現，該函數中並未出現任何和參數 resp 相關的操作，而是一直透過 this.actModel 成員進行解析，究其原因，會發現存在一個 onSuccessBefore 多載函數，用於將 resp 參數解析到 this.actModel 成員上，具體的相關處理 resp 參數內容的程式如程式清單 9-10 所示。

▼ 程式清單 9-10 處理 resp 參數程序呼叫鏈

```
public static <T> T json2Object(String json, Class<T> clazz) {
    return (T) JSON.parseObject(json, clazz);
}
public <T> T parseActModel(String result, Class<T> clazz) {
    return (T) SDJsonUtil.json2Object(result, clazz);
}
public <T> T parseActModel(String result, Class<T> clazz) {
    return (T) SDJsonUtil.json2Object(result, clazz);
}
// com.fanwe.library.adapter.http.callback.SDRequestCallback
@Override
// 解密處理接收資料封包函數
public void onSuccessBefore(SDResponse resp) {
    if (this.mModelClass != null) {
        // 解密函數
        String result = resp.getDecryptedResult();
        if (TextUtils.isEmpty(result)) {
            result = resp.getResult();
        }
        this.actModel = parseActModel(result, this.mModelClass);
    }
}
// com.fanwe.live.appview.main.LiveTabHotView
 private void requestData() {
        CommonInterface.requestIndex(1, this.mSex, 0, this.mCity, new
AppRequestCallback<Index_indexActModel>() {
            /* class com.fanwe.live.appview.main.LiveTabHotView.
AnonymousClass4 */

            /* access modifiers changed from: protected */
            @Override // com.fanwe.library.adapter.http.callback.
SDRequestCallback
            // 將接收到的返回資料進行model資料的給予值
            public void onSuccess(SDResponse resp) {
```

```
            try {
                if (((Index_indexActModel) this.actModel).isOk()) {
                    LiveTabHotView.this.mHeaderView.setData((Index_
indexActModel) this.actModel);
                    LiveTabHotView.this.mHeaderView.setData1((Index_
indexActModel) this.actModel);
                    synchronized (LiveTabHotView.this) {
                        LiveTabHotView.this.mListModel = ((Index_
indexActModel) this.actModel).getList();
                        if (LiveTabHotView.this.mListModel.size() > 8)
{
                            ArrayList arrayList = new ArrayList();
                            for (int sbi = 8; sbi < LiveTabHotView.
this.mListModel.size(); sbi++) {
                                arrayList.add(LiveTabHotView.this.
mListModel.get(sbi));
                            }
                            LiveTabHotView.this.mAdapter.
updateData(arrayList);
                            LiveTabHotView.this.mHeaderView.
updateData(LiveTabHotView.this.mListModel);
                            LiveTabHotView.this.mHeaderView.
setgbxxVis(0);
                        } else {
                            LiveTabHotView.this.mAdapter.
updateData(LiveTabHotView.this.mListModel);
                            LiveTabHotView.this.mHeaderView.
setgbxxVis(8);
                        }
                    }
                }
                LiveTabHotView.this.sbi++;
            } catch (Exception e) {
            }
        }
```

```
        // ...
    });
}
```

這裡依樣畫葫蘆，最終得到簡略房間資訊的 Hook 程式如程式清單 9-11 所示。

◉ 程式清單 9-11 獲取簡略房間資訊的 Hook

```
function hook() {
    Java.perform(function () {
        var JSON = Java.use("com.alibaba.fastjson.JSON")
        var Index_indexActModel = Java.use("com.fanwe.live.model.Index_
indexActModel");
        var gson = Java.use("com.google.gson.Gson").$new();

        var LiveRoomModel = Java.use("com.fanwe.live.model.LiveRoomModel");
        //透過Jadx工具靜態查看smali程式進而確認對應類別名
        //用Objection快速驗證搜索
        //onSuccess所在類別
        Java.use("com.fanwe.live.appview.main.LiveTabHotView$4").onSuccess.
implementation = function (resp) {
            console.log("Entering Room List Parser => ", resp)
            //主動呼叫解密函數
            var result = resp.getDecryptedResult();
            //JSON解析資料
            var resultModel = JSON.parseObject(result, Index_indexActModel.
class);
            var roomList = Java.cast(resultModel, Index_indexActModel).
getList();
            //列印請求得到的房間列表大小並獲取第一個房間的資訊
            console.log("size : ", roomList.size(), roomList.get(0))
            for (var i = 0; i < roomList.size(); i++) {
                var LiveRoomModelInfo = Java.cast(roomList.get(i),
LiveRoomModel);
```

```
            //JSON方式列印資料
            console.log("roominfo: ", i, " ", gson.
toJson(LiveRoomModelInfo));
        }
        return this.onSuccess(resp)
    }
  })
}
```

注意，在程式清單 9-11 中，是透過 hook onSuccess 函數實現獲取簡略房間資訊的效果的，那麼對應的 onSuccess 函數所在類別其實就是 requestIndex 函數的第 4 個參數對應的類型。有一定開發知識或者逆向經驗的讀者一定會敏銳地發現該參數的類型實際上是一個匿名內部類別，泛型的 AppRequestCallback ＜ Index_indexActModel ＞ 格式只是編譯時期需要的資訊，真正在記憶體中對應的匿名內部類別名稱可以透過查看對應的 smali 程式或者透過如圖 9-16 所示的 Hook 方式確定。

最終將 Hook 程式注入處理程序後，即可得到大量簡略房間的資訊，其部分結果如圖 9-18 所示。

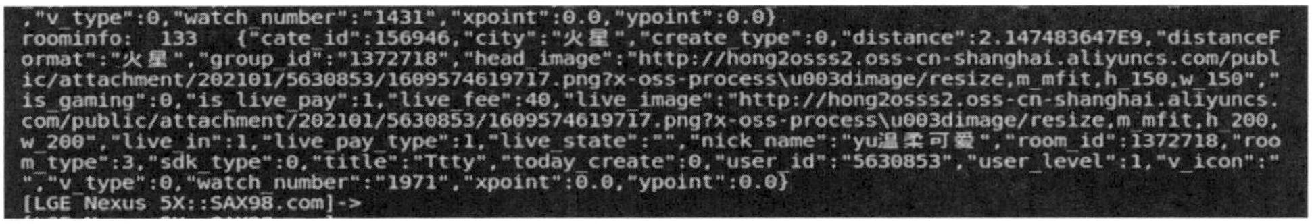

▲ 圖 9-18 獲取簡略房間資訊

在透過 Hook 獲取簡略房間資訊成功後，還需要再加上最終的主動呼叫。這裡為了避免大量的複雜參數建構的問題，直接主動呼叫 requestIndex() 函數的上層函數 requestData 實現，具體主動呼叫程式如程式清單 9-12 所示。

○ 程式清單 9-12 主動呼叫程式

```
function invoke(){
    Java.perform(function(){
        Java.choose("com.fanwe.live.appview.main.LiveTabHotView",{
            onMatch:function(ins){
                console.log("found ins => ",ins)
                // 主動呼叫發送封包函數
                ins.requestData();
            },onComplete:function(){
                console.log("Search completed!")
            }
        })
    })
}
```

9.3.2 複雜函數的主動呼叫

接下來介紹進入直播間時獲取直播間的房間詳情的主動呼叫方式，即 requestRoomInfo 函數的主動呼叫。

同樣，作為三劍客的第一步，這裡先透過 Hook 方式獲取房間的詳情資訊。

為了節省篇幅，將焦點集中於主動呼叫本身，這裡直接略過中間 Hook 實現的分析，舉出最終的 Hook 指令稿，其內容如程式清單 9-13 所示。最終得到的 Hook 結果如圖 9-19 所示。

○ 程式清單 9-13 hookRoomInfo 程式

```
function inspectObject(obj) {
    Java.perform(function () {
        const Class = Java.use("java.lang.Class");
        const obj_class = Java.cast(obj.getClass(), Class);
```

```javascript
        const fields = obj_class.getDeclaredFields();
        console.log("Inspecting " + obj.getClass().toString());
        console.log("\tFields:");
        for (var i in fields){
            //console.log("\t\t" + fields[i].toString());
            var className = obj_class.toString().trim().split(" ")[1] ;
            //console.log("className is => ",className);
            var fieldName = fields[i].toString().split(className.
concat(".")).pop() ;
            console.log(fieldName + " => ",obj[fieldName].value);
        }
        //console.log("\tMethods:");
        //for (var i in methods)
        //console.log("\t\t" + methods[i].toString());
    })
}
function hookROOMinfo() {
    Java.perform(function () {
        var JSON = Java.use("com.alibaba.fastjson.JSON")
        var App_get_videoActModel = Java.use("com.fanwe.live.model.App_get_
videoActModel");

        Java.use("com.fanwe.live.business.LiveBusiness$2").onSuccess.
implementation = function (resp) {
            console.log("Enter LiveBusiness$2 ... ", resp)
            var result = resp.getDecryptedResult();
            var resultVideoModel = JSON.parseObject(result, App_get_
videoActModel.class);
            var roomDetail = Java.cast(resultVideoModel, App_get_
videoActModel);
            inspectObject(roomDetail)
            return this.onSuccess(resp);
        }
    })

}
```

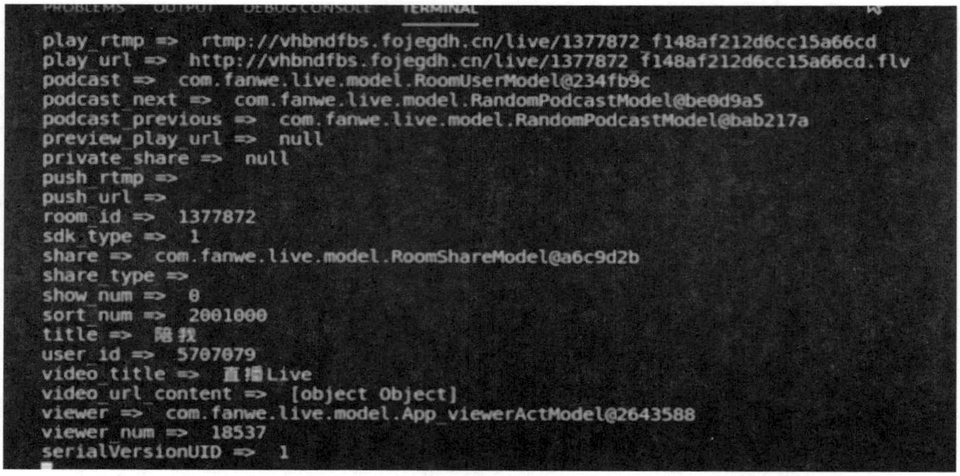

▲ 圖 9-19 Hook 結果

這裡要補充的是在程式清單 9-13 中，inspectObject 函數的作用是透過傳入物件獲取其對應的類別，進而透過 getDeclaredFields() 等 Java 反射的方式獲取實例物件中所有成員的值。

在完成 Hook 工作後，讓我們正式進入正題。

與 requestIndex 函 數 類 似，CommonInterface.requestRoomInfo 函 數 也存在一個參數較簡單的外部呼叫函數：com.fanwe.live.business.LiveBusiness.requestRoomInfo 函數，該函數具體內容如程式清單 9-14 所示。

◉ 程式清單 9-14 requestRoomInfo 函數

```
//com.fanwe.live.business.LiveBusiness
public void requestRoomInfo(String private_key) {
    CommonInterface.requestRoomInfo(getLiveActivity().getRoomId(),
getLiveActivity().isPlayback() ? 1 : 0, private_key, new
AppRequestCallback<App_get_videoActModel>() {
        public String getCancelTag() {
```

```
        //...
    }

    /* access modifiers changed from: protected */
    //成功請求回呼函數
    public void onSuccess(SDResponse sdResponse) {
        LiveInformation.getInstance().setRoomInfo((App_get_
videoActModel) this.actModel);
        if (((App_get_videoActModel) this.actModel).isOk()) {
            LiveBusiness.this.onRequestRoomInfoSuccess((App_get_
videoActModel) this.actModel);
        } else {
            LiveBusiness.this.onRequestRoomInfoError((App_get_
videoActModel) this.actModel);
        }
    }

    /* access modifiers changed from: protected */
    // 失敗請求回呼函數
    public void onError(SDResponse resp) {
        LiveBusiness.super.onError(resp);
        String msg = "request error";
        if (resp.getThrowable() != null) {
            msg = resp.getThrowable().toString();
        }
        LiveBusiness.this.onRequestRoomInfoException(msg);
    }
});
}
```

與 requestIndex 函數不同的是，LiveBusiness.requestRoomInfo() 函數仍舊有一個固定的參數 private_key。這裡透過 Hook 的方式確定該參數的值為 123454。

但是在主動呼叫該函數之前，我們發現在第一步按照程式清單 9-15 中的程式透過 Java.choose() 函數尋找 requestRoomInfo() 函數所在類別 LiveBusiness 的實例時就會折戟沉沙：找不到對應實例。

● 程式清單 9-15　Java.choose() 找不到實例

```
function invoke(){

    Java.perform(function(){
        Java.choose("com.fanwe.live.business.LiveBusiness",{
            onMatch:function(ins){
                console.log("found ins => ",ins)
            },onComplete:function(){
                console.log("Search completed!")
            }
        })
    })
}
```

既然找不到對應的實例，那麼順理成章地想到：是否可以主動建構一個實例去呼叫實例方法呢？

為了主動建構一個 LiveBusiness 物件，這裡首先透過 Objection 得到對應建構函數為 com.fanwe.live.business.LiveBusiness.$init(com.fanwe.live.activity.room.ILiveActivity)，進而發現該類別的建構函數的參數是一個複雜物件，對應靜態結果後，發現該參數甚至是一個 interface（介面）類別，如圖 9-20 所示。

▲ 圖 9-20　獲取建構函數

由於介面類別需要被實例化才可以被當作參數呼叫，因此如果想要主動建構一個 LiveBusiness 物件，必須實例化 ILiveActivity 這個介面。在查閱並結合開發知識後發現，Frida 提供了 Java.registerClass() 這個 API 用於建構一個 Java 類別，最終獲得了一個基礎的主動呼叫函數，其內容如程式清單 9-16 所示。

◯ **程式清單 9-16　主動建構一個實例實現主動呼叫**

```
function invoke2(){
    Java.perform(function(){

        //建構函數：介面com.fanwe.live.business.LiveBusiness(ILiveActivity);
        var ILiveActivity = Java.use("com.fanwe.live.activity.room.
ILiveActivity");
        // 實例化介面
        const ILiveActivityImpl = Java.registerClass({
            name: 'com.fanwe.live.activity.room.ILiveActivityImpl',
            implements: [ILiveActivity],
          });

        var result = Java.use("com.fanwe.live.business.LiveBusiness").$new(
ILiveActivityImpl.$new());
        console.log("result is => ",result.requestRoomInfo("123454"))
    })
}
```

但是將以上指令稿注入處理程序後，手動執行 invoke2 函數就會得到如圖 9-21 所示中的顯示出錯：Missing implementaion for...。而這樣的顯示出錯意味著，在實現一個介面類別時還需要實現該介面類別中所有的介面函數。

▲ 圖 9-21　缺少函數實現的顯示出錯

在發現圖 9-21 中的顯示出錯後,依次在主動呼叫程式中實現提示缺少的函數,最終主動呼叫的程式如程式清單 9-17 所示,最終再次主動呼叫 requestRoomInfo 函數就成功了,其結果如圖 9-22 所示。

⬇ 程式清單 9-17 主動建構一個實例實現主動呼叫

```
//主動new一個實例
function invoke2(){
    Java.perform(function(){

        //建構函數:介面com.fanwe.live.business.LiveBusiness(ILiveActivity);
        var ILiveActivity = Java.use("com.fanwe.live.activity.room.
ILiveActivity");
        //實例化介面
        const ILiveActivityImpl = Java.registerClass({
            name: 'com.fanwe.live.activity.room.ILiveActivityImpl',
            implements: [ILiveActivity],
            methods: {
                //必須實現這些方法,否則顯示出錯。這裡只是簡單地寫了一個抽象函
數,並未具體實現內部邏輯
                openSendMsg(){},
                getCreaterId(){},
                getGroupId(){},
                getRoomId(){ },
                getRoomInfo(){},
                getSdkType(){},
                isAuctioning(){},
                isCreater(){},
                isPlayback(){},
                isPrivate(){}
            }
        });
        var result = Java.use("com.fanwe.live.business.LiveBusiness").$new(
ILiveActivityImpl.$new());
        console.log("result is => ",result.requestRoomInfo("123454"))
```

```
    })
 }
```

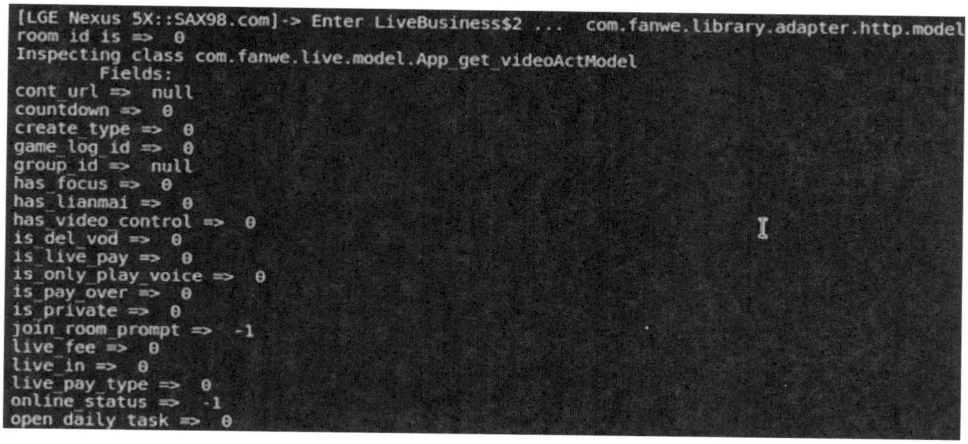

```
[LGE Nexus 5X::SAX98.com]-> Enter LiveBusiness$2 ...  com.fanwe.library.adapter.http.model
room id is =>
Inspecting class com.fanwe.live.model.App_get_videoActModel
        Fields:
cont url => null
countdown => 0
create type => 0
game log id => 0
group id => null
has focus => 0
has lianmai => 0
has video control => 0
is del vod => 0
is live pay => 0
is only play voice => 0
is pay over => 0
is private => 0
join room prompt => -1
live fee => 0
live in => 0
live pay type => 0
online status => -1
open daily task => 0
```

▲ 圖 9-22　主動呼叫結果

雖然圖 9-22 中的主動呼叫成功執行，但是對比圖 9-19 中透過 Hook 得到的資料會發現，圖 9-22 中的資料非常不真實，甚至完全無效。研究發現出現這種情況的原因實際上是在透過 Java.registerClass 這個 Frida 提供的 API 函數實現 ILiveActivity 介面類別中的函數時並未真實地實現函數邏輯，而只是簡單地清空，導致後續主動呼叫時不存在真實的資訊。

鑑於上述原因，筆者採用了另一個獲取實例的方法：透過 Hook 所需實例的另一個實例方法得到對應實例，並保存到外部供主動呼叫時使用，這也正是在前面章節中使用的方法。這裡 Hook 的函數為 LiveBusiness. getLiveQualityData()，最終具體程式如程式清單 9-18 所示。

🔻 程式清單 9-18　Hook 其他函數外部保存供主動呼叫

```
var LiveBusiness = null ;
console.log("LiveBusiness is => ", LiveBusiness)
function hook3(){
```

```
    Java.perform(function(){
        Java.use("com.fanwe.live.business.LiveBusiness").
getLiveQualityData.implementation = function(){
            LiveBusiness = this;
            console.log("now LiveBusiness is => ", LiveBusiness)
        //LiveBusiness.requestRoomInfo("12343"); //立刻主動呼叫
            var result = this.getLiveQualityData()
            return result;
        }
    })
}
//這個方式會崩潰
function invoke3(){
    Java.perform(function(){
        var result = LiveBusiness.requestRoomInfo("12343");
        console.log("result is => ",result)
    })
}
```

如圖 9-23 所示，確實找到了對應的實例，但是在手動呼叫 invoke3() 函數時發現，這種主動呼叫的方式不僅沒有成功，甚至還導致樣本崩潰，而如果在 Hook getLiveQualityData() 函數程式中進行主動呼叫，就會發現是可以成功被呼叫的，因此可以判定在手動呼叫 invoke3() 函數時，對應實例實際上已經被解析掉。

▲ 圖 9-23 找到實例

至此，這個樣本的主動呼叫計畫徹底崩盤。也許有讀者會説，可以繞過需要實例才能進行呼叫的動態函數 LiveBusiness.requestRoomInfo()，轉而主動呼叫無須實例即可呼叫的靜態函數 CommonInterface.requestRoomInfo(int room_id, int is_vod, String private_key,AppRequestCallback＜App_get_videoActModel＞ listener)。但在實現時發現對於該函數的主動呼叫也存在與程式清單 9-17 一樣的問題：該函數的第 4 個參數也需要透過 Java.registerClass() 這個 API 來實例化對應的回呼介面。這裡測試的主動呼叫程式如程式清單 9-19 所示。

⭗ **程式清單 9-19 主動呼叫 CommonInterface.requestRoomInfo() 函數**

```
function invoke4(){
    Java.perform(function(){

        //com.fanwe.live.business.LiveBusiness(ILiveActivity);
        var ILiveActivity = Java.use("com.fanwe.live.activity.room.
ILiveActivity");
        const ILiveActivityImpl = Java.registerClass({
            name: 'com.fanwe.live.activity.room.ILiveActivityImpl',
            implements: [ILiveActivity],
            methods: {
                //主動呼叫離資料越遠，中間實現的細節就越多
                openSendMsg(){},
                getCreaterId(){},
                getGroupId(){},
                getRoomId(){}, //沒有實現正確導致App崩潰
                getRoomInfo(){},
                getSdkType(){},
                isAuctioning(){},
                isCreater(){},
                isPlayback(){},
                isPrivate(){}
            }
        });
```

```
        var LB = Java.use("com.fanwe.live.business.LiveBusiness").$new(ILiv
eActivityImpl.$new());

        var LB2 = Java.use("com.fanwe.live.business.LiveBusiness$2");
        var AppRequestCallback = Java.use('com.fanwe.hybrid.http.
AppRequestCallback');
        Java.use("com.fanwe.live.common.CommonInterface").requestRoomInfo(1
377894,123,"1234",Java.cast(LB2.$new(LB),AppRequestCallback));
    })
}
```

綜上，針對這類實例物件在記憶體中並不持久存在的樣本函數，如果要真正實現函數的主動呼叫，還是透過建構一個實例實現主動呼叫的方式更加可靠，但如果該實例的建構異常複雜，要實現該方法的主動呼叫，難度也就對應地變得更高。更重要的發現在於，如果要實現函數的主動呼叫，就應該始終遵循「離資料越近越好」的原則，需要手動建構的資料越少越好。之所以這裡實現獲取直播間房間資訊很困難，究其原因正是我們的目標函數離資料太遠，導致要實現的細節問題變得異常多。

9.4 本章小結

本章透過對一個違法樣本的 VIP 功能進行繞過以及協定分析，將在前面章節介紹的知識應用於真實的實踐操作中，幫助讀者進一步掌握前面章節的知識。除此之外，還特地單列出了一節用於進一步介紹在使用 Frida 進行函數的主動呼叫時需要注意的問題，並透過筆者的失敗實踐以身說法，希望讀者能從中吸取教訓，並在後續的 Hook 工作中始終謹記「Hook 離資料越近越好」的基本準則。

Android 協定分析之會員制
非法應用破解

由於 Android 應用的開發方式多種多樣，相應地，Android 的協定分析也就存在各種各樣的方法。本章將透過對一個應用的協定分析，盡可能多地介紹一些在協定分析過程中可能會遇到的情況，同時還會介紹一款工具—r0tracer 的使用與具體原理，並透過該工具協助完成樣本的分析過程。

10.1 r0tracer 介紹與原始程式剖析

與 r0capture 一樣，r0tracer（原始程式位址：https://github.com/r0ysue/r0tracer）同樣是基於 Frida、結合多個 Hook 專案開發的一款工具，但是與 r0capture 用於解決應用層抓取封包問題不同，r0tracer 用於在 Android 應用中多功能追蹤 Java 層類別的指令稿，它不僅可以根據黑白名單批次追蹤類別的所有方法，還可以在命中方法後列印出該類別或物件的所有域值、參數、呼叫堆疊和返回值，堪稱精簡版的 Objection 和 Wallbreaker 的結合。

當然，如果僅僅是結合 Objection 和 Wallbreaker，就沒有必要單獨列出一節進行介紹。r0tracer 相對于 Objection 不僅增加了在找不到指定類別時切換 Classloader 的功能，而且增加了延遲時間 Hook 機制，避免因透過 spawn 方式注入應用導致無法立刻找到 App 類別的尷尬處境。同時，相對於 Objection，r0tracer 更是增加了 Objection 無法在指定 Hook 類別的同時對類別的建構函數進行 Hook 的功能。相比於 WallBreaker，r0tracer 還增加了在 Hook 時列印類別中實例域的值的功能。接下來，將從原始程式層面一一介紹以上這些功能。

首先介紹 r0tracer 的基礎功能：查看物件中成員的值。

如程式清單 10-1 所示，在使用時只需傳入實例物件（obj）和最終輸出的日誌參數（input）即可。因為靜態類別的物件只是對應類別的包裝器，還要透過判斷對應參數是否存在 Frida 中定義的 $handle/$h 屬性來避免因為傳入靜態類別導致無法查看域值的情況出現。在獲取傳入實例物件中的成員的方法時，選擇了 Java 提供的反射方法 getDeclaredFields()，最終透過 obj 實例獲取對應成員的值。

● 程式清單 10-1 查看物件域值

```
//查看域值
function inspectObject(obj, input) {
    var isInstance = false;
    var obj_class = null;
    //確定傳進的obj參數是不是靜態類別，這部分程式參考WallBreaker
    if (getHandle(obj) === null) {
        obj_class = obj.class;
    } else {
        var Class = Java.use("java.lang.Class");
```

```
        obj_class = Java.cast(obj.getClass(), Class);
        isInstance = true;
    }
    input = input.concat("Inspecting Fields: => ", isInstance, " => ", obj_
class.toString());
    input = input.concat("\r\n");
    //獲取類別中的成員
    var fields = obj_class.getDeclaredFields();
    for (var i in fields) {
        //判定成員所在類別是不是靜態類別，或者成員是不是靜態成員
        if (isInstance || Boolean(fields[i].toString().indexOf("static ")
>= 0)) {
            //output = output.concat("\t\t static static static " +
fields[i].toString());
            //獲取類別名稱
            var className = obj_class.toString().trim().split(" ")[1];
            //console.Red("className is => ",className);
            //獲取類別成員名稱
            var fieldName = fields[i].toString().split(className.
concat(".")).pop();
            //獲取類別成員類型
            var fieldType = fields[i].toString().split(" ").slice(-2)[0];
            var fieldValue = undefined;
            //獲取實例成員值
            if (!(obj[fieldName] === undefined))
                fieldValue = obj[fieldName].value;
            //拼接結果，保障輸出連續
            input = input.concat(fieldType + " \t" + fieldName + " => ",
fieldValue + " => ", JSON.stringify(fieldValue));
            input = input.concat("\r\n")
        }
    }
    return input;
}


function getHandle(object) {
```

```
    //Frida 12專用屬性
    if (hasOwnProperty(object, '$handle')) {
        if (object.$handle != undefined) {
            return object.$handle;
        }
    }
    //Frida 14以上屬性
    if (hasOwnProperty(object, '$h')) {
        if (object.$h != undefined) {
            return object.$h;
        }
    }
    return null;
}
```

接下來介紹 r0tracer 的兩個 Hook 功能。

首先介紹 r0tracer Hook 指定函數名的所有多載函數（traceMethod 函數）的功能。該函數的具體實現方法如程式清單 10-2 所示，在使用時只需傳入完整的類別名 + 函數名作為 traceMethod 函數參數即可，比如想要 Hook Android 中標準加密類別 Cipher 類別的函數 doFinal 所有多載，只需傳入 doFinal 函數的完整類別名（javax.crypto.Cipher）並拼接上函數名作為參數即可。在 traceMethod 函數內部會進行類別名和函數名的分割，並透過 overloads 這一 Frida 中定義的屬性值獲取所有名稱相同的多載函數並分別進行 Hook。如圖 10-1 所示，最終在任意 doFinal 多載函數被執行時，不僅會呼叫 inspectObject() 函數查看 Cipher 類別中對應的域值，還會列印出對應函數的參數、返回值和呼叫堆疊。

◯ 程式清單 10-2　Trace 所有多載函數

```
//Trace單一類別的所有靜態和實例方法，包括建構方法
function traceMethod(targetClassMethod) {
    var delim = targetClassMethod.lastIndexOf(".");
```

```
    if (delim === -1) return;
    var targetClass = targetClassMethod.slice(0, delim)
    var targetMethod = targetClassMethod.slice(delim + 1,
 targetClassMethod.length)
    // 獲取類別的wrapper
    var hook = Java.use(targetClass);
    var overloadCount = hook[targetMethod].overloads.length; // 獲取函數所
有多載
    console.Red("Tracing Method : " + targetClassMethod + " [" +
overloadCount + " overload(s)]");
    for (var i = 0; i < overloadCount; i++) {
        hook[targetMethod].overloads[i].implementation = function () {
            //初始化輸出，透過concat拼接輸出，以保證一次Hook的連續
            var output = "";
            //畫個橫線
            for (var p = 0; p < 100; p++) {
                output = output.concat("==");
            }
            //列印域值
            if (!isLite) { output = inspectObject(this, output); }
            //進入函數
            output = output.concat("\n*** entered " + targetClassMethod);
            output = output.concat("\r\n")
            //if (arguments.length) console.Black();
            //參數
            var retval = this[targetMethod].apply(this, arguments);
            if (!isLite) {
                for (var j = 0; j < arguments.length; j++) {
                    output = output.concat("arg[" + j + "]: " +
arguments[j] + " => " + JSON.stringify(arguments[j]));
                    output = output.concat("\r\n")
                }
                //呼叫堆疊
                output = output.concat(Java.use("android.util.Log").
getStackTraceString(Java.use("java.lang.Throwable").$new()));
                //返回值
```

```
                    output = output.concat("\nretval: " + retval + " => " +
JSON.stringify(retval));
            }
        // inspectObject(this)
        //離開函數
        output = output.concat("\n*** exiting " + targetClassMethod);
        //最終輸出
        // console.Black(output);
        ...
        printOutput(output);
        return retval;
        }
    }
}
```

▲ 圖 10-1 Hook 指定函數所有多載

除了上述 Hook 單一函數的功能外，r0tracer 還支援 Hook 指定類別中所有函數的功能，與 Objection Hook 指定類別只能處理正常函數不同的是，r0tracer 不僅可以透過呼叫 getDeclaredMethods() 這一 Java 反射函數獲取類別中所有非建構函數，還可以透過呼叫 getDeclaredConstructors() 函數確認指定類別是否存在建構函數，最終透過呼叫上面介紹的 traceMethod 函數完成對所有函數的 Hook 功能，如程式清單 10-3 所示。

🔻 **程式清單 10-3 Hook 指定類別中所有建構方法和動靜態函數**

```
function traceClass(targetClass) {
    //Java.use是獲取特定類別對應的控制碼
    var hook = Java.use(targetClass);
    //利用反射的方式拿到當前類別的所有方法
    var methods = hook.class.getDeclaredMethods();
    //建完物件之後記得將物件釋放掉
    hook.$dispose;
    //將方法名保存到待Hook的陣列中
    var parsedMethods = [];
    var output = "";
    output = output.concat("\tSpec: => \r\n")
    methods.forEach(function (method) {
        output = output.concat(method.toString())
        output = output.concat("\r\n")
        parsedMethods.push(method.toString().replace(targetClass + ".",
"TOKEN").match(/\sTOKEN(.*)\(/)[1]);
    });
    //去掉一些重複的值
    var Targets = uniqBy(parsedMethods, JSON.stringify);
    // targets = [];
    // 透過反射函數getDeclaredConstructors確認是否存在建構函數
    var constructors = hook.class.getDeclaredConstructors();
    if (constructors.length > 0) {
        constructors.forEach(function (constructor) {
            output = output.concat("Tracing ", constructor.toString())
```

```
            output = output.concat("\r\n")
    })
    //Frida透過$init表示建構函數
    Targets = Targets.concat("$init")
}
//對陣列中所有的方法（包括建構函數）進行Hook
Targets.forEach(function (targetMethod) {
    traceMethod(targetClass + "." + targetMethod);
});
// 畫個橫線以進行區分
for (var p = 0; p < 100; p++) {
    output = output.concat("+");
}
console.Green(output);
}
```

除此之外，r0tracer 還支援指定黑白名單的方式進行 Hook。如程式清
單 10-4 所示，在使用這一功能時，只需傳入希望目標類別中包含的字
串作為第一個參數，不希望目標類別中出現的字串作為第二個參數，即
可達到想要的效果。當然，如果指定的目標類別不在當前類別載入器
classLoader 中，還可以傳入非 Null 的任意值作為第三個參數，使得在函
數中執行 Hook 操作時能夠透過 Java.enumerateClassLoaders() 這個 API
函數切換到目標類別所在的 classLoader。如圖 10-2 所示是在想要 Hook
不包含 $ 符號卻包含 javax.crypto.Cipher 字串的相關類別最終的效果。

▲ 圖 10-2 黑白名單 Hook 效果

● 程式清單 10-4 黑白名單進行 Hook

```
function hook(white, black, target = null) {
    console.Red("start")
    //確定是否需要切換classLoader
    if (!(target === null)) {
        console.LightGreen("Begin enumerateClassLoaders ...")
        Java.enumerateClassLoaders({
            onMatch: function (loader) {
                try {
                    //透過遍歷所有loader載入目標類別，確定目標類別所在的
loader
                    if (loader.findClass(target)) {
                        console.Red("Successfully found loader")
                        console.Blue(loader);
                        Java.classFactory.loader = loader;
                        console.Red("Switch Classloader Successfully ! ")
                    }
                }
                catch (error) {
                    console.Red(" continuing :" + error)
                }
            },
            onComplete: function () {
                console.Red("EnumerateClassloader END")
            }
        })
    }
    console.Red("Begin Search Class...")
    var targetClasses = new Array();
    //遍歷記憶體中所有已載入的類別
    Java.enumerateLoadedClasses({
        onMatch: function (className) {
            //黑白名單處理
            if (className.toString().toLowerCase().indexOf(white.
toLowerCase()) >= 0 &&
```

```
                (black == null || black == '' || className.toString().
toLowerCase().indexOf(black.toLowerCase()) < 0)
            ) {
                console.Black("Found Class => " + className)
                targetClasses.push(className);
                //對找到的目標類別執行Hook工作
                traceClass(className);
            }
        }, onComplete: function () {
            console.Black("Search Class Completed!")
        }
    })
    var output = "On Total Tracing :"+String(targetClasses.length)+"
classes :\r\n";
    targetClasses.forEach(function(target){
        output = output.concat(target);
        output = output.concat("\r\n")
    })
    console.Green(output+"Start Tracing ...")
}
```

除了以上介紹的新增功能外，r0tracer 還會有一些操作上的優勢。由於 r0tracer 僅使用 JavaScript 檔案，完全不需要 Python 等語言的包裝，因此可以直接借助 Frida 原生命令列的參數實現延遲時間 Hook、日誌保存等功能。筆者認為，這樣極簡的方式對於一個小工具來說對使用者更加友善，且能夠減少開發程式量，進而更專注於功能本身。

當然，r0tracer 在使用時需要注意，儘量使用 Frida 14 以上版本，如果讀者想要使用 Frida 12 以下的版本，在執行注入時需要加上 --runtime＝v8 選項，將 Frida 引擎從預設的 DUK 切換到 v8 執行，這樣 r0tracer 中的一些語法才能夠得到支援。

接下來，就讓我們用剛介紹的 r0tracer 來完成本章樣本的逆向分析吧！

10.2 付費功能繞過

本次選擇的樣本同樣是一個打著「聊天交友」旗號的違法應用。與前面一致的是，在正式進行協定分析之前，我們先來完成對該應用的付費功能繞過。

樣本 App 經過保護處理，為了方便後續配合動態分析，這裡首先使用 frida-dexdump 進行脫殼供後續靜態分析，但是在使用 Jadx-gui 開啟脫殼的 DEX 檔案後，發現部分 DEX 檔案不被辨識，由於 DEX 格式問題，出現如圖 10-3 所示的 "File not open" 錯誤。

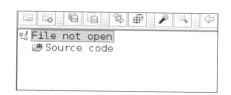

▲ 圖 10-3 Jadx-gui 開啟 DEX 檔案顯示出錯

遇到這種情況時，可以結合 frida_fart 或者其他脫殼工具對應用進行再次脫殼處理，這裡使用 frida_fart 的 Hook 版本進行脫殼後，完美解決了之前 frida-dexdump 脫殼檔案格式不對的問題。

為了方便後續的應用測試工作，這裡註冊了一個測試用的帳號，筆者在測試過程中發現：當使用者登入後，會要求透過付費成為會員後才能進入 App 真實業務頁面，而這部分付費的邏輯，單純從頁面上經過多種方法點擊，是無法直接不付費進入真實首頁面的，如圖 10-4 所示。

▲ 圖 10-4 付費成為會員

站在開發的角度上,當想要從一個頁面跳躍到另一個頁面時,通常可以使用 Intent 跳躍的方式完成,其實現方式如程式清單 10-5 所示。

◐ **程式清單 10-5 頁面跳躍**

```
void startActivity(){
    Intent intent = new Intent(this,MainActivity.class);
    startActivity(intent);
}
```

觀察這部分程式會發現,如果頁面在啟動時沒有做任何驗證,上述方式即可完成針對任意頁面的跳躍。因此,如果想要透過 Frida 指令稿實現頁面直接不付費跳躍至首頁面,只需將程式清單 10-5 中的程式翻譯為 JavaScript 語言即可。幸運的是,Objection 作為一款優秀的協力廠商工具,其中整合了程式清單 10-5 中頁面跳躍方式的命令,因此只需找到 App 首頁面所在的活動類別(透過結合靜態 AndroidManifest.xml 清單檔

案和動態搜索 Activity 並多次測試後,確認 App 首頁面活動類別名稱)
即可透過如下命令完成頁面的跳躍。

```
# android intent launch_activity com.chanson.business.MainActivity
```

由於 App 確實未做業務上的驗證,在執行如上命令後,我們成功繞過了
在登入 App 後需要付費 39 元才能正常使用的硬性要求,如圖 10-5 所示。

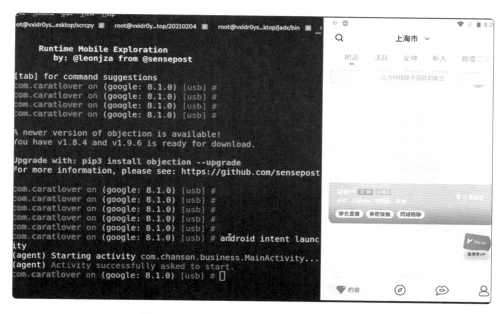

▲ 圖 10-5 利用業務邏輯漏洞實現頁面跳躍

透過後續對 App 業務的研究發現,樣本 App 的核心業務實際上是聊天功
能,但是在測試過程中,我們發現雖然在上一個步驟中透過樣本業務設
計漏洞繞過了付費成為 VIP 才訊息能正常使用的限制,但是實際在與別
人聊天時,在輸入完訊息後,點擊「發送」按鈕還是會出現如圖 10-6 所
示的要求付費成為 VIP 才能正常私聊的提示。

▲ 圖 10-6 要求付費的提示

那麼要如何定位發出提示的視窗並繞過這個限制呢?

毫無疑問,可以使用 hookEvent.js。如圖 10-7 所示,最終透過這個指令稿得到對應的關鍵類別為 com.tencent.qcloud.tim.uikit.modules.chat. layout.input.InputLayout。

```
mOnClickListener name is [com.chanson.business.widget.floatScrollView]
[LGE Nexus 5X::克拉恋人]-- [WatchEvent] onClick: com.tencent.qcloud.tim.uikit.modules.chat.layout.input.InputLayout
[WatchEvent] onClick: com.chanson.business.widget.y
[WatchEvent] onClick: com.tencent.qcloud.tim.uikit.modules.chat.layout.input.InputLayout
[WatchEvent] onClick: com.chanson.business.widget.y
[WatchEvent] onClick: com.tencent.qcloud.tim.uikit.modules.chat.layout.input.InputLayout
[WatchEvent] onClick: com.chanson.business.widget.y
[WatchEvent] onClick: com.tencent.qcloud.tim.uikit.modules.chat.layout.input.InputLayout
[WatchEvent] onClick: com.chanson.business.widget.y
```

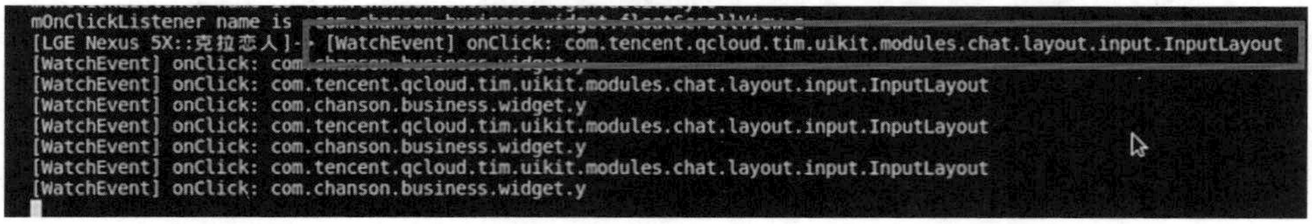

▲ 圖 10-7 hookEvent 執行結果

在定位到對應類別後,透過靜態分析尋找對應的點擊回應函數 onClick(),根據字串一些相關資訊定位後,會發現實際上與「發送資訊」按鈕相關的部分程式如程式清單 10-6 所示。

◆ 程式清單 10-6 訊息發送訊息相關程式

```
public void onClick(View view) {
    if(){
        //...
    } else if (view.getId() == R.id.send_btn && this.mSendEnable) {
            InputLayoutUI.OnSendInterceptListener onSendInterceptListener2
= this.interceptListener;
            if(onSendInterceptListener2 == null ||
            //控制能否發送的函數
            !onSendInterceptListener2.onIntercept()) {
                TextDealHandler textDealHandler2 = this.textDealHandler;
                if (textDealHandler2 != null) {
                    textDealHandler2.textHandle(this.mTextInput.getText().
toString().trim());
                } else {
                    MessageHandler messageHandler = this.mMessageHandler;
                    if (messageHandler != null) {
                        messageHandler.sendMessage(MessageInfoUtil.
buildTextMessage(this.mTextInput.getText().toString().trim()));
                    }
                }
                this.mTextInput.setText("");
            }
        }
    }
}
```

透過靜態分析程式清單 10-6 中的內容後發現，onIntercept 函數實際上才是真實控制資訊能夠發送的關鍵，透過查詢引用等方式找到對應函數的實現，最終發現在 onIntercept 函數中呼叫了 ChatActivity 類別中的 W 函數，由該函數進行能否發送的判定。如程式清單 10-7 所示，最終透過多次動態的 Hook 以及靜態分析確定真實控制是不是 VIP 相關的邏輯，結果最終儲存在 z 變數（程式清單 10-7 中已註釋為「關鍵的判斷條件」）中。

◯ 程式清單 10-7 onIntercept 函數以及 W 函數

```
public final boolean onIntercept() {
    return !this.f10619a.W();
}
// com.chanson.business.message.activity.ChatActivity
private final boolean W() {
        MyInfoBean k;
        BasicUserInfoBean col1;
        MyInfoBean k2;
        MyInfoBean k3;
        BasicUserInfoBean col12;
        BasicUserInfoBean col13;
        if (!V()) {
            return true;
        }
        if (this.f10545d == null) {
            rb.a(rb.f9688c, "資料異常", 0, 2, (Object) null);
            return false;
        } else if (ca()) {
            return false;
        } else {
            CheckTalkBean checkTalkBean = this.f10545d;
            if (checkTalkBean != null) {
                // 關鍵的判斷條件：位置
                boolean z = (Ib.f9521i.h() == 1 && !(((k2 = Ib.f9521i.
k()) == null || (col13 = k2.getCol1()) == null || !col13.isGoddess()) &&
((k3 = Ib.f9521i.k()) == null || (col12 = k3.getCol1()) == null || !col12.
isReal()))) || ((k = Ib.f9521i.k()) != null && (col1 = k.getCol1()) != null
&& col1.isVip() && Ib.f9521i.h() == 2);
                if (!checkTalkBean.getUnlock() && !z) {
                    ChatLayout chatLayout = (ChatLayout) k(R$id.chatLayout);
                    i.a((Object) chatLayout, "chatLayout");
                    chatLayout.getInputLayout().hideSoftInput();
                    x.a(new RunnableC1148a(this), 100);
                    return false;
```

```
            } else if (checkTalkBean.getStatus() == 3 ||
checkTalkBean.getStatus() == 2) {
                rb.a(rb.f9688c, "訊息你已將對方拉黑，無法發送訊息", 0, 2,
(Object) null);

                ...
                return false;
            } else if (checkTalkBean.getStatus() != 1) {
                return true;
            } else {
                rb.a(rb.f9688c, "訊息對方已將你拉黑，無法發送訊息", 0, 2,
(Object) null);

                ...
                return false;
        } else {
            i.a();
            throw null;
        }
    }
}
```

觀察 z 變數的定義會發現其中的邏輯十分複雜，呼叫了幾個類別的函數，
其中就有一個非常明顯地與 VIP 相關的 col1.isVip() 函數，為了快速定位
具體的函數，這裡將組成 z 變數值的所有函數相關完整類別名整理如下：

```
com.chanson.business.g.Ib
com.chanson.business.model.MyInfoBean
com.chanson.business.model.BasicUserInfoBean
```

在複習相關類別後，為了快速定位具體是哪一個函數導致 z 變數的值為
true，可以使用在 10.1 節介紹的 r0tracer 工具中的 traceClass 功能，分別
對上述整理的類別進行 Trace 後，最終確認確實是 BasicUserInfoBean 類
別中的 isVip 函數的返回值為 false 導致 z 變數值為 true，trace 函數的返
回值與對應的呼叫堆疊效果如圖 10-8 所示。

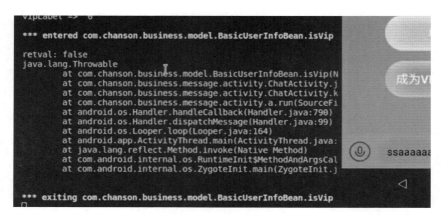

▲ 圖 10-8 traceClass 結果

為了確認是 isVip 訊息函數的返回值最終導致訊息無法發送，這裡使用 Frida 指令稿將該函數的返回值設定為始終返回 true 進行測試，最終對應的程式如程式清單 10-8 所示。在 Hook 訊息指令稿生效後，發現訊息成功發送（見圖 10-9），等待一段時間之後，訊息會成功地顯示「已讀」。

● 程式清單 10-8 hookVIP.js

```
function hookVIP(){
    Java.perform(function(){
        Java.use("com.chanson.business.model.BasicUserInfoBean").isVip.
implementation = function(){
            console.log("Calling isVIP ")
            return true;
        }
    })
}
function main(){
    console.log("Start hook")
    hookVIP()
}
setImmediate(main)
```

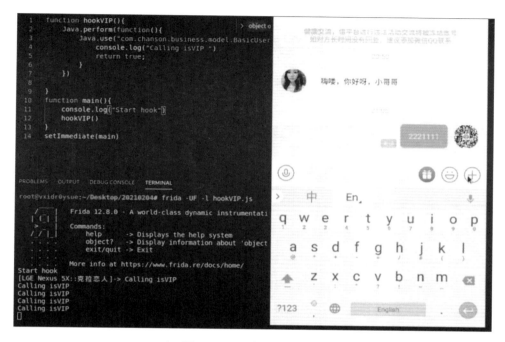

▲ 圖 10-9 訊息成功發送訊息

至此，限制樣本正常使用的障礙已基本移除。如果後續要對協定或者其他方面進行分析，就不再需要浪費精力去解決了。

10.3 協定分析

與第 9 章一樣，為了完成樣本的協定分析，這裡使用 Charles 完成抓取封包工作。

如圖 10-10 所示，在抓取封包的過程中會發現一旦設定了 VPN 代理，A品牌 App 就會提示「網路異常，請檢測網路是否正常」。

▲ 圖 10-10 提示網路異常

與上一章由於存在對抗抓取封包的行為導致樣本無法使用 Charles 進行抓取封包相比，本例樣本並非是因為樣本對抓取封包做了對抗工作，觀察 Charles 中所抓取的封包內容，會發現只有一個 8668 通訊埠的封包一直顯示「正在連接」，具體顯示出錯如圖 10-11 所示，這實際上只是因為非標準通訊埠的資料封包導致 Charles 無法辨識對應通訊埠的資料而已。

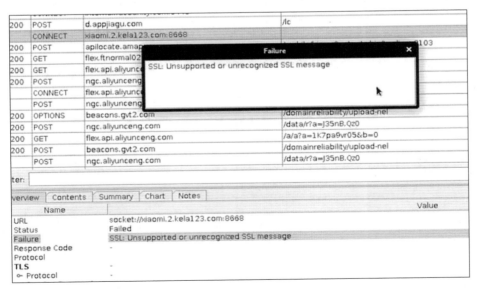

▲ 圖 10-11 8668 通訊埠的封包無法抓取

要解決這個問題，只需要在 Charles 主介面上依次點擊 Porxy → Proxy Settings 並手動將 8668 通訊埠加入 HTTP 協定通訊埠即可，具體增加位置如圖 10-12 所示。

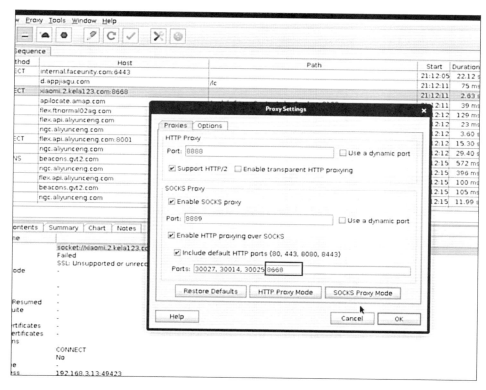

▲ 圖 10-12　增加非標準通訊埠

最終在完成非標準通訊埠的增加後，8668 通訊埠的資料封包成功被 Charles 辨識，最終資料封包 request 封包和 reponse 內容如圖 10-13 所示。

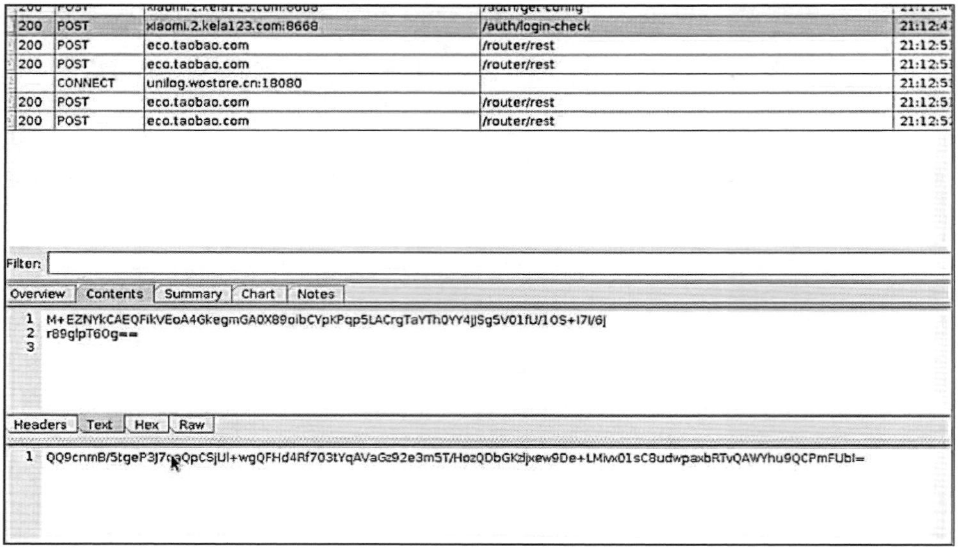

▲ 圖 10-13　8668 通訊埠資料

經過後續的分析發現，實際上 8668 這個通訊埠的通訊內容正是該 App 的關鍵協定，觀察圖 10-13，請求封包和接收封包都處於加密狀態，且疑似 Base64，為了快速定位負責資料封包加密的內容，這裡使用 r0capture 進行抓取封包，利用 r0capture Hook 抓取封包的優勢（堆疊回溯功能）快速定位發送或接收資料封包的過程中所經過的函數流程，最終定位到 8668 通訊埠相關的呼叫堆疊如程式清單 10-9 所示。

◑ 程式清單 10-9　8668 通訊埠資料發送 / 接收呼叫堆疊

```
java.lang.Throwable
    at java.net.SocketOutputStream.socketWrite0(Native Method)
    at java.net.SocketOutputStream.socketWrite(SocketOutputStream.java:109)
    at java.net.SocketOutputStream.write(SocketOutputStream.java:153)
    at h.p.a(SourceFile:5)
    at h.a.a(SourceFile:6)
    at h.t.flush(SourceFile:3)
```

```
at g.a.d.b.a(SourceFile:21)
at g.a.c.b.intercept(SourceFile:26)
at g.a.c.h.a(SourceFile:11)
at g.a.b.a.intercept(SourceFile:7)
at g.a.c.h.a(SourceFile:11)
at g.a.c.h.a(SourceFile:2)
at g.a.a.b.intercept(SourceFile:22)
at g.a.c.h.a(SourceFile:11)
at g.a.c.h.a(SourceFile:2)
at g.a.c.a.intercept(SourceFile:22)
at g.a.c.h.a(SourceFile:11)
at g.a.c.k.intercept(SourceFile:9)
at g.a.c.h.a(SourceFile:11)
at g.a.c.h.a(SourceFile:2)
at com.chanson.common.a.j.intercept(SourceFile:45) // trace這個所在類別
at g.a.c.h.a(SourceFile:11)
at g.a.c.h.a(SourceFile:2)
at g.I.a(SourceFile:28)
at g.I.execute(SourceFile:9)
at j.w.execute(SourceFile:18)
at j.a.a.c.b(SourceFile:5)
at d.a.k.a(SourceFile:77)
at j.a.a.a.b(SourceFile:1)
at d.a.k.a(SourceFile:77)
at d.a.e.e.b.x$b.run(SourceFile:1)
at d.a.q$a.run(SourceFile:2)
at d.a.e.g.k.run(SourceFile:2)
at d.a.e.g.k.call(SourceFile:1)
// ...
at java.lang.Thread.run(Thread.java:764)
```

在得到發送封包函數的呼叫堆疊後，可以透過靜態分析的方式一步一步
分析還原對應的加密流程。這裡為了加快分析的步伐，利用發送資料封
包內容疑似 Base64 編碼的特徵，使用 r0tracer Trace class 的功能去追

蹤 android.os.Base64 類別的呼叫過程，並與程式清單 10-9 中的呼叫堆疊進行對比，最終確定收發送封包的關鍵函數為 com.chanson.common.a.j.intercept()，Base64 相關的關鍵 Trace 記錄如圖 10-14 所示。

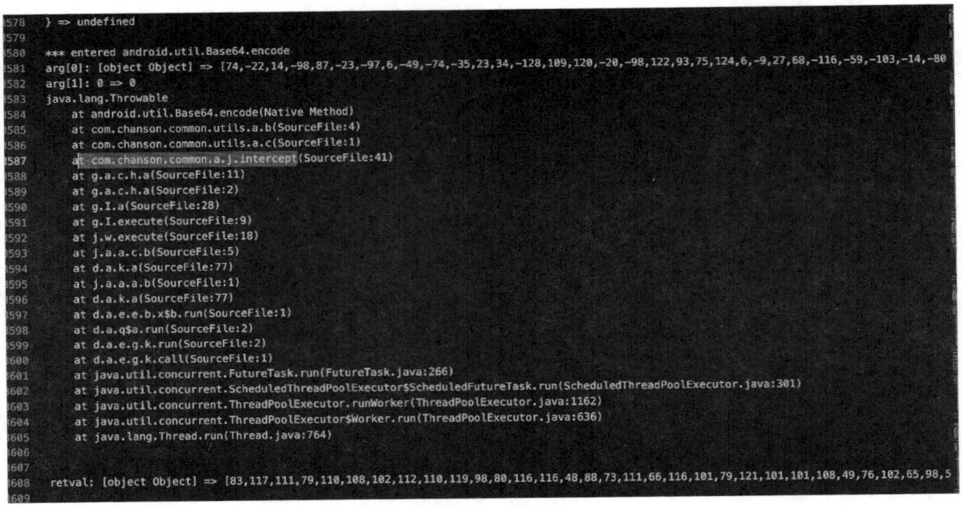

▲ 圖 10-14　Base64 相關的關鍵 Trace 記錄

使用 Jadx 查看該函數的內容並透過靜態分析後發現，在 intercept() 函數中，關鍵類別處理了 JSON 資料，透過 Trace 這個函數後發現，com.chanson.common.utils.a.b.c() 函數執行時，其參數是資料封包內容的明文狀態，如程式清單 10-10 所示。

◉ 程式清單 10-10　intercept 函數

```
public final O intercept(B.a aVar) {
    JSONArray jSONArray;
    J b2 = aVar.b();
    J.a f2 = b2.f();
    Map<String, String> a2 = f.f11633b.a();
    b<Map<String, String>, q> b3 = this.f11642a.b();
    if (b3 != null) {
```

```
        b3.a(a2);
    }
    //...
    f2.a(LocationManager.GPS_PROVIDER, sb.toString());
    String b4 = f.o.b();
    Charset charset = c.f20624a;
    if (b4 != null) {
        //...
        if (f3 != null) {
            byte[] bytes2 = f3.getBytes(charset2);

            byte[] encode2 = Base64.encode(bytes2, 0);

            f2.a("gps_province", n.a(new String(encode2, c.f20624a);
            //...
            f2.a("area", n.a(new String(encode3, c.f20624a), C0533cb.
f5258d, "", false, 4, (Object) null));
            String g2 = f.o.g();
            if (g2 != null) {

            //...
            //關鍵函數
            com.chanson.common.utils.a.b.c(jSONObject.toString());
            f2.b(N.create(C.b(ClipDescription.MIMETYPE_TEXT_PLAIN),
a.c(jSONObject.toJSONString(), "f87210e0ed3079d8")));
        } else {
            throw new kotlin.n("null cannot be cast to non-null type
okhttp3.FormBody");
        }
                }
                return aVar.a(f2.a());
            }
        }
    }
}
```

對比圖 10-14 中的函式呼叫堆疊並結合 r0tracer Trace 功能最終得到關鍵加密函數為 com.chanson.common.utils.a.b()，其對應的加密方式正是 AES/ECB/PKCS5Padding 標準加密模式，其第二個函數 str2 即為金鑰，且金鑰強制寫入為 f87210e0ed3079d8，如程式清單 10-11 所示。

● 程式清單 10-11 AES 資料封包加密

```
public static byte[] b(String str, String str2) {
    if (str == null) {
        return null;
    }
    try {
        Cipher instance = Cipher.getInstance("AES/ECB/PKCS5Padding");
        instance.init(1, new SecretKeySpec(str2.getBytes("utf-8"),
KeyProperties.KEY_ALGORITHM_AES));
        return Base64.encode(instance.doFinal(str.getBytes("utf-8")), 0);
    } catch (Exception e2) {
        e2.printStackTrace();
        return null;
    }
}

public static String c(String str, String str2) {
    return new String(b(str, str2));
}
```

除了透過以上方式確定標準加密模式之外，還可以透過查看加解密沙盒對應應用的 Cipher 檔案與 Charles 中的抓取資料封包進行對比，最終完成該 App 的協定分析工作。這裡為了節省篇幅，就不再介紹了。

10.4 打造智慧聊天機器人

本節將一步一步帶領讀者實現一個智慧聊天機器人。

為了繞過付費限制給其他使用者發送資訊，筆者曾利用 hookEvent. js 訊息指令稿定位到發送訊息的關鍵程式邏輯，但實際上程式清單 10-6 中還會有真正訊息用於發送訊息的嫌疑函數 MessageHandler. sendMessage() 訊息以及用於建構訊息實體的函數 MessageInfoUtil. buildTextMessage()，這兩個函數正是訊息實現發送訊息必不可少的兩部分：訊息發送的訊息內容以及如何發送。

訊息首先來看訊息內容結構的組成。如程式清單 10-12 所示，透過 Jadx 查看 MessageInfoUtil 類別中的 buildTextMessage() 函數內容會發現，該函數主要是根據使用者輸入的字元資料建構成為一個 TIMMessage 實體，並將之作為 MessageInfo 實體的一個成員，同時還設定了訊息一些關於發送人、訊息類型等其他相關資訊，最終將組成的 MessageInfo 物件返回，以訊息完成訊息內容結構的組織。

◯ 程式清單 10-12 建構函數格式

```
public static MessageInfo buildTextMessage(String str) {
    MessageInfo messageInfo = new MessageInfo();
    TIMMessage tIMMessage = new TIMMessage();
    TIMTextElem tIMTextElem = new TIMTextElem();
    tIMTextElem.setText(str);
    tIMMessage.addElement(tIMTextElem);
    setOfflinePushSetting(tIMMessage, str);
    messageInfo.setExtra(str);
    messageInfo.setMsgTime(System.currentTimeMillis() / 1000);
    messageInfo.setElement(tIMTextElem);
```

```
    messageInfo.setSelf(true);
    messageInfo.setTIMMessage(tIMMessage);
    messageInfo.setFromUser(TIMManager.getInstance().getLoginUser());
    messageInfo.setMsgType(0);
    return messageInfo;
}
```

為了確認動態環境下對應的 MessageInfo 實體的情況,這裡使用 r0tracer
對該類別進行 Trace,並將工作日志保存,最終確認在發送資料的那一刻
MessageInfo 實體中域的部分內容如圖 10-15 所示,其中 tttttttt 為發送的
資料內容。

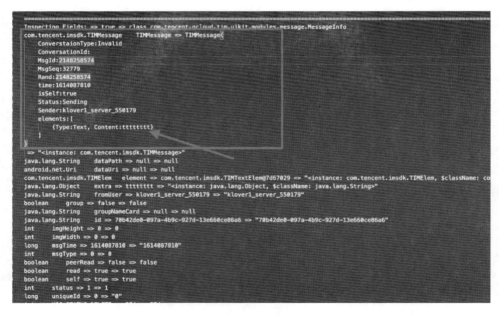

▲ 圖 10-15 MessageInfo 實體域

訊息在完成對訊息結構的分析後,還需要分析 sendMessage() 函數的流
程,由於 MessageHandler 訊息類別只是一個介面類別,因此還需要找到

發送訊息時的真實類別，為此利用 r0tracer 中的日誌內容的呼叫堆疊部分（透過在記錄檔中搜索 sendMessage 字串域與對應呼叫堆疊上下文確定），最終確定真正實現 sendMessage() 函數的實現類別為 com.tencent. qcloud.tim.uikit.modules.chat.base.ChatManagerKit，對應的實現程式如程式清單 10-13 所示。

◆ 程式清單 10-13　sendMessage 函數的實現

```
public void sendMessage(final MessageInfo messageInfo, boolean z, final
IUIKitCallBack iUIKitCallBack) {
    if (!safetyCall()) {
        TUIKitLog.w(TAG, "unSafetyCall");
    } else if (messageInfo != null && messageInfo.getStatus() != 1) {
        messageInfo.setSelf(true);
        messageInfo.setRead(true);
        assembleGroupMessage(messageInfo);
        if (messageInfo.getMsgType() < 256) {
            messageInfo.setStatus(1);
            if (z) {
                this.mCurrentProvider.resendMessageInfo(messageInfo);
            } else {
                this.mCurrentProvider.addMessageInfo(messageInfo);
            }
        }
        String str = TAG;
        TUIKitLog.i(str, "sendMessage:" + ((Object) messageInfo.
getTIMMessage()));
        // 呼叫com.tencent.imsdk.TIMConversation的sendMessage函數
        this.mCurrentConversation.sendMessage(messageInfo.getTIMMessage(),
new TIMValueCallBack<TIMMessage>() {...});
    }
}
// com.tencent.imsdk.TIMConversation
public void sendMessage(@NonNull TIMMessage tIMMessage, @NonNull
TIMValueCallBack<TIMMessage> tIMValueCallBack) {
```

```
if (tIMValueCallBack == null) {
    QLog.e(TAG, "sendMessage ignore, callback is null");
    return;
}
Conversation conversation = this.mConversation;
if (conversation == null) {
    QLog.e(TAG, "sendMessage fail because mConversation is null");
} else {
    conversation.sendMessage(false, tIMMessage, tIMValueCallBack);
}
}
```

分析 ChatManagerKit.sendMessage() 訊息函數發現該函數只是對傳入訊息實體的一些域值進行檢查，最後呼叫 mCurrentConversation.sendMessage() 訊息函數將訊息發送出去，而這個 sendMessage 函數的第一個參數就是 TIMMessage。到這一步，訊息要實現針對單一物件發送訊息已經非常清晰明瞭了：訊息文字訊息封裝成 TIMMessage 類型，並透過 TIMConversation.sendMessage 訊息訊息函數進行訊息的發送工作，要實現給別人發送訊息，只需將上述關鍵程式翻譯為 JavaScript 訊息語言，最終其主動呼叫發送訊息的關鍵 Frida 程式如程式清單 10-14 所示。

⬇ 程式清單 10-14 給單一物件訊息發送訊息

```
var peer = Java.use('java.lang.String').$new("klover1_server_190249"); //發
送身份
var conversation = ins.getConversation(Java.use("com.tencent.imsdk.
TIMConversationType").C2C.value, peer);

var msg = Java.use("com.tencent.imsdk.TIMMessage").$new();
//增加文字內容
var elem = Java.use("com.tencent.imsdk.TIMTextElem").$new();
elem.setText(Java.use("java.lang.String").$new("r0ysue bad bad"));
msg.addElement(elem)
```

```
const callback = Java.registerClass({
    name: 'callback',
    implements: [Java.use("com.tencent.imsdk.TIMValueCallBack")],
    methods: {
        onError(code, desc) {
            console.log("send message failed. code: " + code + " errmsg: "
+ desc);
        },
        onSuccess(msg) {//訊息發送訊息成功
            console.log("SendMsg ok" + msg);
        },
    }
});
conversation.sendMessage(msg, callback.$new())
```

訊息在透過指令稿給單一物件發送訊息成功後，如果要進一步實現批次給別人發送訊息訊息訊息，事實上相比于給單一物件發送訊息，只需要進一步得到所有的 conversation 即可。要做到這一點，我們可以繼續分析該 App 訊息發送訊息的邏輯，但實際上在上述過程中，我們訊息發現該樣本在發送訊息時實際上是使用某大廠所提供的一個協力廠商舊版 SDK 訊息套件，因此要完成批次發送訊息，還可以查閱這個 SDK 套件的開發文件（騰訊雲官方文件位址：https://github.com/tencentyun/qcloud-documents），利用該開發文件說明完成逆向工作。

在對應開發文件的 product/ 行動與通訊 / 即時通訊 /8 用戶端 API/SDK API/ 舊版 SDK API/SDK API（Android）.md 檔案中，最終找到了對應的開發文件。

透過開發文件發現，實際上 TIMManager 類別是整個協力廠商套件的核心類別（見圖 10-16），用於負責 IM SDK 的初始化、登入、建立階段以及管理推送等功能。對應地，筆者在透過 Objection 驗證時，發現這個類別的物件實際上在應用中確實是全域唯一的實例。

```
com.caratlover on (Android: 8.1.0) [usb] # android heap search instances com.tencent.imsdk.
TIMManager
Class instance enumeration complete for com.tencent.imsdk.TIMManager
  Hashcode  Class                                toString()
---------- -----------------------------------  ------------------------------------------
 227890024 com.tencent.imsdk.TIMManager         com.tencent.imsdk.TIMManager@d955368
com.caratlover on (Android: 8.1.0) [usb] #
com.caratlover on (Android: 8.1.0) [usb] #
```

▲ 圖 10-16 TIMManager 全域唯一

進一步查閱文件，發現在 TIMManager 類別中提供了 getConversationList()
函數，用於獲取階段清單，函數的返回數值型別為 java.util.List
＜TIMConversation＞ 訊息，因此要做到批次發送訊息，只需透過
TIMManager 實例獲取所有階段清單即可，訊息透過每一個階段完成訊息
的發送工作。根據以上分析，訊息最終得到的批次發送訊息的指令稿內容
如程式清單 10-15 所示訊息，最終批次發送訊息的結果如圖 10-17 所示。

▲ 圖 10-17 訊息批次發送訊息的結果

◎ 程式清單 10-15 訊息批次發送訊息

```
Java.choose("com.tencent.imsdk.TIMManager", {
    onMatch: function (ins) {
        // 迭代list中的元素
        var iter = ins.getConversationList().listIterator();
        while (iter.hasNext()) {
            console.log(iter.next());
            if (iter.next() != null) {
                var TIMConversation = Java.cast(iter.next(), Java.use("com.
tencent.imsdk.TIMConversation"))
                console.log(TIMConversation.getPeer());
                console.log("try send message...")

                //訊息建構一筆訊息中
                var msg = Java.use("com.tencent.imsdk.TIMMessage").$new();
                //增加文字內容
                var elem = Java.use("com.tencent.imsdk.
TIMTextElem").$new();
                elem.setText("r0ysue222");
                //將elem訊息增加到訊息中
                msg.addElement(elem)

                //if (msg.addElement(elem) != 0) {
                //Log.d(tag, "addElement failed");
                //return;
                //}
                const callback = Java.registerClass({
                    name: 'com.tencent.imsdk.TIMValueCallBackCallback',
                    implements: [Java.use("com.tencent.imsdk.
TIMValueCallBack")],
                    methods: {
                        onError(i, str) { console.log("send message failed.
code: " + i + " errmsg: " + str) },
                        onSuccess(msg) { console.log("SendMsg ok", +msg) }
                    }
```

```
            });
            //訊息發送訊息
            TIMConversation.sendMessage(msg, callback.$new())
            // }
        }
    }
}, onComplete: function () {
        console.log("search compeled")
    }
})
```

至此，訊息批次發送訊息的功能實際上已經完成了。但是如果要打造一個智慧聊天機器人，到這一步還是遠遠不夠的。事實上透過上面的分析讀者會發現，如果想要繼續完善聊天機器人的功能，比如增加獲取、增加、刪除好友，或者建立、加入、邀請、退出群組的功能，其實都是將對應 SDK 開發文件中介紹的 API 進行「翻譯」即可，相信有一定開發基礎的讀者都能夠獨立完成，因此這裡就不再繼續介紹了。

10.5 本章小結

本章介紹了筆者新開發的一款用於 Trace 的工具—r0tracer，可以發現 r0tracer 確實相比於 WallBreaker 查看實例物件中的成員值以及 Objection 中的 Hook 功能做了一些改進。在隨後的實戰講解「會員制」違法應用的協定分析時，也利用 r0tracer 幫助完成了很多工作。另外，在 10.4 節「打造智慧聊天機器人」中，還介紹了逆向的另一種思路—利用開發文件說明完成逆向工作，希望讀者在以後的逆向工作中能夠透過各種方式開啟思路，不要為自己設限，實現真正的逆向「自由」。